西北工业大学精品学术著作
培 育 项 目 资 助 出 版

Computational Methods for Nonlinear Dynamical Systems

Theory and Applications in Aerospace Engineering

Xuechuan Wang　Xiaokui Yue
Honghua Dai　Haoyang Feng
Satya N. Atluri

U0200493

Science Press
Beijing

ELSEVIER

内容简介

非线性动力学系统目前已经成为科学研究和工程应用的重点关注对象。由于缺乏齐次性和叠加性，非线性系统的解往往无法解析获得，只能求助于数值计算方法。然而，面对复杂的强非线性系统，传统数值方法在精度、效率、稳定性等方面常常受到限制，难以满足实际科研工作和工程任务中的仿真计算需求，亟需引入新的思路和方法，推动非线性系统解算的进一步发展。本书根据近十年来相关领域的部分研究成果和作者的研究工作，介绍了非线性动力学系统的一系列全局估计方法和局部计算方法。从经典的渐进方法、有限差分方法和加权残余法开始，对各类典型的非线性动力学系统计算方法进行梳理和总结。在理清已有方法发展脉络的同时，提出了一些新的计算方法研究思路，并借此导出了局部变分迭代法等一系列新型高性能计算方法。本书不仅从理论上对强非线性动力学系统的计算方法进行了归纳、总结和发展，也从具体的航空航天工程问题出发，对方法的实际应用进行了探讨。

本书可供航空航天相关领域科研工作者和工程技术人员阅读，也可供航空航天相关专业的高等院校高年级本科生和研究生学习参考。

ISBN of the Co-Publisher's edition: 978-0-323-99113-1

图书在版编目（CIP）数据

非线性动力学系统的计算方法：航天工程理论及应用 = Computational Methods for Nonlinear Dynamical Systems: Theory and Applications in Aerospace Engineering: 英文 / 汪雪川等著. -- 北京 : 科学出版社 , 2023.6

ISBN 978-7-03-073939-1

Ⅰ. ①非… Ⅱ. ①汪… Ⅲ. ①非线性力学–动力学系统–计算方法–应用–航天工程–英文 Ⅳ. ① V57 ② TP27

中国版本图书馆 CIP 数据核字（2022）第 221127 号

责任编辑：徐杨峰 / 责任校对：谭宏宇
责任印制：黄晓鸣 / 封面设计：殷 靓

科学出版社 出版

北京东黄城根北街 16 号
邮政编码: 100717
http: // www. sciencep. com
南京文脉图文设计制作有限公司排版
广东虎彩云印刷有限公司印刷
科学出版社发行 各地新华书店经销

*

2023 年 6 月第 一 版 开本 : 特 16（787 × 960）
2023 年 6 月第一次印刷 印张 : 15
字数 : 352 000

定价: 140.00 元

（如有印装质量问题，我社负责调换）

Preface

This book aims to introduce many recently developed high-performance computational methods for strongly nonlinear dynamical systems and their applications in practical engineering problems for graduate students and researchers. The study of computational methods covers a broad range of mathematical problems emerging from science and engineering. Leveraging the wide usage of computers, it penetrates most, if not all, modern disciplines. Even sociology and politics are benefiting from computational methods in their state-of-the-art branches, such as social network dynamics. Nevertheless, the most significant developments of computational methods concentrate on applied mathematics and mechanics. The need for simulations in solid mechanics, fluid mechanics, multibody dynamics, and multiphysics has spawned off-the-shelf software that integrates ready-to-use computational methods. New methods are continuously being proposed, improved, and tested to achieve better performance.

The foundation of the computational methods that we use today was laid in the age of Isaac Newton and Leonhard Euler. Using the concepts of differentiation and the Taylor series, the finite difference method was developed. Its early history is marked by the Euler method. Hundreds of variants have been proposed, including the trapezoidal method, the Runge—Kutta method, and the Newmark method. Some of these methods have been used as standard numerical solvers in commercial software (MATLAB, ADAMS, ABAQUS, etc.). Moreover, it was found that the finite difference method can be introduced in a more general way via a weighted residual method, from which the finite element, finite volume, and boundary element methods can be derived. A practical framework of a weighted residual method was built in the 1950s via the development of the finite element method. It quickly stimulated the development of computer-aided engineering and design technology. Alongside that, the asymptotic methods were developed from the early works by Henri Poincaré in his study of three-body problems. The asymptotic methods are very useful when the functional of the problem to be solved is available; therefore they are often applied in weakly nonlinear systems. The best-known asymptotic method for solving nonlinear differential equations is the perturbation method. In earlier times, the use of the perturbation method was limited to astronomical calculations. Scholars such as Laplace and Lagrange used it to investigate the perturbed motion of planets around the sun. Then Urbain Le Verrier successfully predicted the existence of the planet Neptune with this method. After this remarkable event, the application of the perturbation method was gradually broadened to the more general field of nonlinear mechanics.

By contrast, a strongly nonlinear dynamical system possesses complicated behaviors that cannot be found in linearized or weakly nonlinear systems. The difficulties in dealing with strongly nonlinear dynamical systems are twofold: first the simulation of dynamical responses and then the dynamical analysis. Both rely heavily on nonlinear computational methods to tackle the deterioration of convergence, accuracy, and efficiency, which arise from strong nonlinearities. It is a general and interdisciplinary problem to deal with nonlinearity, so we will not discuss it extensively in this book but will focus on dynamical systems that can be modeled or reduced to ordinary differential equations. Although there are numerous finite difference methods for solving ordinary differential equations, they are far from being satisfactory in theoretical analysis and engineering application in strongly nonlinear problems. It is a key point of this book to introduce useful tools to tackle computational difficulties in nonlinear dynamical systems. The solutions for nonlinear dynamical systems are broadly divided into two categories: periodic and transient (nonperiodic). For periodic problems the methodology of a new time domain collocation method is presented. Its equivalence to the high-dimensional harmonic balance method is explained in detail. It shows superiority to other methods in approximating subtle periodic responses of nonlinear dynamical systems. For transient responses the methodology of local variational iteration is presented. It provides possibilities for a new category of numerical methods by combining the principle of a weighted residual method and an asymptotic method. Compared with finite difference methods, the local variational iteration method can achieve much higher computational accuracy and efficiency by using very large step sizes. Applications of these methods are illustrated through examples in orbital mechanics and structural mechanics. The methodologies that are introduced in this book are expected to find applications in a broader range of nonlinear systems.

This book is a collection of recent work involving the collaborative research of the authors from Northwestern Polytechnical University and Texas Tech University. The contents are conveyed in eight chapters. Chapter 1 briefly depicts the outline of computational methods for linear and nonlinear, ordinary and partial differential equations. In Chapters 2–4 the global methods for solving periodic nonlinear dynamical behaviors are presented. Those methods are rooted in weighted residual methods and find their application in the space mission of formation flying. Chapters 5–8 concern local methods for solving transient nonlinear responses with an emphasis on the development of highly accurate and efficient numerical methods that can be used in real-life missions. The applications of those methods are revealed in orbital mechanics and structural dynamics.

We would like to express our deep gratitude to our colleagues and students who generously offered their help and advice in writing this book.

Contents

Chapter 1

Introduction

SUMMARY

Nonlinear dynamical systems are ubiquitous in science and engineering. Although the linearization is often utilized to simplify the original problem, it is not applicable in certain cases. Some dynamical phenomena in nonlinear systems are very distinctive, and cannot be revealed in linear systems [1−4]. Sometimes, the consideration of nonlinearity is so crucial to reveal the true dynamical behaviors that the nonlinear problem must be treated as it is. In this chapter, the principal aim is to introduce the various computational methods for solving nonlinear problems. Hence the attention is focused on the developments of these techniques and their performance, rather than investigating the unique and intriguing features of nonlinear dynamical systems. The contents of this chapter are divided into four parts, including the weighted residual methods (Section 1.1), application of weighted residual methods (Section 1.2), finite difference methods (Section 1.3), and asymptotic methods (Section 1.4).

With the aid of modern computers, the most straightforward approach to investigate the solution trajectory of a nonlinear dynamical system could be the numerical integration methods. There are literally hundreds of different numerical integration methods, of which the majority are finite difference methods such as the Runge-Kutta method [5,6], the Hilber-Hughes-Taylor α method [7], and the Newmark β method [8]. These methods directly use the definition of a differential and the Taylor's theorem to discretize the original differential system and provide piecewise approximations in very small time segments. Depending on whether the approximation in each segment is forward or backward, the methods can normally be classified into explicit and implicit methods [9,10]. In general, the implicit methods are more stable than their explicit counterparts, but the former usually involves the solution of nonlinear algebraic equations.

For nonlinear dynamical systems, long-term responses are often desired. There is the well-known class of methods named Galerkin weak-form methods in the area of numerical analysis. As a special version of the Galerkin's approach, the Harmonic Balance (HB) method [11−13] is well-developed to predict the periodic steady-state solutions for autonomous and non-autonomous nonlinear systems. It is a global approximation method that approximates the steady-state responses with trial functions composed of harmonics or sub-harmonics or super-harmonics. The HB method is a frequency domain method. Through the balance of the harmonics, the fundamental

Computational Methods for Nonlinear Dynamical Systems
DOI: https://doi.org/10.1016/B978-0-323-99113-1.00001-7

frequency ω and the amplitude of the periodic motion can be obtained. However, it is difficult to rewrite nonlinear terms in the form of harmonic series, even with the use of symbolic computation; thus, the resulting algebraic equations are usually very complex. To simplify the symbolic calculations of HB method, the alternate use of the High Dimensional Harmonic Balance (HDHB) method approximates the nonlinear terms using the relationship between its harmonic form and the values of it at a set of collocation points. This approach is much more convenient, but it is also found to be accompanied by the aliasing phenomenon because of the simplification. The HDHB method is later proved to be exactly the same as the Time Domain Collocation (TDC) method [14–16], which is a global collocation method or the spectral collocation method in the time domain.

Besides the periodic responses, the non-periodic responses, including chaotic motions, are more common in nonlinear dynamical systems. Due to the lack of periodicity, the accurate prediction of long term non-periodic responses for the whole domain is difficult. Normally, the approximation is made only in a finite time interval $t_{i+1} - t_i$. Since the response is non-periodic, the trial function can be any appropriate function besides the harmonics. Take the Radial Basis Functions (RBF) for example. By assuming the trial function in terms of RBFs, a RBF Collocation method [17,18] is proposed to solve the initial value and boundary value problems of nonlinear systems. Naturally, to achieve long-term prediction of non-periodic motion of nonlinear systems, the local methods are much more preferable than their global counterparts. By dividing the entire domain into finite but small segments and approximating the solution on each segment separately, the local methods can be more efficient and reliable [19,20].

To approximate the solution of a nonlinear dynamical system with arbitrary boundary conditions, the methods based on the weighted residual principle, which is more general than the idea of the global Galerkin's method, can be used. In the works of Atluri [19] and Dong, etc. [20], these methods are illustrated clearly and systematically, including the Finite Element Method (FEM), the Finite Element Method (FEM), the Boundary Element Method (BEM) and the Meshless Local Petrov Galerkin (MLPG) method, etc. These methods are developed using different test functions and trial functions. Among them, the collocation method is the simplest, using the Dirac Delta functions as the test functions. The selection of trial functions is flexible. Depending on the problem, one can use harmonics, polynomials, Radial Basis Functions (RBFs) and Moving Least Square (MLS) functions. But no matter which trial function is used, it will eventually lead to a system of nonlinear algebraic equations (NAEs). Therefore, the problem ultimately becomes how to solve the NAEs. For that, many NAE solvers are proposed. The most popular Newton-like methods mostly involve the inverse of the Jacobian matrix, and can be very sensitive to the initial guess of the solution. Many efforts have been made to solve the NAEs without inverting the Jacobian and without being sensitive to the initial guess. Among there are the scalar-homotopy iterative algorithms developed by Liu, etc. [21], and Liu and Atluri [22].

In the area of solving strongly nonlinear dynamical problems, the analytical asymptotic methods are also commonly used, such as the Modified Lindstedt-Poincare Method [23], the Homotopy Perturbation Method [24,25], the Variational Iteration Method (VIM) [26,27], the Picard Iteration Method (PIM) [28], the Adomian Decomposition Method (ADM) [29], and so on. These methods start from the solution of a linearized problem and iteratively correct the initial guess so that it approaches the real solution of the nonlinear problem. It implies is that the solution of the nonlinear problem can be obtained by constructing an iterative formula involving a functional of the original equations, instead of transforming the nonlinear differential equations into NAEs to be solved.

1.1 The weighted residual methods

Before treating nonlinear dynamical systems, the static problem of a beam on elastic foundation [19] is considered in this section. With this problem as an example, the various computational methods that are rooted in the idea of weighted residuals are introduced.

1.1.1 Problem description

The deflection of a beam on elastic foundation is simply described by a 4th order ordinary differential equation in dimensionless form

$$y'''' + y - f(x) = 0, \quad x \in \Omega, \tag{1.1}$$

where y is the normalized vertical displacement (deflection), f is the normalized distributed load on the beam, and $\Omega = \{x|0 < x < 1\}$ is the domain of interest. The boundary conditions are expressed as follows:

$$y = \bar{y} \text{ at } S, \; y' = \bar{y}' \text{ at } S', \; y'' = \bar{y}'' \text{ at } S'', \; y''' = \bar{y}''' \text{ at } S''',$$

in which S, S', S'', and S''' are the boundaries where y (displacement), y' (rotation), y'' (moment), and y''' (shear) are prescribed, respectively.

Depending on the boundary conditions, the problem can be either well-posed or ill-posed. For the well-posed problem, the prescriptions of $y \,\&\, y'''$ and $y' \,\&\, y''$ are mutually disjoint, i.e.

$$S \cup S''' = S' \cup S'' = \partial\Omega, \quad S \cap S''' = S' \cap S'' = \varnothing, \tag{1.2}$$

where $\partial\Omega$ being the boundary points $x = 0, 1$. The well-posed problem is physically consistent with the solid mechanics used to describe the beam, in which the shear force reaction y''' can be obtained once the deflection y is prescribed, and the moment reaction y'' can be obtained once the rotation y' is prescribed. Thus, the well-posed problem is "natural", meaning that a unique solution exists and is not very sensitive to small variations of boundary values, although the solution does change continuously with the boundary condition. A problem that is not well-posed, i.e. Eq. (1.2) does not hold, is termed ill-posed. The ill-posed problems are very common in various engineering applications such as structural health monitoring, system control, and medical imaging.

1.1.2 Primal methods

1.1.2.1 Global unsymmetric weak form I

To solve Eq. (1.1), a trial function u is used to approximate the true solution. It results to a residual error in the governing equation:

$$R = u'''' + u - f \neq 0, \quad x \in \Omega, \tag{1.3}$$

Although it is usually difficult to make the residual error R diminish in general nonlinear problems, it is relatively easier and often feasible to make the weighted residuals equal to zero. With a test function v, the weighted residual form of the governing equation is written as follows, which is

$$\int_\Omega vR\mathrm{d}x = \int_\Omega v(u'''' + u - f)\mathrm{d}x = 0. \tag{1.4}$$

Conventionally, the weighted residual form is also referred to as the weak form.

Since Eq. (1.4) involves the fourth-order derivative of the trial function u, it is required that the third order derivative of u should be continuous, or briefly, that u should be C^3 functions. Some commonly used C^3 functions include harmonics, polynomials, radial basis functions (RBFs), and moving least squares functions (MLSFs) [20]. As to the test function v, there is no requirement on its continuity. The selection of trial and test functions is very flexible, leading to various combinations. Among them, two cases are of certain importance in the application to nonlinear dynamical problems and will be further explored in Section 1.2.

Case 1:

test functions are selected as Dirac Delta functions, leading to the collocation method.

Case 2:

test functions and trial functions are both selected as the same orthogonal basis functions, leading to the Galerkin method.

The collocation method is the simplest and the most straightforward among the various branches of weighted residual method. Based on the types of trial functions, the collocation methods are divided into different classes. Three kinds of commonly used trial functions are listed herein

Harmonics:

$$u = \sum_{n=1}^{N} a_n \sin(n\pi x) + \sum_{n=0}^{N} b_n \cos(n\pi x) = b_0 + \sum_{n=1}^{N} [a_n \sin(n\pi x) + b_n \cos(n\pi x)] \quad (1.5)$$

Polynomials (first kind of Chebyshev polynomials):

$$u = \sum_{n=0}^{N} T_n(x), \text{ where } T_0(x) = 1, \; T_1(x) = x, \; \ldots, \quad (1.6)$$
$$T_{n+1}(x) = 2xT_n(x) - T_{n-1}(x), \; -1 \le x \le 1.$$

Radial Basis Function (Gaussian):

$$u = \sum_{n=1}^{N} a_n \phi(x - x_n) = \sum_{n=1}^{N} a_n e^{-(x-x_n)^2}, \text{ where } x_n \; (n = 1, \ldots, N) \text{ are the supporting nodes.} \quad (1.7)$$

Note that there are numerous types of trial functions [30–32]. Each of them possesses unique properties and behaves differently in application. From the perspective of computational accuracy and efficiency, the orthogonal functions are often preferred [33], in that they can bring great convenience in numerical computation as well as symbolic operation using their orthogonality property.

No matter which kind of trial function is used, we can always write the trial function in the following form:

$$u = \boldsymbol{\Phi}(x)\boldsymbol{\alpha}, \quad (1.8)$$

where each column of the row vector $\boldsymbol{\Phi}(x)$ represent each of the basis functions, and the column vector $\boldsymbol{\alpha}$ is the vector of undetermined coefficients.

As is stated above, when the Dirac Delta functions are used as test functions, i.e. $v = \delta(x - x_I)$ for a group of pre-selected points x_I ($I = 1, \ldots, N$) along the beam, the weighted residual formulation Eq. (1.4) leads to the point collocation method [34]:

$$u''''(x_I) + u(x_I) = [\boldsymbol{\Phi}''''(x_I) + \boldsymbol{\Phi}(x_I)]\boldsymbol{\alpha} = f(x_I) \quad (1.9)$$

The boundary conditions can be incorporated in the formulation by collocating points on the boundaries:

$$u(x_I) = \boldsymbol{\Phi}(x_I)\boldsymbol{\alpha} = \bar{y}, \text{ for } x_I \in S; \quad (1.10)$$
$$u'(x_I) = \boldsymbol{\Phi}'(x_I)\boldsymbol{\alpha} = \bar{y}', \text{ for } x_I \in S';$$

$$u''(x_I) = \boldsymbol{\Phi}''(x_I)\boldsymbol{\alpha} = \overline{y}'', \text{ for } x_I \in S'';$$
$$u'''(x_I) = \boldsymbol{\Phi}'''(x_I)\boldsymbol{\alpha} = \overline{y}''', \text{ for } x_I \in S'''.$$

By solving the algebraic equations consisting of Eqs. (1.9) and (1.10), the coefficients $\boldsymbol{\alpha}$ can be obtained, thus an approximated solution is provided by Eq. (1.8).

Other than the Dirac Delta function, we can also use Heaviside functions as test functions in Eq. (1.4), leading to the finite volume method [20].

1.1.2.2 Global symmetric weak form

In Eq. (1.4) the continuity requirements for the admissible trial function u and the admissible test function v are different, hence the weak form in Eq. (1.4) is unsymmetric. For some methods, such as the Galerkin-based finite element method, the symmetric weak form is preferred. By integrating Eq. (1.4) by parts twice, the symmetric weak form is derived as following:

$$[n_x u''' v]_{\partial\Omega} - [n_x u'' v']_{\partial\Omega} + \int_{\Omega} (u'' v'' + uv - fv)\mathrm{d}x = 0. \tag{1.11}$$

For Eq. (1.11), both the trial and test function are required to be C^1 continuous. Therefore, it allows us to use simple functions, instead of complex C^3 continuous functions, to approximate the solution. As aforementioned, the trial and test function can be chosen as harmonics, polynomials and RBFs, etc., leading to the Galerkin method. However, these functions (harmonics, polynomials, RBFs) often lead to a dense, ill-conditioned system of algebraic equations. Suppose that the trial function is

$$u = \sum_{n=0}^{N} [a_n \sin(n\pi x) + b_n \cos(n\pi x)]. \tag{1.12}$$

The basis functions in Eq. (1.12) are defined in the whole domain and are mostly non-zero for an arbitrary x in the domain Ω. These kind of basis functions are classified as global approximation functions. In practice, it is more favorable to use local functions (which are non-zero only over elements of the global domain) instead of global ones. For instance, we define the trial function as element based Hermite Interpolation as following

$$u = \sum_{n=1}^{N} h_n(x), \text{ with the Hermite Interpolation function } h_n(x) \text{ defined in } \Omega_n. \tag{1.13}$$

Then only one local basis function $h_n(x)$ need to be considered for $x \in \Omega_n$, where Ω_n $(n = 1, \ldots, N)$ are nonoverlapping subdomains in Ω. By adopting the local trial functions, the weak form Eq. (1.11) leads to the finite element method [35–37], where the resulting system of algebraic equations is sparse.

1.1.2.3 Global unsymmetric weak form II

Further integrating Eq. (1.11) by parts for twice yields another unsymmetric weak form:

$$[n_x u''' v]_{\partial\Omega} - [n_x u'' v']_{\partial\Omega} + [n_x u' v'']_{\partial\Omega} - [n_x u v''']_{\partial\Omega} + \int_{\Omega} (v'''' + v) u \mathrm{d}x - \int_{\Omega} f v \mathrm{d}x = 0\theta, \tag{1.14}$$

in which there is no continuity requirement for the trial function. If one can find the Green's function for

$$v'''' + v = \delta(x - \eta), \tag{1.15}$$

i.e. the fundamental solution of an infinite beam on an elastic foundation, Eq. (1.14) will reduce to the boundary integral equation

$$C(\eta)u(\eta) - \int_{\Omega} f v \mathrm{d}x + (n_x u''' v)_{\partial\Omega} - (n_x u'' v')_{\partial\Omega} + (n_x u' v'')_{\partial\Omega} - (n_x u v''')_{\partial\Omega} = 0, \tag{1.16}$$

where

$$C(\eta) = \begin{cases} 1, & \eta \notin \partial\Omega \\ 0.5, & \eta \in \partial\Omega \end{cases}$$

In Eq. (1.16) the trial function only appears in the boundary terms, thus it leads to the boundary element method [38,39].

1.1.2.4 Local weak forms

In the previously developed methods, the global weak form is considered in the whole domain Ω. By dividing Ω into subdomains Ω_I (over-lapping or non-over-lapping), local weak forms are defined as follows:

$$\int_{\Omega_I} vR\mathrm{d}x = \int_{\Omega_I} v(u'''' + u - f)\mathrm{d}x = 0. \tag{1.17}$$

The corresponding local boundary conditions are set by the continuity requirement of u on the local boundaries $\partial\Omega_I$.

If Ω_I are non-over-lapping subdomains, the local boundary conditions can be expressed as follows:

$$\begin{gathered} u\left(\partial\Omega_I^+\right) = u\left(\partial\Omega_{I+1}^-\right),\ u'\left(\partial\Omega_I^+\right) = u'\left(\partial\Omega_{I+1}^-\right), \\ u''\left(\partial\Omega_I^+\right) = u''\left(\partial\Omega_{I+1}^-\right),\ u'''\left(\partial\Omega_I^+\right) = u'''\left(\partial\Omega_{I+1}^-\right). \end{gathered} \tag{1.18}$$

In the local weak forms, the trial and test functions can be much simpler than those in the global weak forms Eq. (1.3). As a subdomain is much smaller than the original domain, it is relatively very easy to obtain highly accurate local solutions, even when the trial functions are composed of a small group of basis functions. In addition to that, the local weak forms are also more efficient in computation, because the resulting system of algebraic equations is partitioned into blocks.

The local weak forms can also be rewritten into symmetric and unsymmetric forms, therefore it enables us to develop local collocation, local finite volume, local Galerkin, local finite element, and local boundary element method [40–43].

Herein, it should be emphasized that the local approximation methods are different from the global approximation methods using local trial functions. In the former, the weak forms are satisfied in each local subdomain, and the continuity of the solution is ensured by the local boundary conditions; while in the latter, the weak form is satisfied in the whole domain, and the local trial function needs to be selected carefully to satisfy the continuity requirements a-priori.

1.1.3 Mixed methods

In the primal problem description, the governing Eq. (1.3) is of 4th order. This brings much inconvenience to the computation, due to the fact the trial function has to meet high order continuity requirements. It not only makes the trial function quite complex, but also easily leads to ill-conditioned matrix by including higher order derivatives of the trial function. To decrease the order of the governing Eq. (1.3), extra variables are introduced. By adding the moment m, Eq. (1.3) is rewritten as

$$\begin{cases} u'' - m = 1 \\ m'' + u - f = 0 \end{cases}, \tag{1.19}$$

or equivalently in matrix form as

$$\boldsymbol{A}\boldsymbol{P}'' + \boldsymbol{B}\boldsymbol{P} - \boldsymbol{g} = 0, \tag{1.20}$$

where

$$\boldsymbol{A} = \begin{bmatrix} 1 & 0 \\ 0 & 1 \end{bmatrix}, \boldsymbol{B} = \begin{bmatrix} 0 & -1 \\ 1 & 0 \end{bmatrix}, \boldsymbol{g} = \begin{bmatrix} 0 \\ f \end{bmatrix}, \boldsymbol{P} = \begin{bmatrix} u \\ m \end{bmatrix}.$$

Considering a vector test function $\boldsymbol{V}$, the weak form of Eq. (1.20) can be obtained

$$\int_{\Omega} (\boldsymbol{A}\boldsymbol{P}'' + \boldsymbol{B}\boldsymbol{P} - \boldsymbol{g})^{\mathrm{T}} \boldsymbol{V} \mathrm{d}x = 0. \tag{1.21}$$

The required continuity of the trial function is reduced from C^3 to C^1.

By introducing displacement u, rotation θ, moment m, and shear q as independent variables, the original governing equation is expressed as

$$\begin{cases} u' = \theta \\ \theta' = m \\ m' = q \\ q' + u - f = 0 \end{cases}. \tag{1.22}$$

The corresponding matrix form is

$$\boldsymbol{CZ}' + \boldsymbol{DZ} - \boldsymbol{h} = 0, \tag{1.23}$$

where

$$\boldsymbol{C} = \begin{bmatrix} 1 & 0 & 0 & 0 \\ 0 & 1 & 0 & 0 \\ 0 & 0 & 1 & 0 \\ 0 & 0 & 0 & 1 \end{bmatrix}, \ \boldsymbol{B} = \begin{bmatrix} 0 & -1 & 0 & 0 \\ 0 & 0 & -1 & 0 \\ 0 & 0 & 0 & -1 \\ 1 & 0 & 0 & 0 \end{bmatrix}, \ \boldsymbol{g} = \begin{bmatrix} 0 \\ 0 \\ 0 \\ f \end{bmatrix}, \ \boldsymbol{Z} = \begin{bmatrix} u \\ \theta \\ m \\ q \end{bmatrix}.$$

The weighted residual weak form can be written as follows:

$$\int_{\Omega} (\boldsymbol{CZ}' + \boldsymbol{DZ} - \boldsymbol{h})^{\mathrm{T}} \boldsymbol{W} \mathrm{d}x = 0. \tag{1.24}$$

Now the continuity requirement of trial functions is reduced to C^0, that is to say, the solution of the problem considered can be approximated by linear interpolation.

Based on the mixed forms, one can develop various mixed weighted residual methods [35,44], in contrast to the primal methods that were introduced in Subsection 1.1.2.

1.2 Application of weighted residual methods

Two classical problems in nonlinear dynamics are considered in this section: the Duffing oscillator and the Van Der Pol oscillator. The governing equations of them are both expressed as 2nd order nonlinear differential equations.

Duffing Oscillator:

$$x'' + \mu x' + \varepsilon x + \xi x^3 = f(t), \tag{1.25}$$

Van der Pol Oscillator:

$$x'' - \mu(1 - x^2)x' + x = f(t), \tag{1.26}$$

where x is a variable dependent on time t, while μ, ε, ξ are system parameters and $f(t)$ is an external force.

Due to the existence of nonlinear terms (cubic nonlinear term and nonlinear damping term respectively) in Eqs. (1.25) and (1.26) and, it is barely possible to obtain analytical solutions to these equations in closed forms. Therefore, numerical methods and approximate approaches are often used to reveal the nonlinear dynamical behaviors. In this section, the weighted residual methods in the time domain are employed, from which one can accurately approximate the true solution and investigate some characteristics of the nonlinear dynamical systems in a semi-analytical way.

1.2.1 Transient motions

The transient motion of a nonlinear system is usually irregular and fast-changing. It occurs before the system settles down to a steady state, such as a fixed point or a limit cycle. To capture the transient motion of a nonlinear system for long term is difficult. Normally, it is achieved by dividing the time domain into many small intervals and approximating the solution in each interval successively.

1.2.1.1 Collocation method

Denoting the local time interval as Ω_I, the local weak form of Eq. (1.25) is written as

$$\int_{\Omega_I} vR\mathrm{d}t = \int_{\Omega_I} v(u'' + \mu u' + \varepsilon u + \xi u^3 - f)\mathrm{d}t = 0, \tag{1.27}$$

where v is test function, u is trial function, and R stands for the residual error.

Let the test function v be a series of Dirac Delta functions: $v = \delta(t - t_m)$, $t_m \in \Omega_I, m = 1, 2, \ldots, M$. Eq. (1.27) gives rise to a series of collocation equations

$$u''(t_m) + \mu u'(t_m) + \varepsilon u(t_m) + \xi u(t_m)^3 - f(t_m) = 0, \tag{1.28}$$

The local initial conditions are satisfied by imposing

$$u(t_0) = x(t_0), \text{ and } u'(t_0) = x'(t_0), \tag{1.29}$$

where t_0 is the initial time in the local interval.

Suppose the trial function u is composed of a set of basis functions (harmonics, polynomials, RBFs, etc.)

$$u(t) = \sum_{n=1}^{N} \alpha_n \phi_n(t) = \boldsymbol{\Phi}(t)\boldsymbol{\alpha}. \tag{1.30}$$

Substituting it into Eq. (1.28) we have

$$[\boldsymbol{\Phi}''(t_m) + \mu\boldsymbol{\Phi}'(t_m) + \varepsilon\boldsymbol{\Phi}(t_m)]\boldsymbol{\alpha} + \xi[\boldsymbol{\Phi}(t_m)\boldsymbol{\alpha}]^3 - f(t_m) = 0, \tag{1.31}$$

where $m = 1, 2, ..., M$. By solving the system of nonlinear algebraic equations given in Eq. (1.31), the coefficients $\boldsymbol{\alpha}$ can be determined.

Similar to the preceding process introduced to solve Duffing equation, the approximated solution of the Van der Pol oscillator can be obtained by solving

$$[\boldsymbol{\Phi}''(t_m) + \boldsymbol{\Phi}(t_m)]\boldsymbol{\alpha} - \mu\{1 - [\boldsymbol{\Phi}(t_m)\boldsymbol{\alpha}]^2\}[\boldsymbol{\Phi}'(t_m)\boldsymbol{\alpha}] - f(t_m) = 0 \tag{1.32}$$

in each local time interval Ω_I. The number of collocation points M can be equal to or more than the number of basis functions N.

1.2.1.2 Galerkin method

Let the test and the trial function in Eq. (3.3) be the same set of orthogonal functions $\overline{\boldsymbol{\Phi}}(t)$ in the local time domain Ω_I,

$$v(t) = u(t) = \sum_{n=1}^{N} \alpha_n \overline{\phi}_n(t) = \overline{\boldsymbol{\Phi}}(t)\boldsymbol{\alpha}. \tag{1.33}$$

The local weak form becomes

$$\int_{\Omega_I} \overline{\boldsymbol{\Phi}}(t)\boldsymbol{\alpha}\left\{\left[\overline{\boldsymbol{\Phi}}''(t) + \mu\overline{\boldsymbol{\Phi}}'(t) + \varepsilon\overline{\boldsymbol{\Phi}}(t)\right]\boldsymbol{\alpha} + \xi\left[\overline{\boldsymbol{\Phi}}(t)\boldsymbol{\alpha}\right]^3 - f(t)\right\}\mathrm{d}t = 0. \tag{1.34}$$

Using the orthogonal property of $\overline{\boldsymbol{\Phi}}(t)$, implying

$$\int_{\Omega_I} \overline{\phi}_i\overline{\phi}_j = \begin{cases} 1, & i = j \\ 0, & i \neq j \end{cases}, \tag{1.35}$$

Eq. (1.34) leads to a system of equations

$$\int_{\Omega_I} \overline{\phi}_n(t)\left\{\left[\overline{\boldsymbol{\Phi}}''(t) + \mu\overline{\boldsymbol{\Phi}}'(t) + \varepsilon\overline{\boldsymbol{\Phi}}(t)\right]\alpha + \xi\left[\overline{\boldsymbol{\Phi}}(t)\alpha\right]^3 - f(t)\right\}\mathrm{d}t = 0, \tag{1.36}$$

where $n = 1, 2, ..., N$. For the existence of cubic nonlinear term, the integration in Eq. (1.36) could be troublesome when N is relatively large. In order to explicitly express Eq. (1.36) as algebraic equations of the coefficients $\boldsymbol{\alpha}$, Mathematica can be used to handle with the complex symbolic manipulation.

For demonstration, the Chebyshev polynomials of the first kind $T_n(\tau)$ are used as basis functions [45,46]. Because the orthogonality of $T_n(\tau)$ is valid for $-1 \leq \tau \leq 1$, time t is replaced by the rescaled time τ, where

$$\tau = 2(t - t_0)/(t_f - t_0) - 1. \tag{1.37}$$

The local weak form of the Galerkin method using the Chebyshev polynomials of the first kind is expressed as

$$\int_{\Omega_I} T_n(\tau)\{[\boldsymbol{T}''(\tau)+\mu\boldsymbol{T}'(\tau)+\varepsilon\boldsymbol{T}(\tau)]\boldsymbol{\alpha}+\xi[\boldsymbol{T}(\tau)\boldsymbol{\alpha}]^3-f[(\tau+1)(t_f-t_0)/2+t_0]\}\mathrm{d}\tau=0, \tag{1.38}$$

in which $\boldsymbol{T}(\tau)=[T_1(t), T_2(t), \ldots, T_N(t)]$ and $(\bullet)'$ denotes $\mathrm{d}(\bullet)/\mathrm{d}t$.

The final algebraic form of Galerkin method is much more complex than collocation method. Normally, the explicit expression of Eq. (1.36) cannot be written in matrix and vector form. Thus, it is often laborious to solve Eq. (1.36), even with the help of Mathematica and MATLAB. However, the solution of Galerkin method is usually more accurate than that of collocation method when the same number of basis functions are used.

1.2.2 Periodic motions

In nonlinear dynamics, the global motion is related to the limit cycle oscillation (LCO). As one of the most commonly observed phenomena in nonlinear dynamical systems, the existence of LCOs as well as their shape and period information is of great interest in both theoretical studies and engineering applications.

1.2.2.1 Collocation method

To approximate the periodic solution, the trial function is selected as periodic basis functions such as harmonics with unknown frequency ω.

$$u(t)=b_0+\sum_{n=1}^{N}(a_n\sin n\omega t+b_n\cos n\omega t). \tag{1.39}$$

The weak form of Eq. (1.25) is written as

$$\int_{\Omega_T} vR\mathrm{d}t=\int_{\Omega_T} v(u''+\mu u'+\varepsilon u+\xi u^3-f)\mathrm{d}t=0, \tag{1.40}$$

with v being Dirac Delta function, and Ω_T being a time interval, of which the length is the period T of the limit cycle oscillation. Substituting Eq. (1.39) into Eq. (1.40), we get a system of nonlinear algebraic equations (NAEs) in matrix form

$$[\omega^2\mathbf{E}(\omega)\boldsymbol{A}^2+\mu\omega\mathbf{E}(\omega)\boldsymbol{A}+\varepsilon\mathbf{E}(\omega)]\mathbf{Q}+\xi[\mathbf{E}(\omega)\mathbf{Q}]^3-\boldsymbol{F}=\mathbf{0}, \tag{1.41}$$

where

$$
\boldsymbol{E}(\omega)=\begin{bmatrix}
1 & \sin\omega t_1 & \cos\omega t_1 & \cdots & \sin N\omega t_1 & \cos N\omega t_1\\
1 & \sin\omega t_2 & \cos\omega t_2 & \cdots & \sin N\omega t_2 & \cos N\omega t_2\\
\vdots & \vdots & \vdots & \vdots & \vdots & \vdots\\
1 & \sin\omega t_M & \cos\omega t_M & \cdots & \sin N\omega t_M & \cos N\omega t_M
\end{bmatrix}_{(2N+1)\times M},
$$

$$
\mathbf{Q}=\begin{bmatrix} b_0 & a_1 & b_1 & \cdots & a_N & b_N \end{bmatrix}^{\mathrm{T}}
$$

$$
\boldsymbol{A}=\begin{bmatrix}
0 & & & & & \\
 & & -1 & & & \\
 & 1 & & & & \\
 & & & \ddots & & \\
 & & & & & -N\\
 & & & & N &
\end{bmatrix}_{(2N+1)\times(2N+1)},
$$

$$
\boldsymbol{F}=\begin{bmatrix} f(t_1) & f(t_2) & \cdots & f(t_M) \end{bmatrix}^{\mathrm{T}}.
$$

The coefficients $\mathbf{Q}$ and frequency ω are unknowns to be determined.

To solve the NAEs in Eq. (1.41), Newton-Raphson (NR) iteration method is mostly used because of its simplicity and high convergence speed. However, the NR method is sensitive to the initial values used to initialize the iteration, and it needs to compute the Jacobian matrix and its inverse. In the case where proper initial values can not be obtained, or the Jacobian matrix is ill-conditioned, the NR method may fail to work. To remedy that, one may use the vector/scalar homotopy method as an alternative [21,22,47], which is insensitive to initial values and free of inverting the Jacobian matrix.

1.2.2.2 Galerkin method (harmonic balance method)

Let both the trial and test function be composed of the same harmonics:

$$
v(t)=u(t)=b_0+\sum_{n=1}^{N}(a_n\sin n\omega t+b_n\cos n\omega t). \tag{1.42}
$$

The weak form Eq. (1.40) can be discretized as

$$
\begin{cases}
\int_{\Omega_I}(u''+\mu u'+\varepsilon u+\xi u^3-f)\mathrm{d}t=0\\
\int_{\Omega_I}(\sin n\omega t)(u''+\mu u'+\varepsilon u+\xi u^3-f)\mathrm{d}t=0\ ,\quad n=1,2,\ldots,N.\\
\int_{\Omega_I}(\cos n\omega t)(u''+\mu u'+\varepsilon u+\xi u^3-f)\mathrm{d}t=0
\end{cases} \tag{1.43}
$$

To solve Eq. (1.43), the coefficient a_1 may be prescribed as zero. In this way, there are $2N+1$ equations for $2N+1$ unknowns, thus the system is solvable.

The weak form of the Galerkin method is described in Eq. (1.43) The corresponding NAEs can also be obtained by substituting the trial function in Eq. (1.42) into

$$u'' + \mu u' + \varepsilon u + \xi u^3 - f = 0, \tag{1.44}$$

and eliminating the coefficients of each harmonics $\sin n\omega t$ and $\cos n\omega t$ separately. This approach is exactly the harmonic balance method [11,12,48].

It should be noted that the collocation method and the harmonic balance method may produce approximate solutions that are physically meaningless. To relieve that, more collocation points or harmonics may be added to the approximations. Normally, the more harmonics are included in the trial function, the more accurate approximation can be obtained. However, as the number of harmonics increases, the resulting system of NAEs becomes more complex and thus more computational resource will be spent.

Compared to the purely numerical methods, the advantages of weighted residual methods in predicting the global motion includes:

(i) The periodic solution is approximated semi-analytically, in which the frequency and the shape of the LCOs are explicitly provided.

(ii) Both the stable and unstable LCOs can be predicted by weighted residual methods, while the numerical integration can only reveal the stable ones.

(iii) It is more convenient to use weighted residual methods to study the evolution of LCOs for varying system parameters, than using numerical methods.

1.3 Finite difference methods

Consider a second-order nonlinear dynamic system, which is recast into a system of first-order ordinary differential equations (ODEs):

$$\begin{cases} x_1' = g_1(x_1, x_2, f, t) = x_2 \\ x_2' = g_2(x_1, x_2, f, t) \end{cases}. \tag{1.45}$$

This system can be further rewritten as matrix form

$$\boldsymbol{x}' = \boldsymbol{g}(\boldsymbol{x}, f, t), \tag{1.46}$$

where $\boldsymbol{x} = [x_1, x_2]^{\mathrm{T}}$ and f is the external force applied to the system. For a specified set of initial conditions, Eq. (1.46) can be solved with various implicit and explicit numerical integration methods.

1.3.1 Explicit methods

In explicit methods, the future unknown state is directly expressed in terms of the currently-unknown state with an explicit formula. Euler method is the simplest explicit method

$$\boldsymbol{x}(t+\Delta t)=\boldsymbol{x}(t)+\Delta t\boldsymbol{g}\left[\boldsymbol{x}(t),f(t),t\right], \tag{1.47}$$

which is a first order Taylor series expansion in the time domain.

Another explicit method is the second order central difference method:

$$\begin{aligned} x_2\left(t+\frac{\Delta t}{2}\right) &= x_2\left(t-\frac{\Delta t}{2}\right)+\Delta t g_2(x,f,t) \\ x_1(t+\Delta t) &= x_1(t)+\Delta t x_2\left(t+\frac{\Delta t}{2}\right). \end{aligned} \tag{1.48}$$

This method is presented in Belytschko's paper [49] and has been widely used for transient finite element analyses of large-scale nonlinear structures, such as crash simulation of automobiles.

The family of Runge-Kutta (RK) methods are considered as the most widely used explicit methods for numerical integration. The first order RK method is simply the forward Euler method in Eq. (1.47) Among the various RK methods, the classical 4th -order RK method is the most commonly used. It evaluates the future unknown state by adding four increments to the currently-known system state:

$$\boldsymbol{x}(t+\Delta t)=\boldsymbol{x}(t)+\frac{\Delta t}{6}(\boldsymbol{k}_1+2\boldsymbol{k}_2+2\boldsymbol{k}_3+\boldsymbol{k}_4), \tag{1.49}$$

where

$$\begin{cases} \boldsymbol{k}_1=\boldsymbol{F}[t,\boldsymbol{x}(t)] \\ \boldsymbol{k}_2=\boldsymbol{F}\left[t+\dfrac{\Delta t}{2},\boldsymbol{x}(t)+\dfrac{\Delta t}{2}\boldsymbol{k}_1\right] \\ \boldsymbol{k}_3=\boldsymbol{F}\left[t+\dfrac{\Delta t}{2},\boldsymbol{x}(t)+\dfrac{\Delta t}{2}\boldsymbol{k}_2\right] \\ \boldsymbol{k}_4=\boldsymbol{F}[t+\Delta t,\boldsymbol{x}(t)+\Delta t\boldsymbol{k}_3] \end{cases}$$

Against to the fixed step size in RK4, the adaptive step-size 4th-order RK methods are developed by Fehlberg [5] and are now known as the Runge-Kutta-Fehlberg (RKF) methods. Several higher order adaptive RKF methods [6] are widely used for very high accuracy applications such as orbit propagation problems [50].

1.3.2 Implicit methods

Implicit methods approximate the unknown future state in a backward manner, which normally results in a set of linear or nonlinear algebraic equations. The backward Euler method is an illustration of this concept:

$$\boldsymbol{x}(t+\Delta t)=\boldsymbol{x}(t)+\Delta t\boldsymbol{g}\left[\boldsymbol{x}(t+\Delta t),f(t+\Delta t),t+\Delta t\right]. \tag{1.50}$$

Newmark introduces the Newmark-β method [8] based on the extended mean value theorem, which is among the most widely-used implicit methods for numerically evaluating the dynamical response of engineering structures.

A generalization of the Newmark-β method is introduced by Hilber [7]:

$$\begin{cases} x_1(t+\Delta t)=x_1(t)+\Delta t x_2(t)+\dfrac{1}{2}\Delta t^2[(1-2\beta)a(t)+2\beta a(t+\Delta t)] \\ x_2(t+\Delta t)=x_2(t)+\Delta t[(1-\gamma)a(t)+\gamma a(t+\Delta t)] \\ a(t+\Delta t)=(1+a)x_2'(t+\Delta t)-\alpha x_2'(t) \end{cases}, \tag{1.51}$$

with $\gamma=\frac{1-2\alpha}{2}$, $\beta=\left(\frac{1-2\alpha}{2}\right)^2$, $\alpha\in\left[-\frac{1}{3},0\right]$. This method is known as Hilbert-Hughes-Taylor or HHT-α method.

For $\alpha=0$ the HHT-α method collapses to the well-known Newmark method.

For the above numerical methods, the step size plays an important role in the computational accuracy. Normally the step size must be relatively small to obtain an accurate solution for nonlinear problem. If an implicit method is used, it will be necessary to solve a set of linear/nonlinear algebraic equations, which bring in extra computational burden.

1.4 Asymptotic methods

One way to obtain analytical approximated solutions of nonlinear problems in time domain is using asymptotic method [51,52]. It is often used to study the local behavior of a system in the neighborhood of a nominal solution. The most well-known asymptotic method for solving nonlinear differential equations should be the perturbation method [2,53].

1.4.1 Perturbation method

This method assumes that the nonlinear terms in the equations are associated with small parameters. In that case, the solution can be obtained by adding small corrections to the nominal solution that is readily solved for. To explain the details of perturbation method, a second order differential equation is taken for instance.

$$x'' + \varepsilon x + \mu x^3 = 0. \tag{1.52}$$

Since μ is a small parameter, the solution of linear equation $x'' + \varepsilon x = 0$ is considered close enough to the true solution of Eq. (1.52), and is denoted as x_0. Obviously, a nonzero residual occurs by substituting x_0 into Eq. (1.52)

$$R_0 = x_0'' + \varepsilon x_0 + \mu x_0^3 = \mu x_0^3 = O(\mu). \tag{1.53}$$

To eliminate the residual R_0, a small correction x_1 is added to x_0, leading to

$$R_1 = (x_0'' x_1'') + \varepsilon(x_0 + x_1) + \mu(x_0 + x_1)^3 = x_1'' + \varepsilon x_1 + \mu x_0^3 + \mu\left(3x_0^2 x_1 + 3x_0 x_1^2 + x_1^3\right). \tag{1.54}$$

Let $x_1'' + \varepsilon x_1 + \mu x_0^3 = 0$, we have

$$R_1 = \mu\left(3x_0^2 x_1 + 3x_0 x_1^2 + x_1^3\right) = O\left(\mu^2\right).$$

Then by adding another correction x_2 to the previously obtained solution, the residual of Eq. (1.52) is obtained as

$$\begin{aligned} R_2 &= (x_0'' + x_1'' + x_2'') + \varepsilon(x_0 + x_1 + x_2) + \mu(x_0 + x_1 + x_2)^3 \\ &= x_2'' + \varepsilon x_2 + 3\mu x_0^2 x_1 + \mu\left(3x_0 x_1^2 + x_1^3 + 3x_0^2 x_2 + 6x_0 x_1 x_2 + 3x_1^2 x_2 + 3x_0 x_2^2 + 3x_1 x_2^2 + x_2^3\right) \end{aligned} \tag{1.55}$$

Let $x_2'' + \varepsilon x_2 + 3\mu x_0^2 x_1 = 0$, R_2 becomes

$$R_2 = \mu(3x_0 x_1^2 + x_1^3 + 3x_0^2 x_2 + 6x_0 x_1 x_2 + 3x_1^2 x_2 + 3x_0 x_2^2 + 3x_1 x_2^2 + x_2^3) = O\left(\mu^3\right). \tag{1.56}$$

Repeating this process leads to the solution of Eq. (1.52) in series form

$$x(t) = x_0(t) + x_1(t) + x_2(t) + \ldots, \tag{1.57}$$

where $x_0, x_1, x_2, \ldots$ are obtained by solving

$$\begin{cases} x_0'' + \varepsilon x_0 = 0 \\ x_1'' + \varepsilon x_1 + \mu x_0^3 = 0 \\ x_2'' + \varepsilon x_2 + 3\mu x_0^2 x_1 = 0 \\ \vdots \end{cases} \tag{1.58}$$

The preceding description of the perturbation method looks concise and clear. However, by proceeding in this way, a difficulty may often be encountered in the form of "secular terms", where the convergence of the solution Eq. (1.57) is destroyed by infinitely growing terms as $t \to \infty$. In the following, Eq. (1.52) is used as an example to illustrate the appearance of secular terms.

By solving Eq. (1.58), the nominal solution x_0 is obtained first

$$x_0 = C_1 \cos(\sqrt{\varepsilon} t) + C_2 \sin(\sqrt{\varepsilon} t). \tag{1.59}$$

Hence the second equation becomes

$$x_1'' + \varepsilon x_1 + \frac{\mu}{4}\Big[3(C_1 + C_1 C_2{}^2)\cos(\sqrt{\varepsilon} t) + (C_1{}^3 - 3C_1 C_2^2)\cos(3\sqrt{\varepsilon} t) + 3(C_1{}^2 C_2 + C_2{}^3)\sin(\sqrt{\varepsilon} t) + (3C_1{}^2 C_2 - C_2^3)\sin(3\sqrt{\varepsilon} t)\Big].$$

Since it is a linear differential equation, its solution x_1 can be obtained by the superposition of multiple simpler solutions. Specifically, we focus on the solution $x_{1,1}$ of

$$x_{1,1}'' + \varepsilon x_{1,1} + \cos(\sqrt{\varepsilon} t) = 0, \tag{1.60}$$

which is solved as

$$x_{1,1} = C_1 \cos(\sqrt{\varepsilon} t) + C_2 \sin(\sqrt{\varepsilon} t) + \frac{-\cos(\sqrt{\varepsilon} t)\cos(2\sqrt{\varepsilon} t) - 2t\sqrt{\varepsilon}\sin(\sqrt{\varepsilon} t) - \sin(\sqrt{\varepsilon} t)\sin(2\sqrt{\varepsilon} t)}{4\varepsilon}. \tag{1.61}$$

It can be seen the secular term appears as $-2t\sqrt{\varepsilon}\sin(\sqrt{\varepsilon} t)$.

The reason why secular terms appear is as follows. By solving the first equation, the frequency of x_0 is prescribed as $\sqrt{\varepsilon}$, while the true periodic solution of the nonlinear system, if there is any, may not be of the same frequency. A particularly convincing example is given by the expanding a simple periodic function $\sin(\sqrt{\varepsilon} + \delta)t$:

$$\sin(\sqrt{\varepsilon} + \delta)t = \sin\sqrt{\varepsilon} t + \delta t \cos\sqrt{\varepsilon} t - \frac{\delta^2 t^2}{2!}\sin\sqrt{\varepsilon} t - \cdots, \tag{1.62}$$

where δ is a small variation of the frequency.

To eliminate the secular terms, a rescaled time $\tau = \omega t$ is used in Eq. (1.52), with ω being the frequency of the true solution, which is unknown. Hence Eq. (1.52) becomes

$$\omega^2 x'' + \varepsilon x + \mu x^3 = 0,$$

and Eq. (1.58) is rewritten as follows:

$$\begin{cases} \omega_0^2 x_0'' + \varepsilon x_0 = 0 \\ \omega_0^2 x_1'' + \omega_1^2 x_0'' + \varepsilon x_1 + \mu x_0^3 = 0 \\ \omega_0^2 x_2'' + \omega_2^2 x_0'' + \omega_1^2 x_1'' + \varepsilon x_2 + 3\mu x_0^2 x_1 = 0 \\ \vdots \end{cases} \tag{1.63}$$

Since $x_i(\tau + 2\pi) = x_i(\tau)$, we have the following conditions to help solving Eq. (1.63), namely

$$x_0(0) = A,\ x_{i+1}(0) = 0,\ \dot{x}_i(0) = 0,\ i = 0, 1, 2, \cdots.$$

The frequency is then determined by $\omega^2 = {\omega_0}^2 + {\omega_1}^2 + {\omega_2}^2 + \cdots$. This is the modified perturbation method.

At last, it should be noted that the convergence of the perturbation method is not guaranteed. Although a process of solving the given example is demonstrated above, it is more like as a guideline rather than a presentation of rules. In practice, the application of perturbation method could be very flexible and experience is valuable.

1.4.2 Adomian decomposition method

This method is essentially another version of perturbation method, but it introduces the notion of Adomian polynomials and presents the iterative formulas for solving nonlinear systems in a particular form [29,54–56]. A brief description of the Adomian decomposition method (ADM) is made as the following.

Considering a one-dimensional nonlinear dynamical system $Lx + Rx + Nx = f(t)$, where L is the highest-ordered derivative, R and N are the linear and nonlinear operators on x, it can be rewritten as

$$Lx = f(t) - Rx - Nx. \tag{1.64}$$

The solution is supposed to be found as a series of functions

$$x = \sum_{n=0}^{\infty} x_n. \tag{1.65}$$

Substituting Eq. (1.65) into Eq. (1.64), we have

$$Lx = f(t) - Rx - N(x_0 + x_1 + x_2 + \cdots). \tag{1.66}$$

The nonlinear term can be equivalently expanded into Taylor series

$$N(x) = N(x_0) + N'(x_0)(x_1 + x_2 + \cdots) + \frac{N''(x_0)}{2!}(x_1 + x_2 + \cdots)^2 + \cdots. \tag{1.67}$$

After rearrangement, it is expressed as

$$
\begin{aligned}
N(x) = N(x_0) + N'(x_0)x_1 + \left[N'(x_0)x_2 + \frac{N''(x_0)}{2!}{x_1}^2\right] \\
+ \left[N'(x_0)x_3 + N''(x_0)x_1x_2 + \frac{N'''(x_0)}{3!}{x_1}^3\right] + \cdots.
\end{aligned}
\tag{1.68}
$$

From Eq. (1.68), the Adomian polynomials are defined as

$$
\begin{aligned}
&A_0 = N(x_0),\ A_1 = N'(x_0)x_1,\ A_2 = N'(x_0)x_2 + \frac{N''(x_0)}{2!}x_1^2, \\
&A_3 = N'(x_0)x_3 + N''(x_0)x_1x_2 + \frac{N'''(x_0)}{3!}x_1^3.
\end{aligned}
$$

Therefore, Eq. (1.66) can be decomposed as

$$
\begin{cases} Lx_0 = f(t) - Rx_0 \\ Lx_{n+1} = -Rx_{n+1} - A_n \end{cases}
\tag{1.69}
$$

where $n = 0, 1, 2, \cdots$. By solving Eq. (1.69), analytical solutions can be obtained for the nonlinear system. However, the convergence of the Adomian series is not guaranteed. Although it is often declared that the ADM is not restricted to small assumptions or weak nonlinearity, mostly it works only when the nonlinear term is relatively not very significant.

It is well known that a higher order differential equation can always be transformed into a system of first order differential equations. For example, the equation

$$
\frac{\mathrm{d}^n x}{\mathrm{d}\tau^n} = f\left(\frac{\mathrm{d}^n x}{\mathrm{d}\tau^n}, \cdots \frac{\mathrm{d}x}{\mathrm{d}\tau}, x, \tau\right)
\tag{1.70}
$$

can be rewritten as a system of equations like

$$
\frac{\mathrm{d}x_0}{\mathrm{d}\tau} = x_1,\ \frac{\mathrm{d}x_1}{\mathrm{d}\tau} = x_2,\ \ldots,\ \frac{\mathrm{d}x_{n-1}}{\mathrm{d}\tau} = f\left(\frac{\mathrm{d}x_{n-1}}{\mathrm{d}\tau}, \ldots x_1, x_0, \tau\right)
\tag{1.71}
$$

by introducing the variables $x_0, \ldots x_{n-1}$. Therefore, the nonlinear ordinary differential equations can be expressed in a general form as

$$
\frac{\mathrm{d}\boldsymbol{x}}{\mathrm{d}\tau} = \boldsymbol{F}(\boldsymbol{x}, \tau),\ \text{where } \boldsymbol{x} = (x_1, x_2, \ldots)^{\mathrm{T}},\ \boldsymbol{F} = (f_1, f_2, \ldots)^{\mathrm{T}},\ \tau \in [t_0, t],
\tag{1.72}
$$

Where, here onwards, a bold symbol indicates a vector or a matrix, and for brevity, the differential operator $\mathrm{d}/\mathrm{d}\tau$ is denoted as $\boldsymbol{L}$ in the following sections. In the preceding equation, $\boldsymbol{F}(\boldsymbol{x}, \tau)$ is a nonlinear function of the state vector $\boldsymbol{x}$ and the independent variable τ.

This equation can be recast as

$$\boldsymbol{x} = \boldsymbol{x}(t_0) + \boldsymbol{L}^{-1}\boldsymbol{F}(\boldsymbol{x}, \tau), \tag{1.73}$$

where $\boldsymbol{x}(t_0)$ is the initial condition.

The innovative part of ADM is that it approximates both the solution and the nonlinear part as sequences of functions and introduces the Adomian polynomials so that the original problem will be solved progressively.

$$\boldsymbol{x} = \sum_{n=0}^{\infty} \overline{\boldsymbol{x}}_n,\ \boldsymbol{F}(\boldsymbol{x}, \tau) = \sum_{n=0}^{\infty} \boldsymbol{A}_n(\overline{\boldsymbol{x}}_0, \overline{\boldsymbol{x}}_1, \ldots \overline{\boldsymbol{x}}_n). \tag{1.74}$$

This enables us to rewrite Eq. (1.73) as

$$\sum_{n=0}^{\infty} \overline{\boldsymbol{x}}_n = \boldsymbol{x}(t_0) + \boldsymbol{L}^{-1} \sum_{n=0}^{\infty} \boldsymbol{A}_n(\overline{\boldsymbol{x}}_0, \overline{\boldsymbol{x}}_1, \ldots \overline{\boldsymbol{x}}_n). \tag{1.75}$$

The solution can thus be derived recursively in the following way

$$\overline{\boldsymbol{x}}_1 = \boldsymbol{x}(t_0) + \boldsymbol{L}^{-1}\boldsymbol{A}_0, \tag{1.76}$$

$$\overline{\boldsymbol{x}}_{n+1} = \boldsymbol{L}^{-1}\boldsymbol{A}_n. \tag{1.77}$$

The Adomian polynomials $\boldsymbol{A}_n$ are generated by simply rearranging the Taylor series expansion of $\boldsymbol{F}(\boldsymbol{x}, \tau)$ about $\overline{\boldsymbol{x}}_0 = \boldsymbol{x}(t_0)$. They are expressed as

$$\boldsymbol{A}_0 = \boldsymbol{F}(\overline{\boldsymbol{x}}_0),\ \boldsymbol{A}_1 = \boldsymbol{F}'(\overline{\boldsymbol{x}}_0)\overline{\boldsymbol{x}}_1,\ \boldsymbol{A}_2 = \boldsymbol{F}'(\overline{\boldsymbol{x}}_0)\overline{\boldsymbol{x}}_2 + \boldsymbol{F}''(\overline{\boldsymbol{x}}_0)\frac{\overline{\boldsymbol{x}}_1^{\,2}}{2!},\ \ldots, \tag{1.78}$$

where

$$\boldsymbol{F}'(\overline{\boldsymbol{x}}_0) = \frac{\partial \boldsymbol{F}}{\partial \overline{\boldsymbol{x}}_0} = \begin{bmatrix} \dfrac{\partial F_1}{\partial \overline{x}_{1,0}} & \dfrac{\partial F_1}{\partial \overline{x}_{2,0}} & \cdots \\ \dfrac{\partial F_2}{\partial \overline{x}_{1,0}} & \dfrac{\partial F_2}{\partial \overline{x}_{2,0}} & \\ \vdots & & \ddots \end{bmatrix},$$

$$\boldsymbol{F}''(\overline{\boldsymbol{x}}_0) = \frac{\partial^2 \boldsymbol{F}}{\partial \overline{\boldsymbol{x}}_0^{\,2}} = \begin{bmatrix} \dfrac{\partial^2 F_1}{\partial \overline{x}_{1,0}\partial \overline{x}_{1,0}} & \dfrac{\partial^2 F_1}{\partial \overline{x}_{1,0}\partial \overline{x}_{2,0}} & \cdots & \dfrac{\partial^2 F_1}{\partial \overline{x}_{2,0}\partial \overline{x}_{1,0}} & \dfrac{\partial^2 F_1}{\partial \overline{x}_{2,0}\partial \overline{x}_{2,0}} & \cdots \\ \dfrac{\partial^2 F_2}{\partial \overline{x}_{1,0}\partial \overline{x}_{1,0}} & \dfrac{\partial^2 F_2}{\partial \overline{x}_{1,0}\partial \overline{x}_{2,0}} & \cdots & \dfrac{\partial^2 F_2}{\partial \overline{x}_{2,0}\partial \overline{x}_{1,0}} & \dfrac{\partial^2 F_2}{\partial \overline{x}_{2,0}\partial \overline{x}_{2,0}} & \\ \vdots & & \ddots & \vdots & & \ddots \end{bmatrix}$$

1.4.3 Picard iteration method

Picard iteration method takes a very simple form. However, its application is very limited, since it needs to integrate the nonlinear terms in each iteration step, which could be very difficult to implement. Compared to Adomian's method, the Picard iteration lacks the ease of computation and the ability to solve a wide class of equations. But by combining with other computational techniques, it is still possible to make advances. For example, the MCPI method [46] combines the Chebyshev polynomials with the Picard's method and is applied to the two-body gravitational integration problem. It is shown that the integration process of Picard's method becomes very simple and the method achieves high accuracy and efficiency.

Consider an initial value problem (IVP) governed by a system of first-order differential equations

$$\boldsymbol{L}\boldsymbol{x}=\boldsymbol{F}(\boldsymbol{x},\tau),\ \tau\in[t_0,t],\ \boldsymbol{x}(t_0)=[x_1(t_0),x_2(t_0),\ldots]^{\mathrm{T}}. \tag{1.79}$$

It is equal to the associated integral equations

$$\boldsymbol{x}(t)=\boldsymbol{x}(t_0)+\int_{t_0}^{t}\boldsymbol{F}[\tau,\boldsymbol{x}(\tau)]\mathrm{d}\tau,\ \tau\in[t_0,t]. \tag{1.80}$$

The PIM solves this problem in a recursive way by constructing a series of approximating functions. The process works as

(1) Give an initial guess of the solution $\boldsymbol{x}_0(\tau)$ that satisfies the initial condition $\boldsymbol{x}_0(t_0)=\boldsymbol{x}(t_0)$.

(2) Substitute it into the recursive formula that holds

$$\boldsymbol{x}_{n+1}(t)=\boldsymbol{x}(t_0)+\int_{t_0}^{t}\boldsymbol{F}[\tau,\boldsymbol{x}_n(\tau)]\mathrm{d}\tau, \tag{1.81}$$

for $n\geq 0$.

The Picard iteration method and its applications in modern engineering problems will be further mentioned in Chapters 5–7.

References

[1] J. Guckenheimer, P. Holmes, Nonlinear Oscillations, Dynamical Systems, and Bifurcations of Vector Fields, Springer-Verlag, 1983.

[2] C. Hayashi, Nonlinear Oscillations in Physical Systems, McGraw-Hill, 1964.

[3] R.E. Mickens, An Introduction to Nonlinear Oscillations, Cambridge University Press, Cambridge, 1981.

[4] N. Minorsky, Introduction to Nonlinear Mechanics, J. W. Edwards, Ann Arbor, MI, 1947.

[5] E. Fehlberg, Low-Order Classical Runge-Kutta Formulas With Step Size Control and Their Application to Some Heat Transfer Problems, Technical Report, NASA, 1969.

[6] S. Filippi, J. Gräf, New Runge–Kutta–Nyström formula-pairs of order 8 (7), 9 (8), 10 (9) and 11 (10) for differential equations of the form $y''=f(x, y)$, Journal of Computational and Applied Mathematics 14 (3) (1986) 361–370.

[7] H.M. Hilber, T.J. Hughes, R.L. Taylor, Improved numerical dissipation for time integration algorithms in structural dynamics, Earthquake Engineering & Structural Dynamics 5 (3) (1977) 283–292.

[8] N.M. Newmark, A method of computation for structural dynamics, Journal of the Engineering Mechanics Division 85 (3) (1959) 67–94.

[9] M. Dokainish, K. Subbaraj, A survey of direct time-integration methods in computational structural dynamics-I, explicit methods, Computers & Structures 32 (6) (1989) 1371–1386.

[10] K. Subbaraj, M. Dokainish, A survey of direct time-integration methods in computational structural dynamics-II, implicit methods, Computers & Structures 32 (6) (1989) 1387–1401.

[11] L.P. Liu, E.H. Dowell, The secondary bifurcation of an aeroelastic airfoil motion: effect of high harmonics, Nonlinear Dynamics 37 (1) (2004) 31–49.

[12] L.P. Liu, E.H. Dowell, Harmonic balance approach for an airfoil with a freeplay control surface, AIAA Journal 43 (4) (2005) 802–815.

[13] J.P. Thomas, K.C. Hall, E.H. Dowell, A harmonic balance approach for modeling nonlinear aeroelastic behavior of wings in transonic viscous flow, AIAA Paper 1924 (2003) 1–6.

[14] H.H. Dai, J.K. Paik, S.N. Atluri, The global nonlinear Galerkin method for the analysis of elastic large deflections of plates under combined loads: a scalar homotopy method for the direct solution of nonlinear algebraic equations, CMC: Computers Materials and Continua 23 (2011) 69–99.

[15] H.H. Dai, M. Schnoor, S.N. Alturi, A simple collocation scheme for obtaining the periodic solutions of the Duffing equation, and its equivalence to the high dimensional harmonic balance method: subharmonic oscillations, Computer Modeling in Engineering & Sciences 84 (5) (2012) 459–497.

[16] H.H. Dai, X.K. Yue, J.P. Yuan, S.N. Atluri, A time domain collocation method for studying the aeroelasticity of a two dimensional airfoil with a structural nonlinearity, Journal of Computational Physics 270 (2014) 214–237.

[17] T.A. Elgohary, L. Dong, J.L. Junkins, S.N. Atluri, A simple, fast, and accurate time integrator for strongly nonlinear dynamical systems, Computer Modeling in Engineering & Sciences 100 (3) (2014) 249–275.

[18] T.A. Elgohary, J.L. Junkins, S.N. Atluri, An RBF-collocation algorithm for orbit propagation, in: Advances in Astronautical Sciences: AAS/AIAA Space Flight Mechanics Meeting, 2015.

[19] S.N. Atluri, Methods of Computer Modeling in Engineering & the Sciences, Volume I, Tech Science Press, Forsyth, 2005.

[20] L. Dong, A. Alotaibi, S.A. Mohiuddine, S.N. Atluri, Computational methods in engineering: a variety of primal & mixed methods, with global & local interpolations, for well-posed or Ill-posed BCs, Computer Modeling in Engineering & Sciences 99 (1) (2014) 1–85.

[21] C.S. Liu, W. Yeih, C.L. Kuo, S.N. Atluri, A scalar homotopy method for solving an over/under determined system of non-linear algebraic equations, Computer Modeling in Engineering and Sciences 53 (1) (2009) 47–71.

[22] C.S. Liu, S.N. Atluri, A globally optimal iterative algorithm using the best decent vector $\dot{\mathbf{x}} = \lambda[\alpha_c\mathbf{F} + \mathbf{B}^T\mathbf{F}]$, with the critical value α_c, for solving a system of nonlinear algebraic equations $\mathbf{F(x)} = \mathbf{0}$, CMES: Computer Modeling in Engineering & Sciences 84 (6) (2012) 575–601.

[23] Y.K. Cheung, S.H. Chen, S.L. Lau, A modified Lindstedt-Poincare method for certain strongly nonlinear oscillators, International Journal of Non-Linear Mechanics 26 (¾) (1991) 367–378.

[24] J.H. He, Homotopy perturbation technique, Computer Methods in Applied Mechanics and Engineering 178 (1999) 257.

[25] J. He, Homotopy perturbation method for solving boundary value problems, Physics Letters A 350 (1–2) (2006) 87–88.

[26] J. He, Variational iteration method: a kind of non-linear analytical technique: some examples, International Journal of Non-linear Mechanics 34 (1999) 699–708.

[27] J.H. He, Variational iteration method for autonomous ordinary differential systems, Applied Mathematics and Computation 114 (2–3) (2000) 115–123.

[28] T. Fukushima, Picard iteration method, Chebyshev polynomial approximation, and global numerical integration of dynamical motions, The Astronomical Journal 113 (5) (1997) 1909–1914.
[29] G. Adomian, A review of the decomposition method in applied mathematics, Journal of Mathematical Analysis and Applications 135 (1988) 501–544.
[30] M.D. Buhmann, Radial Basis Functions: Theory and Implementations, volume 5, Cambridge University Press, Cambridge, 2003.
[31] Q. Chen, Y. Zhang, S. Liao, F. Wan, Newton-Kantorovich/pseudospectral solution to perturbed astrodynamic two-point boundary-value problems, Journal of Guidance, Control, and Dynamics 36 (2) (2013) 485–498.
[32] W. Chen, Z. Fu, C.S. Chen, Recent Advances on Radial Basis Function Collocation Methods, Springer, Berlin, 2013.
[33] M. Abramowitz, I. Stegun, Handbook of Mathematical Functions with Formulas, Graphs, and Mathematical Tables, US. Department of Commerce, 1968.
[34] N. Bellomo, B. Lods, R. Revelli, L. Ridolfi, Generalized Collocation Methods, Solutions to Nonlinear Problems, Birkhäuser, Boston, 2008.
[35] L. Dong, S.N. Atluri, A simple procedure to develop efficient & stable hybrid/mixed elements, and voronoi cell finite elements for macro-& micromechanics, CMC: Computers Materials & Continua 24 (1) (2011) 61–141.
[36] A.P. Zielinski, I. Herrera, Trefftz method: fitting boundary conditions, International Journal for Numerical Methods in Engineering 24 (1987) 871–891.
[37] O.C. Zienkiewicz, P.B. Morice, The Finite Element Method in Engineering Science, McGraw-Hill, London, 1971.
[38] S. Atluri, The Meshless Method (MLPG) for Domain & BIE Discretizations, Tech Science Press, 2004.
[39] H. Okada, H. Rajiyah, S.N. Atluri, A novel displacement gradient boundary element method for elastic stress analysis with high accuracy, Journal of Applied Mechanics 55 (1988) 786–794.
[40] S.N. Atluri, H.G. Kim, J.Y. Cho, A critical assessment of the truly meshless local Petrov-Galerkin (MLPG), and local boundary integral equation (LBIE) methods, Computational Mechanics 24 (1999) 348–372.
[41] S.N. Atluri, J. Sladek, V. Sladek, T. Zhu, The local boundary integral equation (LBIE) and it's meshless implementation for linear elasticity, Computational Mechanics 25 (2000) 180–198.
[42] S.N. Atluri, S.P. Shen, The Meshless Local Petrov-Galerkin (MLPG) Method, Tech Science Press, 2002.
[43] T. Zhang, Y. He, L. Dong, S. Li, A. Alotaibi, S.N. Atluri, Meshless local petrov-galerkin mixed collocation method for solving Cauchy inverse problems of steady-state heat transfer, CMES: Computer Modeling in Engineering & Sciences 97 (6) (2014) 509–553.
[44] S.W. Lee, T.H.H. Pian, Improvement of plate and shell finite elements by mixed formulations, AIAA Journal 16 (1978) 29–34.
[45] X. Bai, J.L. Junkins, Modified Chebyshev-Picard iteration methods for solution of boundary value problems, The Journal of Astronautical Sciences 58 (4) (2011) 615–642.
[46] R.M. Woollands, A.B. Younes, J.L. Junkins, New solutions for the perturbed lambert problem using regularization and Picard iteration, Journal of Guidance, Control, and Dynamics 38 (9) (2015) 1548–1562.
[47] C.S. Liu, S.N. Atluri, An iterative algorithm for solving a system of nonlinear algebraic equations, $\mathbf{F}(\mathbf{x}) = \mathbf{0}$, using the system of ODEs with an optimum α in $\dot{\mathbf{x}} = \lambda[\alpha\mathbf{F} + (1-\alpha)\mathbf{B}^T\mathbf{F}]$; $B_{ij} = \partial F_i/\partial x_j$, Computer Modeling in Engineering and Sciences 73 (4) (2011) 395–431.
[48] L. Liu, E.H. Dowell, K. Hall, A novel harmonic balance analysis for the van der pol oscillator, International Journal of Non-Linear Mechanics 42 (1) (2007) 2–12.
[49] T. Belytschko, J.I. Lin, T. Chen-Shyh, Explicit algorithms for the nonlinear dynamics of shells, Computer Methods in Applied Mechanics and Engineering 42 (2) (1984) 225–251.
[50] K. Fox, Numerical integration of the equations of motion of celestial mechanics, Celestial Mechanics 33 (2) (1984) 127–142.

[51] N.N. Bogoliubov, Y.A. Mitropolsky, Asymptotic Methods in the Theory of Non-linear Oscillations, Hindustan Publishing Corporation, India, 1961.
[52] J.H. He, Some asymptotic methods for strongly nonlinear equations, International Journal of Modern Physics B 20 (10) (2006) 1141–1199.
[53] A.H. Nayfeh, Introduction to Perturbation Methods, John Wiley, New York, 1981.
[54] M. Al-Sawalha, M.S.M. Noorani, I. Hashim, On accuracy of Adomian decomposition method for hyperchaotic Rössler systems, Chaos Solitons & Fractals 40 (2009) 1801–1807.
[55] S. Ghosh, A. Roy, D. Roy, An adaption of Adomian decomposition for numeric-analytic integration of strongly nonlinear and chaotic oscillators, Computer Methods in Applied Mechanics and Engineering 196 (2007) 1133–1153.
[56] A.M. Wazwaz, A new algorithm for calculating Adomian polynomials for nonlinear operators, Applied Mathematics and Computation 111 (2000) 53–69.

Chapter 2

Harmonic Balance Method and Time Domain Collocation Method

SUMMARY

As discussed in Chapter 1, a variety of spatial discretization methods can be used to reduce linear/nonlinear partial or differential equations in spatial coordinates to linear/nonlinear algebraic equations (L/NAEs). The earliest such methods are the finite-difference methods. More recent methods are based on the general concept of setting the weighted residual error in the differential equations in the spatial domain to zero. Such methods include, for example, the Galerkin method, the collocation method, the finite volume method, the primal Galerkin finite element method, the hybrid/mixed finite element methods, the boundary-element method, and Meshless Local Petrov Galerkin (MLGP) methods discovered by Atluri and coworkers since 1998 [1−4].

If the partial/ordinary differential equations in both space and time coordinates are spatially discretized by any of the methods mentioned above, one obtains semi-discrete linear/nonlinear coupled ordinary differential equations in time. Or, as in coupled nonlinear Duffing Oscillators, one may directly encounter coupled nonlinear ordinary differential equations (ODEs) in the time variables. These ODEs in time have to be solved for very long times, given some initial conditions at $t = 0$. Also, often times, these ODEs exhibit periodic solutions, and hence it may be sufficient to obtain the solution only in a time interval which corresponds to the period of the periodic solution.

In this chapter we take Duffing equation as example to illustrate that the subharmonic oscillations (when the period of the forcing function is nearly three-times the natural frequency of the linear system) can be captured using the time-collocation method over a period, assuming harmonic as well as subharmonic Fourier series as the trial functions. We show that the present simple notion of collocation of the error in the nonlinear ODE, with the assumed trial functions, is entirely equivalent to the so-called High Dimensional Harmonic Balance Method (HDHB) or the Time Domain Harmonic Balance Method, introduced earlier by Hall, Thomas, and Clark [5]; Thomas, Dowell, and Hall [6].

Closed form solutions to the Duffing equation

$$\ddot{x} + \xi\dot{x} + \alpha x + \beta x^3 = F\cos\omega t \tag{2.1}$$

(where ξ is the coefficient of damping, $\sqrt{\alpha}$ is the natural frequency of the linear system, ω is the frequency of the external force, β is the coefficient of the cubic nonlinearity, F

Computational Methods for Nonlinear Dynamical Systems
DOI: https://doi.org/10.1016/B978-0-323-99113-1.00002-9

is the magnitude of the external force, x is the amplitude of motion, t is the time, and $(\dot{})$ denotes a time-differentiation), are largely unknown in all but a few simple cases due to its nonlinear character. This relatively innocent looking differential equation, however, possesses a great variety of periodic solutions. The solution of Duffing's equation (2.1) has both periodic and transient solutions. However, most of the research is devoted to the periodic solutions. In practice, experimentalists often observe the motions to be periodic after the transients die out. In this chapter, we focus our attention on the periodic solutions.

Levenson [7] first pointed out that the Duffing equation with $\xi = 0$ may possess periodic solutions with frequency equal to $1/n$ of the frequency of the impressed force for any integer n. Moriguchi and Nakamura [8] verified this argument by numerical trials and found that for a sufficiently small ξ, subharmonic resonances of any fractional order exist. They vanish as ξ increases or β approaches zero. In this chapter, other than the harmonic oscillation, the $1/3$ subharmonic oscillation, whose fundamental frequency is one-third that of the applied force, when ω in Eq. (2.1) is in the vicinity of 3 times $\sqrt{\alpha}$, is investigated because the nonlinear characteristic of Eq. (2.1) is cubic. The perturbation methods (the Lindstedt-Poincare method, the averaging method, the Krylov-Bogoliubov-Mitropolsky (KBM) method and the multiple scale method [9]) were constructed to find approximated solution of weakly nonlinear dynamical systems. These methods, however, require the existence of a small parameter in the equation, which is not available for many cases. In this chapter, we consider a strong nonlinearity when β in Eq. (2.1) is larger than α.

The Galerkin method in the time-domain, applied within an appropriate period of the periodic solution (otherwise known as the harmonic balance method), can be used to tackle this type of problem. It presumes a Fourier series expansion for the desired periodic solution and then obtains the nonlinear algebraic equations of the coefficients by balancing each harmonic. The two harmonic approximation (i.e., a two-term approximation in the Fourier series in time) was used to investigate the property of the Duffing equation [10–13] and then extended to find a higher fidelity approximation for the periodic solutions [14,15]. However, this method is practically confined to a small number of harmonics, due to the need for a large number of symbolic operations.

This limitation is conquered by developing a high dimensional harmonic balance method (HDHB) [5,16], which has been successfully applied in aeroelastic problems, time delay problem, Duffing oscillator, Van der Pol's oscillator, etc. It was regarded as a variation of the harmonic balance method that can avoid many symbolic operations. In this chapter we show that the HDHB is not a kind of harmonic balance method; it is essentially a version of the simple collocation method presented in this chapter. In addition, the HDHB produced additional meaningless solutions [17], which made the HDHB method sometimes not practically useful. In using the collocation method, we provide appropriate initial values by a simple approach such that only physically meaningful solutions are calculated.

In Section 2.1 the simple point collocation method is presented and applied to find periodic solutions of harmonic and $1/3$ subharmonic oscillations. The nonlinear algebraic equations are obtained through the use of collocation in the time domain, within a period of oscillation. In Section 2.2 we explore the relationship between the collocation method and the HDHB and demonstrate that the HDHB approach is actually a transformed collocation method. Section 2.3 provides initial values to the NAEs solver, and an undamped system is analyzed by the proposed scheme. Meanwhile, the amplitude-frequency response relations for a damped Duffing equation with various values of damping, nonlinearity and force amplitude are explored.

2.1 Time collocation in a period of oscillation

In this section, we apply the collocation method in the time domain within a period of oscillation, for the periodic solutions of both harmonic and subharmonic oscillations, for the Duffing equation:

$$\ddot{x} + \xi\dot{x} + x + \beta x^3 = F\cos\omega t \tag{2.2}$$

The harmonic solution of Eq. (2.2) is sought in the form:

$$x(t) = A_0 + \sum_{n=1}^{N} A_n \cos n\omega t + B_n \sin n\omega t \tag{2.3}$$

The assumed form of $x(t)$ can be simplified by considering the symmetrical property of the nonlinear restoring force. First, Hayashi [13] pointed out that under circumstances when the nonlinearity is symmetric, i.e., $f(x)$ is odd in x, A_0 can be discarded. Second, it was demonstrated by Urabe [18] both numerically and theoretically that the even harmonic components in Eq. (2.3) are zero. The approximate solution is simplified to:

$$x(t) = \sum_{n=1}^{N} A_n \cos(2n-1)\omega t + B_n \sin(2n-1)\omega t \tag{2.4}$$

where N is the number of harmonics used in the desired approximation. $x(t)$ in Eq. (2.4) is called the N harmonic approximation (or labeled as the N-th order approximation in the present chapter) of the harmonic solution.

In using the collocation method in the time domain, within a period of oscillation, we obtain the residual-error function $R(t)$ by substituting the approximate solution, Eq. (2.4), into the following equation:

$$R(t) = \ddot{x} + \xi\dot{x} + x + \beta x^3 - F\cos\omega t \neq 0. \tag{2.5}$$

Upon enforcing $R(t)$ to be zero at $2N$ equidistant points t_j over the domain $[0, 2\pi/\omega]$, we obtain a system of $2N$ nonlinear algebraic equations:

$$R_j(A_1, A_2, \ldots, A_N; B_1, B_2, \ldots, B_N) := \ddot{x}(t_j) + \xi\dot{x}(t_j) + x(t_j) + \beta x^3(t_j) - F\cos\omega t_j = 0_j, \tag{2.6}$$

where

$$x(t_j) = \sum_{n=1}^{N} A_n \cos(2n-1)\omega t_j + B_n \sin(2n-1)\omega t_j, \tag{2.7a}$$

$$\dot{x}(t_j) = \sum_{n=1}^{N} -(2n-1)\omega A_n \sin(2n-1)\omega t_j + (2n-1)\omega B_n \cos(2n-1)\omega t_j, \tag{2.7b}$$

$$\ddot{x}(t_j) = \sum_{n=1}^{N} -(2n-1)^2\omega^2 A_n \cos(2n-1)\omega t_j - (2n-1)^2\omega^2 B_n \sin(2n-1)\omega t_j, \tag{2.7c}$$

where j is an index value ranging from 1 to $2N$. Eq. (2.6) is the collocation-resulting system of NAEs for the harmonic solution.

Finally, the coefficients in Eq. (2.6) can be solved by nonlinear algebraic equations (NAEs) solvers, e.g., the Newton-Raphson method and the Jacobian matrix inverse-free algorithms [19–22]. Herein the more familiar Newton-Raphson method is employed. We emphasize that the Jacobian matrix $\boldsymbol{B}$ of Eq. (2.6) can be readily derived upon differentiating R_j with respect to A_i and B_i.

$$\boldsymbol{B} = \left[\frac{\partial R_j}{\partial A_i}, \frac{\partial R_j}{\partial B_i}\right]_{2N \times 2N}, \tag{2.8}$$

where

$$\frac{\partial R_j}{\partial A_i} = -(2i-1)^2\omega^2 \cos(2i-1)\omega t_j - \xi(2i-1)\omega \sin(2i-1)\omega t_j$$
$$+ \cos(2i-1)\omega t + 3\beta x^2(t_j)\cos(2i-1)\omega t$$

$$\frac{\partial R_j}{\partial B_i} = -(2i-1)^2\omega^2 \sin(2i-1)\omega t_j + \xi(2i-1)\omega \cos(2i-1)\omega t_j$$
$$+ \sin(2i-1)\omega t + 3\beta x^2(t_j)\sin(2i-1)\omega t$$

In order to capture the subharmonic behavior, a different approximate solution must be defined. Similarly, the N-th order approximation of the 1/3 subharmonic solution can be assumed a as

$$x(t) = \sum_{n=1}^{N} a_n \cos\frac{1}{3}(2n-1)\omega t + b_n \sin\frac{1}{3}(2n-1)\omega t. \tag{2.9}$$

After collocation, the resulting NAEs are

$$R_j(a_1, a_2, \ldots, a_N; b_1, b_2, \ldots, b_N) := \ddot{x}(t_j) + \xi\dot{x}(t_j) + x(t_j) + \beta x^3(t_j) - F\cos\omega t_j = 0_j, \tag{2.10}$$

where $j = 1, \ldots, 2N$. Eq. (2.10) is the collocation-resulting system of NAEs for the 1/3 subharmonic solutions. A critical difference now to capture the subharmonic solutions, is that the collocation should be performed over $[0, 6\pi/\omega]$, since the 1/3 subharmonic solution has a period which is three times that of the harmonic solution. The collocation-resulting NAEs may then be solved as above.

2.2 Relationship between collocation and harmonic balance

In this section, we explore the relation between the present simple collocation method and the High Dimensional Harmonic Balance method (HDHB) to give a better understanding of the HDHB.

For comparison, we choose the same model [17] as follows:

$$m\ddot{x} + d\dot{x} + kx + \alpha x^3 = F \sin \omega t. \tag{2.11}$$

All the parameters in the above Duffing equation are kept in order to identify the source of the terms in the NAEs.

2.2.1 Harmonic balance method

Traditionally, to employ the standard harmonic balance method (HB), the solution of x is sought in the form of a truncated Fourier series expansion:

$$x(t) = x_0 + \sum_{n=1}^{N} [x_{2n-1} \cos n\omega t + x_{2n} \sin n\omega t], \tag{2.12}$$

where N is the number of harmonics used in the truncated Fourier series, and x_n, $n = 0, 1, \ldots, 2N$, are the unknown coefficients to be determined in the harmonic balance method. We differentiate $x(t)$ with respect to t, leading to

$$\dot{x}(t) = \sum_{n=1}^{N} [-n\omega x_{2n-1} \sin n\omega t + n\omega x_{2n} \cos n\omega t], \tag{2.13a}$$

$$\ddot{x}(t) = \sum_{n=1}^{N} \left[-(n\omega)^2 x_{2n-1} \cos n\omega t - (n\omega)^2 x_{2n} \sin n\omega t\right]. \tag{2.13b}$$

Considering the cubic nonlinearity in Eq. (2.11), the nonlinear term can be expressed in terms of the truncated Fourier series with $3N$ harmonics:

$$x^3(t) = r_0 + \sum_{n=1}^{3N} [r_{2n-1} \cos n\omega t + r_{2n} \sin n\omega t]. \tag{2.14}$$

where $r_0, r_1, ..., r_{6N}$ are obtained by the following formulas:

$$r_0 = \frac{1}{2\pi}\int_0^{2\pi}\left\{x_0 + \sum_{k=1}^{N}[x_{2k-1}\cos k\theta + x_{2k}\sin k\theta]\right\}^3 d\theta, \tag{2.15a}$$

$$r_{2n-1} = \frac{1}{\pi}\int_0^{2\pi}\left\{x_0 + \sum_{k=1}^{N}[x_{2k-1}\cos k\theta + x_{2k}\sin k\theta]\right\}^3 \cos n\theta d\theta, \tag{2.15b}$$

$$r_{2n} = \frac{1}{\pi}\int_0^{2\pi}\left\{x_0 + \sum_{k=1}^{N}[x_{2k-1}\cos k\theta + x_{2k}\sin k\theta]\right\}^3 \sin n\theta d\theta. \tag{2.15c}$$

where $n = 1, 2, ..., 3N$, $\theta = \omega t$, and k is a dummy index.

In the harmonic balance method, one should balance the harmonics 1, $\cos n\omega t$, and $\sin n\omega t$, $n = 1, 2, ..., N$, to obtain the simultaneous $2N + 1$ nonlinear algebraic equations. All the higher order harmonics $[n \geq N + 1]$ in the nonlinear term are omitted. Thus only the first N harmonics are retained, that is

$$x_{HB}^3(t) = r_0 + \sum_{k=1}^{N}[r_{2n-1}\cos n\omega t + r_{2n}\sin n\omega t]. \tag{2.16}$$

Therefore, only $r_0, r_1, ..., r_{2N}$ are needed in the harmonic balance method.

Next, substituting Eqs. (2.12)–(2.13b) and (2.16) into Eq. (2.11) and collecting the terms associated with each harmonic 1, $\cos n\theta$, and $\sin n\theta$, $n = 1, ..., N$, we finally obtain a system of NAEs in a vector form:

$$\left(m\omega^2\boldsymbol{A}^2 + d\omega\boldsymbol{A} + k\boldsymbol{I}\right)\boldsymbol{Q}_x + \alpha\boldsymbol{R}_x = F\boldsymbol{H}, \tag{2.17}$$

where $\boldsymbol{I}$ is a $2N + 1$ dimension identity matrix, and

$$\boldsymbol{Q}_x = \begin{bmatrix} x_0 \\ x_1 \\ \vdots \\ x_{2N} \end{bmatrix}, \boldsymbol{R}_x = \begin{bmatrix} r_0 \\ r_1 \\ \vdots \\ r_{2N} \end{bmatrix}, \boldsymbol{H} = \begin{bmatrix} 0 \\ 0 \\ 1 \\ 0 \\ \vdots \\ 0 \end{bmatrix},$$

$$A = \begin{bmatrix} 0 & 0 & 0 & \cdots & 0 \\ 0 & \boldsymbol{J}_1 & 0 & \cdots & 0 \\ 0 & 0 & \boldsymbol{J}_2 & \cdots & 0 \\ \vdots & \vdots & \vdots & \ddots & \vdots \\ 0 & 0 & 0 & \cdots & \boldsymbol{J}_N \end{bmatrix}, \boldsymbol{J}_n = n\begin{bmatrix} 0 & \omega \\ -\omega & 0 \end{bmatrix}.$$

One should note that r_n, $n = 0, 1, \ldots, 2N$, are analytically expressed in terms of the coefficients x_n, $n = 0, 1, \ldots, 2N$, which makes the HB algebraically expensive for application. If many harmonics are used or the nonlinearity is more complicated than the cubic nonlinearity, are considered, the expressions for the nonlinear terms, Eqs. (2.15) become much more complicated.

2.2.2 High dimensional harmonic balance method

In order to eliminate needs for analytical expressions arising from the nonlinear term of the standard harmonic balance method, Thomas, Dowell, and Hall [16]; Hall, Thomas, and Clark [5] developed the high dimensional harmonic balance method (HDHB). The key aspect is that instead of working in terms of Fourier coefficient variables x_n as in the HB method, the coefficient variables are instead recast in the time domain and stored at $2N + 1$ equally spaced sub-time levels $x(t_i)$ over a period of one cycle of motion. The objective of the HDHB is to express the $\mathbf{Q}_x$, $\boldsymbol{R}_x$ in x_n [See Eq. (2.17)] by $\tilde{\mathbf{Q}}_x$, $\tilde{\boldsymbol{R}}_x$ in $x(t_n)$.

In the HDHB, the $2N + 1$ harmonic balance Fourier coefficient solution variables are related to the time domain solution at $2N + 1$ equally spaced sub-time levels over a period of oscillation via a constant Fourier transformation matrix. That is

$$\mathbf{Q}_x = \boldsymbol{E}\tilde{\mathbf{Q}}_x \tag{2.18}$$

where

$$\tilde{\mathbf{Q}}_x = \begin{bmatrix} x(t_0) \\ x(t_1) \\ x(t_2) \\ \vdots \\ x(t_{2N}) \end{bmatrix}, \quad \mathbf{Q}_x = \begin{bmatrix} x_0 \\ x_1 \\ x_2 \\ \vdots \\ x_{2N} \end{bmatrix}, \tag{2.19}$$

with $t_i = 2\pi i/(2N + 1)\omega, (i = 0, 1, 2, \ldots, 2N)$, and the transform matrix is

$$\boldsymbol{E} = \frac{2}{2N+1}\begin{bmatrix} \frac{1}{2} & \frac{1}{2} & \cdots & \frac{1}{2} \\ \cos\theta_0 & \cos\theta_1 & \cdots & \cos\theta_{2N} \\ \sin\theta_0 & \sin\theta_1 & \cdots & \sin\theta_{2N} \\ \cos 2\theta_0 & \cos 2\theta_1 & \cdots & \cos 2\theta_{2N} \\ \sin 2\theta_0 & \sin 2\theta_1 & \cdots & \sin 2\theta_{2N} \\ \vdots & \vdots & \ddots & \vdots \\ \cos N\theta_0 & \cos N\theta_1 & \cdots & \cos N\theta_{2N} \\ \sin N\theta_0 & \sin N\theta_1 & \cdots & \sin N\theta_{2N} \end{bmatrix}, \tag{2.20}$$

where $\theta_i = \omega t_i = 2\pi i/2N + 1, (i = 0, 1, 2, \ldots, 2N)$. One should note that θ_i is the corresponding phase point of t_i.

Furthermore, the time domain solutions at the $2N+1$ equally spaced sub-time levels can be expressed in terms of the harmonic balance Fourier coefficient solution using the inverse of the Fourier transformation matrix, i.e.

$$\tilde{\mathbf{Q}}_x = \boldsymbol{E}^{-1}\mathbf{Q}_x, \tag{2.21}$$

where

$$\boldsymbol{E}^{-1} = \begin{bmatrix} 1 & \cos\theta_0 & \sin\theta_0 & \cdots & \cos N\theta_0 & \sin N\theta_0 \\ 1 & \cos\theta_1 & \sin\theta_1 & \cdots & \cos N\theta_1 & \sin N\theta_1 \\ \vdots & \vdots & \vdots & \ddots & \vdots & \vdots \\ 1 & \cos\theta_{2N} & \sin\theta_{2N} & \cdots & \cos N\theta_{2N} & \sin N\theta_{2N} \end{bmatrix}. \tag{2.22}$$

Similarly, $\boldsymbol{H} = \boldsymbol{E}\tilde{\boldsymbol{H}}$, where

$$\tilde{\boldsymbol{H}} = \begin{bmatrix} \sin\theta_0 \\ \sin\theta_1 \\ \vdots \\ \sin\theta_{2N} \end{bmatrix}. \tag{2.23}$$

So far, $\mathbf{Q}_x$ and $\boldsymbol{H}$ have been transformed by the transformation matrix. Now, we turn to process the nonlinear term $\boldsymbol{R}_x$. We define the $\tilde{\boldsymbol{R}}_x$ as

$$\tilde{\boldsymbol{R}}_x = \begin{bmatrix} x^3(t_0) \\ x^3(t_1) \\ \vdots \\ x^3(t_{2N}) \end{bmatrix}. \tag{2.24}$$

In the studies by Thomas, Dowell, and Hall [6] Hall, Thomas, and Clark [5]; Liu, Dowell, Thomas, Attar, and Hall [17], they use the relation $\boldsymbol{R}_x = \boldsymbol{E}\tilde{\boldsymbol{R}}_x$ without further discussion. However, this relation is not strictly true, as seen below.

We consider the relation between $\boldsymbol{E}^{-1}\boldsymbol{R}_x$ and $\tilde{\boldsymbol{R}}_x$ instead.

$$\begin{aligned} \boldsymbol{E}^{-1}\boldsymbol{R}_x &= \begin{bmatrix} 1 & \cos\theta_0 & \sin\theta_0 & \cdots & \cos N\theta_0 & \sin N\theta_0 \\ 1 & \cos\theta_1 & \sin\theta_1 & \cdots & \cos N\theta_1 & \sin N\theta_1 \\ \vdots & \vdots & \vdots & \ddots & \vdots & \vdots \\ 1 & \cos\theta_{2N} & \sin\theta_{2N} & \cdots & \cos N\theta_{2N} & \sin N\theta_{2N} \end{bmatrix} \begin{bmatrix} r_0 \\ r_1 \\ \vdots \\ r_{2N} \end{bmatrix} \\ &= \begin{bmatrix} r_0 + \sum_{n=1}^{N} [r_{2n-1}\cos n\theta_0 + r_{2n}\sin n\theta_0] \\ r_0 + \sum_{n=1}^{N} [r_{2n-1}\cos n\theta_1 + r_{2n}\sin n\theta_1] \\ \vdots \\ r_0 + \sum_{n=1}^{N} [r_{2n-1}\cos n\theta_{2N} + r_{2n}\sin n\theta_{2N}] \end{bmatrix} \\ &= \begin{bmatrix} x_{HB}^3(t_0) \\ x_{HB}^3(t_1) \\ \vdots \\ x_{HB}^3(t_{2N}) \end{bmatrix}, \end{aligned}$$

Considering Eq. (2.14),

$$\tilde{\boldsymbol{R}}_x = \begin{bmatrix} x^3(t_0) \\ x^3(t_1) \\ \vdots \\ x^3(t_{2N}) \end{bmatrix} = \begin{bmatrix} r_0 + \sum_{n=1}^{3N}[r_{2n-1}\cos n\theta_0 + r_{2n}\sin n\theta_0] \\ r_0 + \sum_{n=1}^{3N}[r_{2n-1}\cos n\theta_1 + r_{2n}\sin n\theta_1] \\ \vdots \\ r_0 + \sum_{n=1}^{3N}[r_{2n-1}\cos n\theta_{2N} + r_{2n}\sin n\theta_{2N}] \end{bmatrix}.$$

It is clear that $\boldsymbol{E}^{-1}\boldsymbol{R}_x$ and $\tilde{\boldsymbol{R}}_x$ are not equal.

Once the approximate relation: $\boldsymbol{E}^{-1}\boldsymbol{R}_x = \tilde{\boldsymbol{R}}_x$ is applied, using $\boldsymbol{Q}_x = \boldsymbol{E}\tilde{\boldsymbol{Q}}_x$, $\boldsymbol{H}_x = \boldsymbol{E}\tilde{\boldsymbol{H}}_x$, Eq. (2.17) is then rewritten as

$$\left(m\omega^2\boldsymbol{A}^2 + d\omega\boldsymbol{A} + k\boldsymbol{I}\right)\boldsymbol{E}\tilde{\boldsymbol{Q}}_x + \alpha\boldsymbol{E}\tilde{\boldsymbol{R}}_x = F\boldsymbol{E}\tilde{\boldsymbol{H}}. \tag{2.25}$$

It is seen that by using the approximation $\boldsymbol{R}_x = \boldsymbol{E}^{-1}\tilde{\boldsymbol{R}}_x$ in Eq. (2.25), the HDHB absorbs the higher harmonics in the nonlinear term $\tilde{\boldsymbol{R}}_x$. This may be one source of non-physical solutions generated by the HDHB method.

Multiplying both sides of the above equation by $\boldsymbol{E}^{-1}$ yields:

$$\left(m\omega^2\boldsymbol{D}^2 + d\omega\boldsymbol{D} + k\boldsymbol{I}\right)\tilde{\boldsymbol{Q}}_x + \alpha\tilde{\boldsymbol{R}}_x = F\tilde{\boldsymbol{H}}, \tag{2.26}$$

where $\boldsymbol{D} = \boldsymbol{E}^{-1}\boldsymbol{A}\boldsymbol{E}$. The Eq. (2.26) is referred to as the HDHB solution system.

We emphasize that the HDHB is distinct from the harmonic balance method only in the nonlinear term, where the HDHB includes higher order harmonic terms $(n = N + 1, \ldots, 3N)$.

In this section, the HDHB is derived baesd on an approximation from the standard harmonic balance method. The HDHB and the harmonic balance method are not equivalent. Interestingly, the HDHB can be derived strictly from the point collocation method presented in Section 2.1.

2.2.3 Equivalence between HDHB and collocation

Herein, we derive the HDHB from the collocation method to demonstrate their equivalence. In section 2.1, the Duffing equation and the trial function used are not uniform to those in this section. Thus, we need to reformulate the collocation method herein. Using the approximate solution, Eq. (2.12), we first write the residual-error function of the Eq. (2.11) as

$$R(t) = m\ddot{x} + d\dot{x} + kx + \alpha x^3 - F\sin\omega t \neq 0. \tag{2.27}$$

Upon enforcing $R(t)$ to be zero at $2N+1$ equidistant points t_i over the domain $[0, 2\pi/\omega]$, we obtain a system of $2N+1$ nonlinear algebraic equations:

$$R_i(x_0, x_1, ..., x_{2N}) := m\ddot{x}(t_i) + d\dot{x}(t_i) + kx(t_i) + \alpha x^3(t_i) - F\sin\omega t_i = 0_i. \tag{2.28}$$

Later on, we explain the time domain transformation or the Fourier transformation in the view of collocation. Now, we consider each term in the above equation separately.

For comparison, the trial solution of the collocation method is the same as in Eq. (2.12). Collocating $x(t)$ in Eq. (2.12) at points t_i, we have

$$x(t_i) = x_0 + \sum_{n=1}^{N} [x_{2n-1}\cos n\omega t_i + x_{2n}\sin n\omega t_i]. \tag{2.29}$$

Considering $\theta_i = \omega t_i$, Eq. (2.29) can be rewritten in a matrix form

$$\begin{bmatrix} x(t_0) \\ x(t_1) \\ \vdots \\ x(t_{2N}) \end{bmatrix} = \begin{bmatrix} 1 & \cos\theta_0 & \sin\theta_0 & \cdots & \cos N\theta_0 & \sin N\theta_0 \\ 1 & \cos\theta_1 & \sin\theta_1 & \cdots & \cos N\theta_1 & \sin N\theta_1 \\ \vdots & \vdots & \vdots & \ddots & \vdots & \vdots \\ 1 & \cos\theta_{2N} & \sin\theta_{2N} & \cdots & \cos N\theta_{2N} & \sin N\theta_{2N} \end{bmatrix} \begin{bmatrix} x_0 \\ x_1 \\ \vdots \\ x_{2N} \end{bmatrix} \tag{2.30}$$

Therefore

$$\tilde{\mathbf{Q}}_x = \begin{bmatrix} x(t_0) \\ x(t_1) \\ \vdots \\ x(t_{2N}) \end{bmatrix} = \boldsymbol{E}^{-1}\mathbf{Q}_x. \tag{2.31}$$

In comparison with Eq. (2.21), we see that the Fourier transformation matrix $\boldsymbol{E}$ can be interpreted as the collocation-resulting matrix in Eq. (2.30).

Similarly, collocating $\dot{x}(t)$ at $2N+1$ equidistant time points t_i, we have

$$\dot{x}(t_i) = \sum_{n=1}^{N} [-n\omega x_{2n-1}\sin n\omega t_i + n\omega x_{2n}\cos n\omega t_i]. \tag{2.32}$$

The above equation can be written in a matrix form

$$\begin{bmatrix} \dot{x}(t_0) \\ \dot{x}(t_1) \\ \vdots \\ \dot{x}(t_{2N}) \end{bmatrix} = \omega \begin{bmatrix} 0 & -\sin\theta_0 & \cos\theta_0 & \cdots & -N\sin N\theta_0 & N\cos N\theta_0 \\ 0 & -\sin\theta_1 & \cos\theta_1 & \cdots & -N\sin N\theta_1 & N\cos N\theta_1 \\ \vdots & \vdots & \vdots & \ddots & \vdots & \vdots \\ 0 & -\sin\theta_{2N} & \cos\theta_{2N} & \cdots & -N\sin N\theta_{2N} & N\cos N\theta_{2N} \end{bmatrix} \begin{bmatrix} x_0 \\ x_1 \\ \vdots \\ x_{2N} \end{bmatrix}. \tag{2.33}$$

We observe that the square matrix in the above equation can be expressed by two existing matrices:

$$\begin{bmatrix} 0 & -\sin\theta_0 & \cos\theta_0 & \cdots & -N\sin N\theta_0 & N\cos N\theta_0 \\ 0 & -\sin\theta_1 & \cos\theta_1 & \cdots & -N\sin N\theta_1 & N\cos N\theta_1 \\ \vdots & \vdots & \vdots & \ddots & \vdots & \vdots \\ 0 & -\sin\theta_{2N} & \cos\theta_{2N} & \cdots & -N\sin N\theta_{2N} & N\cos N\theta_{2N} \end{bmatrix}$$
$$= \begin{bmatrix} 1 & \cos\theta_0 & \sin\theta_0 & \cdots & \cos N\theta_0 & \sin N\theta_0 \\ 1 & \cos\theta_1 & \sin\theta_1 & \cdots & \cos N\theta_1 & \sin N\theta_1 \\ \vdots & \vdots & \vdots & \ddots & \vdots & \vdots \\ 1 & \cos\theta_{2N} & \sin\theta_{2N} & \cdots & \cos N\theta_{2N} & \sin N\theta_{2N} \end{bmatrix} \begin{bmatrix} 0 & 0 & 0 & \cdots & 0 \\ 0 & \boldsymbol{J}_1 & 0 & \cdots & 0 \\ 0 & 0 & \boldsymbol{J}_2 & \cdots & 0 \\ \vdots & \vdots & \vdots & \ddots & \vdots \\ 0 & 0 & 0 & \cdots & \boldsymbol{J}_N \end{bmatrix}$$
$$= \boldsymbol{E}^{-1}\boldsymbol{A}.$$

Thus, we have

$$\begin{bmatrix} \dot{x}(t_0) \\ \dot{x}(t_1) \\ \vdots \\ \dot{x}(t_{2N}) \end{bmatrix} = \omega \boldsymbol{E}^{-1}\boldsymbol{A}\boldsymbol{Q}_x. \tag{2.34}$$

In the same manner, collocating $\ddot{x}(t)$ at $2N+1$ equidistant time points t_i, we have

$$\ddot{x}(t_i) = \sum_{n=1}^{N} \left[-n^2\omega^2 x_{2n-1}\cos n\omega t_i - n^2\omega^2 x_{2n}\sin n\omega t_i \right]. \tag{2.35}$$

Eq. (2.35) is written in a matrix form

$$\begin{bmatrix} \ddot{x}(t_0) \\ \ddot{x}(t_1) \\ \vdots \\ \ddot{x}(t_{2N}) \end{bmatrix} = \omega^2 \begin{bmatrix} 0 & -\cos\theta_0 & -\sin\theta_0 & \cdots & -N^2\cos N\theta_0 & -N^2\sin N\theta_0 \\ 0 & -\cos\theta_1 & -\sin\theta_1 & \cdots & -N^2\cos N\theta_1 & -N^2\sin N\theta_1 \\ \vdots & \vdots & \vdots & \ddots & \vdots & \vdots \\ 0 & -\cos\theta_{2N} & -\sin\theta_{2N} & \cdots & -N^2\cos N\theta_{2N} & -N^2\sin N\theta_{2N} \end{bmatrix} \begin{bmatrix} x_0 \\ x_1 \\ \vdots \\ x_{2N} \end{bmatrix}. \tag{2.36}$$

Note that the square matrix in the above equation is equal to $\boldsymbol{E}^{-1}\boldsymbol{A}^2$. Therefore,

$$\begin{bmatrix} \ddot{x}(t_0) \\ \ddot{x}(t_1) \\ \vdots \\ \ddot{x}(t_{2N}) \end{bmatrix} = \omega^2 \boldsymbol{E}^{-1}\boldsymbol{A}^2\boldsymbol{Q}_x. \tag{2.37}$$

Now, substituting Eqs. (2.31,2.34,2.37) and (2.24) into the collocation-resulting algebraic Eq. (2.28), we obtain

$$\boldsymbol{E}^{-1}\left(m\omega^2\boldsymbol{A}^2 + d\omega\boldsymbol{A} + k\boldsymbol{I}\right)\mathbf{Q}_x + \alpha\tilde{\boldsymbol{R}}_x = F\tilde{\boldsymbol{H}}. \tag{2.38}$$

By using Eq. (2.31), i.e. $\mathbf{Q}_x = \boldsymbol{E}\tilde{\mathbf{Q}}_x$, the above equation can be written as

$$\left(m\omega^2\boldsymbol{D}^2 + d\omega\boldsymbol{D} + k\boldsymbol{I}\right)\tilde{\mathbf{Q}}_x + \alpha\tilde{\boldsymbol{R}}_x = F\tilde{\boldsymbol{H}}. \tag{2.39}$$

Eq. (2.39) is the transformed collocation system. No approximation is adopted during the derivation. We see that Eq. (2.39) is the same as Eq. (2.26). Therefore, we have demonstrated the equivalence of the collocation method and the high dimensional harmonic balance method (HDHB). We come to the conclusion that the HDHB approach is no more than a cumbersome version of the presently proposed simple collocation method.

In summary: (a) The collocation method is simpler. It does not call for the Fourier transformation and works in terms of Fourier coefficient variables. Section 2.1 shows that the collocation algebraic system and its Jacobian matrix can be obtained easily without intense symbolic operation. (b) The HDHB is a transformed collocation method. It can be derived from the collocation method rigorously.

The reason for the occurrence of the non-physical solution by HDHB can be understood by treating it as a collocation method. In the previous studies, they were not aware of the fact that the HDHB is essentially a collocation method. Thus, they mostly compare the HDHB1 or HDHB2 with HB1 and HB2. As is known that the harmonic balance method (time Galerkin method) works relatively well with few harmonics. As the number N of the harmonics is increased in the trial solution, Eq. (2.12), it may not be sufficient to collocate the residual-error, Eq. (2.27), only at $2N+1$ points in a period [2]. One may have to use M collocation points, $M > 2N+1$, to obtain a reasonable solution. As $M \to \infty$, one may develop a method of least-squared error, wherein one seeks to minimize $\int_0^T R^2(t)dt$ [T is the period of the periodic solution] with respect to the coefficients x_n, $(n = 0, 1, ..., 2N)$ of Eq. (2.12). This will be pursued in a future study.

2.3 Initialization of Newton-Raphson method

In Section 2.1, the collocation method has been formulated. The algebraic systems arising from the harmonic oscillation and 1/3 subharmonic oscillation are given in Eq. (2.6) and (2.10), respectively. In order to solve the resulting NAEs, one has to give initial values for the Newton iterative process to start. It is known that the system has multiple solutions, viz, multiple steady states. Hence it is expected to provide the deterministic initial values to direct the solutions to the system of NAEs to the desired solution. In this section, we provide the initial values for the higher harmonic approximation. The initial values for undamped and damped systems are considered separately.

2.3.1 Initial values for undamped system

In this subsection, we consider the undamped system:

$$\ddot{x} + x + \beta x^3 = F\cos\omega t. \tag{2.40}$$

2.3.1.1 Initial values for the iterative solution of NAEs for capturing the 1/3 subharmonic solution of the undamped system

In the case of undamped system, the trial function in Eq. (2.9) can be simplified further. All the sine terms turn out to be zero in the course of the calculation. This is because the damping is absent. Further rigorous demonstrations can be found in Stoker [10], Urabe [18–23], and Urabe and Reiter [24]. For brevity, we therefore omit the sines at the onset and seek the subharmonic solution in the following form

$$x(t) = \sum_{n=1}^{N} a_n \cos\frac{1}{3}(2n-1)\omega t. \tag{2.41}$$

To find the starting values for the Newton iterative process, we simply consider the approximation with $N = 2$:

$$x(t) = a_1^{(2)}\cos\frac{1}{3}\omega t + a_2^{(2)}\cos\omega t. \tag{2.42}$$

The superscript (2) is introduced, on the other hand, to distinguish from the coefficients a_1, a_2 in the N-th order approximation in Eq. (2.41), and on the other hand to denote the order of harmonic approximation. For brevity, however, we omit the superscript unless needed.

Substitution of Eq. (2.42) in Eq. (2.40) and equating the coefficients of $\cos 1/3\omega t$ and $\cos\omega t$, leads to two simultaneous nonlinear algebraic equations

$$a_1\left[36 - 4\omega^2 + 27\left(a_1^2 + a_1 a_2 + 2a_2^2\right)\beta\right] = 0 \tag{2.43a}$$

$$\left(a_1^3 + 6a_1^2 a_2 + 3a_2^3\right)\beta + 4a_2\left(1 - \omega^2\right) - 4F = 0 \tag{2.43b}$$

From the Eq. (2.43a) we have two possibilities:

$$a_1 = 0 \text{ or } 36 - 4\omega^2 + 27\left(a_1^2 + a_1 a_2 + 2a_2^2\right)\beta = 0.$$

Each possibility leads Eqs. (2.43) to a different system:

$$a_1 = 0 \tag{2.44a}$$

$$3a_2^3\beta + 4a_2\left(1 - \omega^2\right) = 4F \tag{2.44b}$$

and

$$\omega^2 = 9 + \frac{27}{4}\left(a_1^2 + a_1 a_2 + 2a_2^2\right)\beta \tag{2.45a}$$

$$\left(a_1^3 + 6a_1^2 a_2 + 3a_2^3\right)\beta + 4a_2\left(1 - \omega^2\right) = 4F \tag{2.45b}$$

We can see that a_2 in Eq. (2.44b) actually reduces to $A_1^{(1)}$ [Eq. (2.4)], since 1/3 subharmonic component is zero. Similar to the definition of lower-case letters, the capital A_1 coefficient is in reference to Eq. (2.4) where the subharmonics are not yet included in the trial function. The superscript (1) denotes the order of approximation.

For a hard spring system, i.e. $\beta > 0$, it can be immediately seen from Eq. (2.45a) that the frequency ω of the impressed force must be greater than 3 to ensure the existence of real roots for Eqs. (2.45) Here 3 refers to three times the natural frequency of the linear system.

To initialize the Newton iterative process, we compute the second order approximation as the initial values of the N-th order approximation.

As stated above we solve Eqs. (2.45) to obtain coefficients of the second order subharmonic approximation. We set the initial values of the coefficients of the N-th order approximation in Eq. (2.41) as

$$a_1 = a_1^{(2)}, a_2 = a_2^{(2)}, a_3 = a_4 = \cdots = a_N = 0.$$

Starting from the initial values, we can solve the NAEs resulting from the application of collocation in the time domain within an appropriate period of the periodic solution, similar to Eq. (2.10), by the Newton-Raphson method.

It should be noted that there might be multiple sets of solutions for Eqs. (2.45) at a certain frequency. Each set of initial values, viz, the coefficients of the low order approximation, may direct the NAEs to its corresponding high order approximation as will be verified later.

2.3.1.2 Initial values for the iterative solution of NAEs for capturing the harmonic solution of the undamped system

Similar to the N-th order approximation of the 1/3 subharmonic solution in Eq. (2.41), the N-th order approximation of the harmonic solution can be sought in the form

$$x(t) = \sum_{n=1}^{N} A_n \cos(2n-1)\omega t. \tag{2.46}$$

In Subsection 2.3.1.1, we have obtained Eqs. (2.44) which are the NAEs for the second order 1/3 subharmonic solution. Since a_1 is 0, a_2 in Eq. (2.44b) actually turns out to be $A_1^{(1)}$. Therefore, the N-th order approximation can start by letting

$$A_1 = A_1^{(1)}, A_2 = A_3 = \cdots = A_N = 0. \tag{2.47}$$

The first order harmonic approximation is verified reasonably accurately in the example in Section 2.4. Once the initial values are obtained, we can solve the system of NAEs by Newton-Raphson method.

2.3.2 Initial values for damped system

In Subsection 2.3.1, we provided initial values for the harmonic and subharmonic solutions of an undamped system. The solution is developed in Fourier series as Eq. (2.9) for 1/3 subharmonic solution or Eq. (2.4) for harmonic solution. The N harmonic, i.e. N-th order, approximations of Eqs. (2.9) and (2.4) are supposed to be close to Eqs. (2.41) and (2.46) therefore, the initial values can be supplied by the low harmonic approximation of the undamped Duffing equation [18].

However, this is not applicable to the system with a relatively large damping. On the one hand, one may ask how small should the damping be so as to be safe to use the undamped initial values. On the other hand, Urabe's scheme fails to provide reasonable initial values for a strongly damped system. In our scheme, we seek the initial values by solving the lowest two harmonic approximation.

2.3.2.1 Initial values for the iterative solution of NAEs for capturing the 1/3 subharmonic solution of the damped system

We assume the second order 1/3 subharmonic solution as follows

$$x(t) = a_1 \cos\frac{1}{3}\omega t + b_1 \sin\frac{1}{3}\omega t + a_2 \cos\omega t + b_2 \sin\omega t. \tag{2.48}$$

Substitution of Eq. (2.48) into the Duffing equation as well as collecting the coefficients of $\cos 1/3\omega t$, $\sin 1/3\omega t$, $\cos\omega t$, and $\sin\omega t$, leads to a system of four simultaneous NAEs, which is given in Appendix A.

Hence, this system of simultaneous NAEs determines the coefficients of the second order 1/3 subharmonic approximation. For any given problem (ξ, β, F, and ω specified), no matter how strong the damping is, we can calculate the initial values by solving Eq. (2.53) in Appendix A. Multiple sets of solutions can be obtained easily by Mathematica. In a physical view, the multiple solutions correspond to various steady state motions.

Therefore, the N-th order approximation can start with

$$\begin{aligned} a_1 &= a_1^{(2)}, b_1 = b_1^{(2)}, \\ a_2 &= a_2^{(2)}, b_2 = b_2^{(2)}, \\ a_3 &= a_4 = \cdots = a_N = 0, \\ b_3 &= b_4 = \cdots = b_N = 0. \end{aligned}$$

Consequently, the system of $2N$ nonlinear algebraic equations in Eq. (2.10) is solved for the 1/3 subharmonic solution.

2.3.2.2 Initial values for the iterative solution of NAEs for capturing the harmonic solution of the damped system

Similarly, the second order approximation for the harmonic oscillation is

$$x(t) = A_1 \cos \omega t + B_1 \sin \omega t + A_2 \cos 3\omega t + B_2 \sin 3\omega t. \tag{2.49}$$

Substitution of Eq. (2.49) into the Duffing equation (2.6) and then collecting coefficients of $\cos\omega t$, $\sin \omega t$, $\cos 3\omega t$ and $\sin 3\omega t$, leads to a system of NAEs in Appendix B.

This system of NAEs determines the coefficients of the second order approximation. Hence, the N-th order approximation can start with

$$\begin{aligned} A_1 &= A_1^{(2)}, B_1 = B_1^{(2)}, \\ A_2 &= A_2^{(2)}, B_2 = B_2^{(2)}, \\ A_3 &= A_4 = \cdots = A_N = 0, \\ B_3 &= B_4 = \cdots = B_N = 0. \end{aligned}$$

Consequently, the collocation-resulting NAEs can be solved. Then, we can obtain the N-th order harmonic solution after inserting the determined coefficients into trial function in Eq. (2.4).

2.4 Numerical examples

2.4.1 Undamped Duffing equation

We apply the proposed method of collocation in the time domain, within a period of oscillation, to solve the undamped Duffing equation. Ludeke and Cornett [25] studied the undamped Duffing equation having the form

$$\frac{d^2x}{d\tau^2} + 2x + 2x^3 = 10 \cos\Omega\tau \tag{2.50}$$

with an analog computer. We solve this problem by the present scheme. Firstly, making a transformation:

$$\tau = \frac{t}{\sqrt{2}}, \Omega = \sqrt{2}\omega,$$

we have

$$\ddot{x} + x + x^3 = 5\cos\omega t, \tag{2.51}$$

where $\dot{x}$ denotes $\mathrm{d}x/\mathrm{d}t$. For the harmonic solution, we solve Eq. (2.44) to obtain the first order approximation as the initial values for a specified ω_g. For the subharmonic solution, Eqs. (2.45) is solved for the second order approximation.

We can sweep ω, starting from ω_g, back and forth to find the frequency response curve of the considered problem. Throughout this chapter, the solution of the previous frequency is used as the initial values of its immediate subsequent frequency. Thus, the specified ω_g is named the generating frequency. It is not hard to choose a proper ω_g. We will illustrate this in the examples.

For the undamped case, we can plot the graphs of a_1 vs ω, a_2 vs ω, and $A_1^{(1)}$ vs ω, which provide the information of the onset of the subharmonic oscillation, the bifurcation point and the pure subharmonic frequency. Fig. 2.1 shows the general pattern of the response curves. The solid and dashed curves in Fig. 2.1(A) and Fig. 2.1(B) indicate which branch must be considered simultaneously. They also indicate the onset of the occurrence of the 1/3 subharmonic response. $a_1 = 0$, viz, point A in Fig. 2.1(A) determines the bifurcation frequency. $a_2 = 0$, viz, point B in Fig. 2.1(B) is the frequency where the pure 1/3 subharmonic oscillation occurs. Fig. 2.1(C) is the amplitude frequency response curve of the harmonic response of the first order approximation.

In this case, we choose the generating frequency $\omega_g = 4$. It shows from Fig. 2.1(B)(C) that all five sets of solutions exist at ω_g. This means we can find five steady state motions at a single generating frequency, by sweeping ω, from where, we obtain all five branches. The initial values from Eqs. (2.44) and (2.45) are listed in Table 2.1.

The comparison of the initial values in Table 2.1 with the high order solutions in Table 2.2 and Table 2.3 confirm that the initial values are relatively close to the high order solutions. Essentially, the low order approximation and its corresponding N-th solution are qualitatively the same (on the same branch of the response curve). In the tables, SUBSOL and HARSOL denote subharmonic solution and harmonic solution respectively.

Table 2.2 shows that the 1/3 subharmonic and harmonic components dominate others, which illustrates the validity of using the second order approximation as the initials to its high order solution. Table 2.3 shows that the first harmonic is much larger than the higher order components, which also confirms the validity of applying

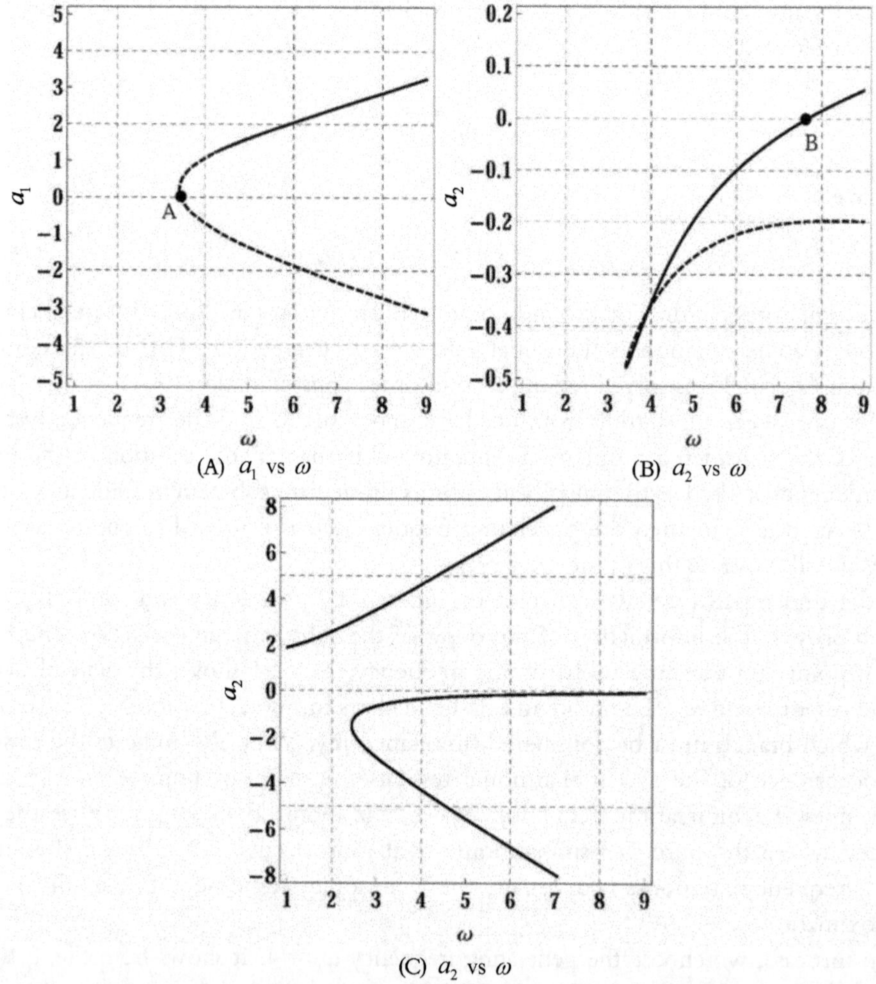

Figure 2.1 The second order approximation of 1/3 subharmonic solution of the undamped Duffing equation: (A) 1/3 subharmonic amplitude varying with frequency ω; (B) fundamental harmonic amplitude varying with frequency ω; (C) fundamental harmonic amplitude a_2 varying with frequency ω, in this case $a_1 = 0$, a_2 represents $A_1^{(1)}$.

$A_1^{(1)}$ as the initial value. We can conclude by comparing Table 2.1 and Table 2.3, that the first, second and third columns of Table 2.3 correspond to the upper, unstable and lower branches, respectively.

By sweeping ω back and forth, starting at ω_g, over all branches, we finally obtain the response curves for both harmonic and subharmonic oscillations. Fig. 2.2 plots the peak amplitude $|x|$ vs frequency curve. Both harmonic and subharmonic responses are provided. Unless otherwise specified, the stop criterion of the Newton-Raphson

Table 2.1 Initial values at $\omega_g = 4$.

	$\cos 1/3\omega t$	$\cos\omega t$
SUBSOL1	−0.7194743839857893	−0.3605474805023799
SUBSOL2	1.0808818787916024	−0.3561553451108572
HARSOL1	0	4.630311850542268
HARSOL2	0	−4.295095097328164
HARSOL3	0	−0.3352167532141037

Table 2.2 Coefficients of the subharmonic solutions for $N = 8$, at $\omega_g = 4$.

	SUBSOL1	SUBSOL2
$\cos 1/3\omega t$	−0.716782379738396	1.079100277428763
$\cos\omega t$	−0.360622588276577	−0.356287363474015
$\cos 5/3\omega t$	−0.004933880272422	−0.005011933668311
$\cos 7/3\omega t$	−0.000866908478371	0.001202617368613
$\cos 3\omega t$	−0.000100868293579	−0.000056982080267
$\cos 11/3\omega t$	−0.000004288223870	−0.000005549305123
$\cos 13/3\omega t$	−0.000000470514428	0.000000435376073
$\cos 5\omega t$	−0.000000038364455	0.000000006711708

Table 2.3 Coefficients of the harmonic solutions for $N = 12$, at $\omega_g = 4$.

	HARSOL1	HARSOL2	HARSOL3
$\cos\omega t$	4.521893823447083	−4.205552387932138	−0.335217130152840
$\cos 3\omega t$	0.207195704577293	−0.160416386994274	−0.000065931786928
$\cos 5\omega t$	0.009041646033017	−0.005939980362846	−0.000000013934922
$\cos 7\omega t$	0.000395158065012	−0.000219574108227	−0.000000000002897
$\cos 9\omega t$	0.000017272530534	−0.000008115364584	−0.000000000000001
$\cos 11\omega t$	0.000000755010323	−0.000000299931252	−0.000000000000000
$\cos 13\omega t$	0.000000033002979	−0.000000011084898	0.000000000000000
$\cos 15\omega t$	0.000000001442629	−0.000000000409676	0.000000000000000
$\cos 17\omega t$	0.000000000063060	−0.000000000015141	0.000000000000000
$\cos 19\omega t$	0.000000000002757	−0.000000000000560	0.000000000000000
$\cos 21\omega t$	0.000000000000121	−0.000000000000021	0.000000000000000
$\cos 23\omega t$	0.000000000000005	−0.000000000000001	−0.000000000000000

solver is $\varepsilon = 10^{-10}$ throughout the paper. Since damping is absent in the current problem, both harmonic and subharmonic responses will go to infinity with the increase of the impressed frequency.

Fig. 2.3 provides the response curves of the harmonic solution. The amplitude of each harmonic is plotted. For the upper branch of $|A_1|$, i.e. the amplitude of the first

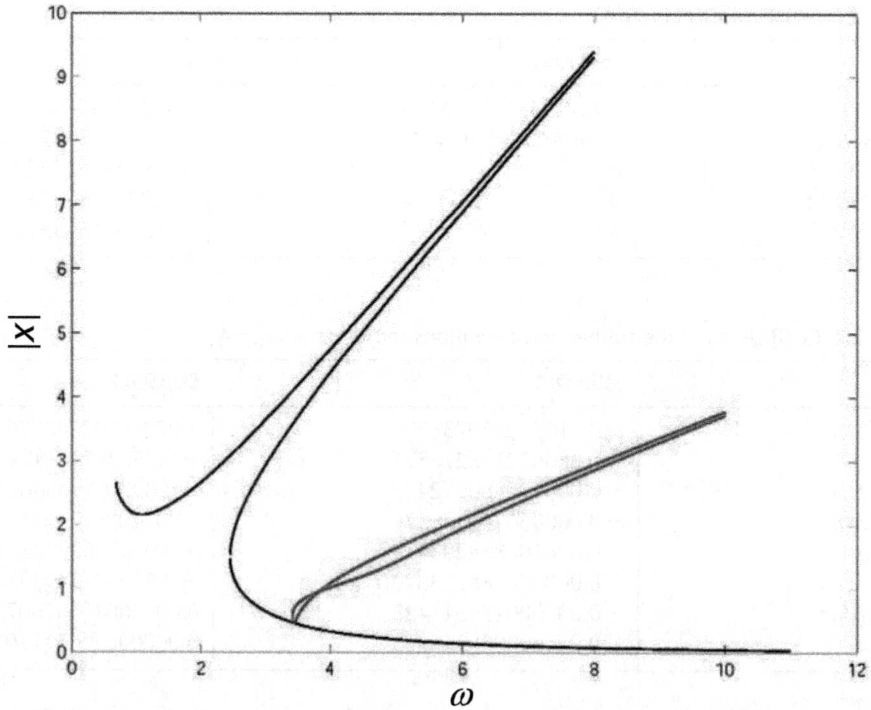

Figure 2.2 The peak amplitude $|x|$ vs frequency curve for the Duffing equation: $\ddot{x}+x+x^3=5\cos\omega t$. The *black* curve represents the harmonic response; the *red* curve represents the 1/3 subharmonic response.

harmonic. It dominates the oscillation from about $\omega=1$ to infinity. The middle branch is an unstable one which is practically unrealizable. The third harmonic component is comparable with the lower branch of the first harmonic component at $\omega>4$. It can be seen that the fifth component is very weak far away from $\omega<1$. It should be noted that the third and fifth harmonics are significant where $\omega<1$.

Fig. 2.4 provides the response curves of the subharmonic solution. The amplitudes of the 1/3, 1 and 5/3 harmonic components are given. It indicates that the 5/3 harmonic component is very weak compared with the 1/3 and the first harmonic components. The subharmonic oscillation emerges from about $\omega=3.5$, which agrees with the above statement that the 1/3 subharmonic oscillation starts at the frequency being greater than three times of the natural frequency.

It indicates that the pure subharmonic oscillation may occur at the frequency where A_1 is zero. It is about $\omega=7.5$ from Fig. 2.4. This is also predicted by the second order approximation in Eq. (2.45) with $a_2=0$, ω as unknown, giving $\omega=7.66384$.

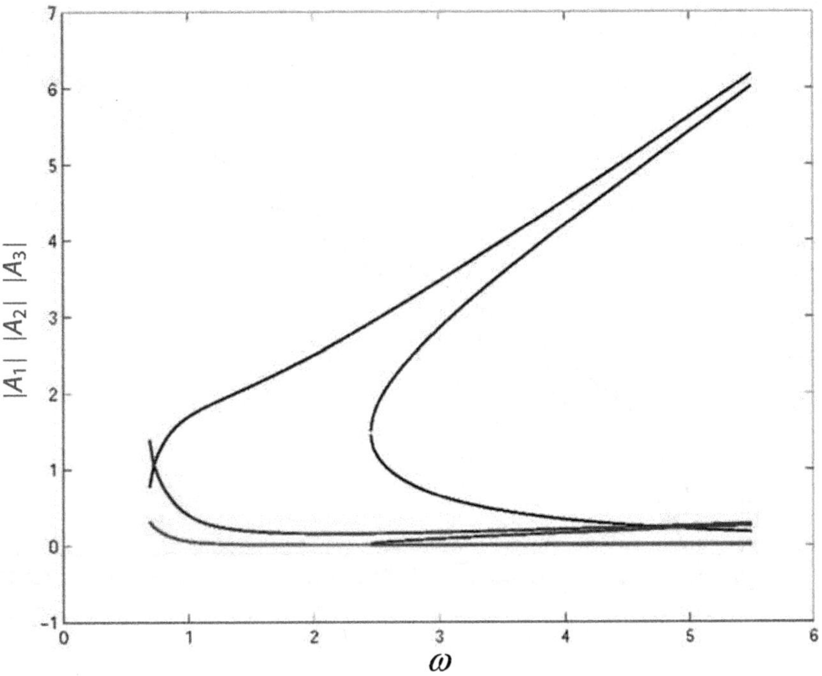

Figure 2.3 The harmonic amplitude vs frequency curves of the harmonic solution for the Duffing equation $\ddot{x}+x+x^3=5\cos\omega t$: the *black* curve represents the first harmonic amplitude $|A_1|$ vs ω; the *blue* curve represents the third harmonic amplitude $|A_2|$ vs ω; the *red* curve represents the fifth harmonic amplitude A_3 vs ω.

2.4.2 Damped Duffing equation

In this section, we investigate the effect of each parameter in the Duffing equation

$$\ddot{x}+\xi\dot{x}+x+\beta x^3=F\cos\omega t. \tag{2.52}$$

For doing so, we compute the amplitude frequency curves for various ξ, β, and F. As before, we exclusively focus on the harmonic and 1/3 subharmonic responses.

Eqs. (2.9) are the N-th order approximations to the harmonic and 1/3 subharmonic solutions. Eqs. (2.48) and (2.49) are used to generate the initial values for the higher order approximations of harmonic and subharmonic solutions respectively. The generating frequency ω_g is chosen according to the considered case.

2.4.2.1 The effect of damping ξ

Fig. 2.5 gives the amplitude-frequency curves with various damping. It indicates that a smaller damping ξ stretches the response curve. The smaller the damping is, the longer the tip of the upper harmonic response and the subharmonic response will be. When

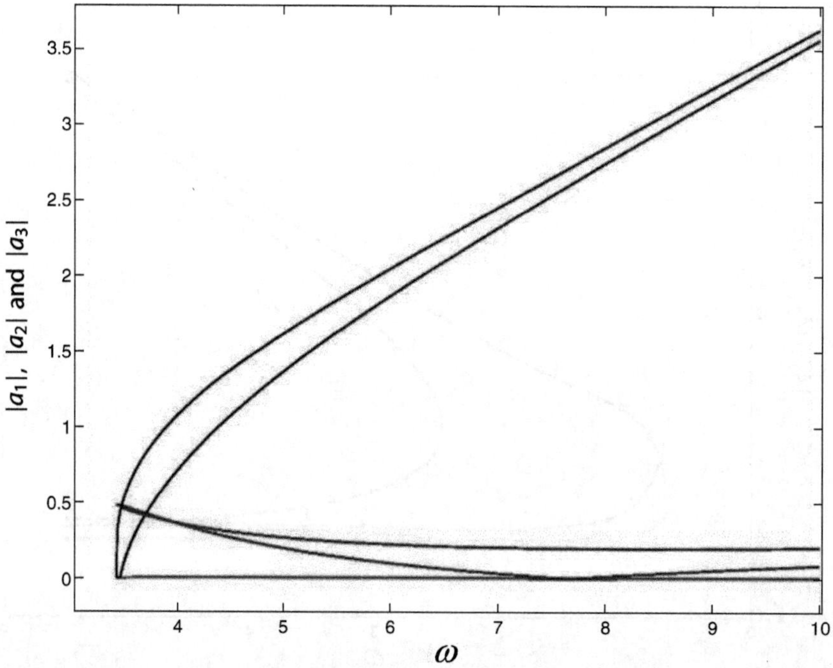

Figure 2.4 The harmonic amplitude vs frequency curves of the 1/3 subharmonic solution for the Duffing equation $\ddot{x} + x + x^3 = 5\cos\omega t$: the black curve represents the 1/3 subharmonic amplitude $|a_1|$ vs ω; the blue curve represents the harmonic amplitude $|a_2|$ vs ω; the red curve represents the 5/3 ultra-subharmonic amplitude $|a_3|$ vs ω.

$\xi = 0$, the upper harmonic response and the subharmonic response go to infinity as stated in the undamped case. It also indicates that the damping does not bend the curve, which means that the response curves for different damping have the same backbone. Fig. (2.5A) also reveals that the damping almost does not influence the response curve except elongating the tip area.

Fig. (2.5B) is the zoom-in of the subharmonic part in Fig. (2.5A) It shows that a larger ξ narrows the occurrence domain of the subharmonic solution. It should be noted that for $\xi = 0.1$, 0.2 in this case, the subharmonic solution does not exist. Hence, there exists a certain damping value, greater than which the subharmonic solution disappears.

2.4.2.2 The effect of nonlinearity β

Fig. 2.6 shows the amplitude vs frequency curves of the harmonic and 1/3 subharmonic solutions for different values of nonlinearity. It shows in Fig. 2.6(A) and Fig. 2.6(B) that a positive nonlinearity has the effect of bending the response curve to the right. The larger the nonlinearity is, the more the curve bends. It applies to both harmonic and subharmonic response curves.

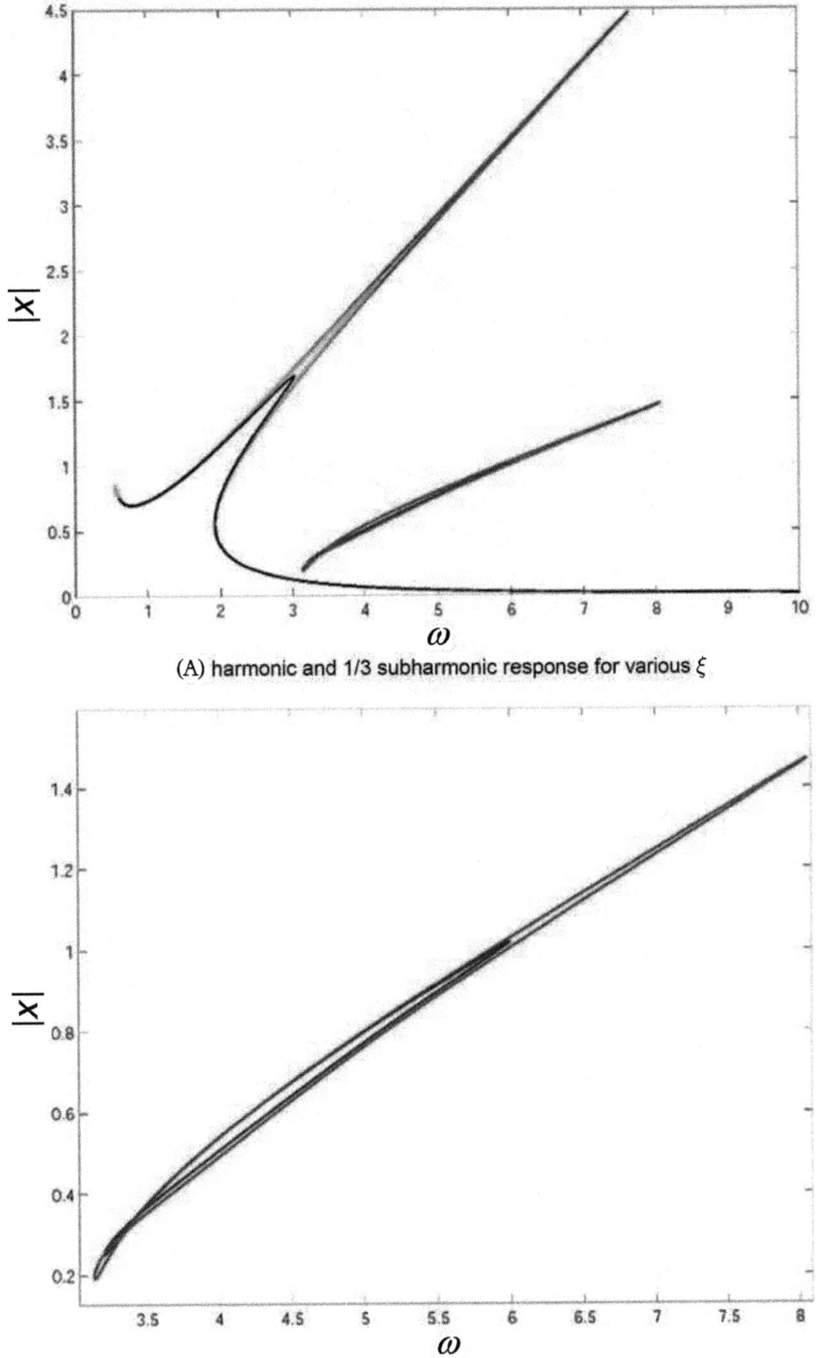

|x|
ω
(A) harmonic and 1/3 subharmonic response for various ξ
|x|
ω

(B) 1/3 subharmonic response

Figure 2.5 The amplitude vs frequency curves of the harmonic and 1/3 subharmonic solutions for the Duffing equation $\ddot{x}+\xi\dot{x}+x+4x^3=\cos\omega t$, where red curve: $\xi=0.03$; blue curve: $\xi=0.05$; green curve: $\xi=0.1$; black curve: $\xi=0.2$. Note that when $\xi=0.1$ and 0.2, 1/3 subharmonic response does not occur.

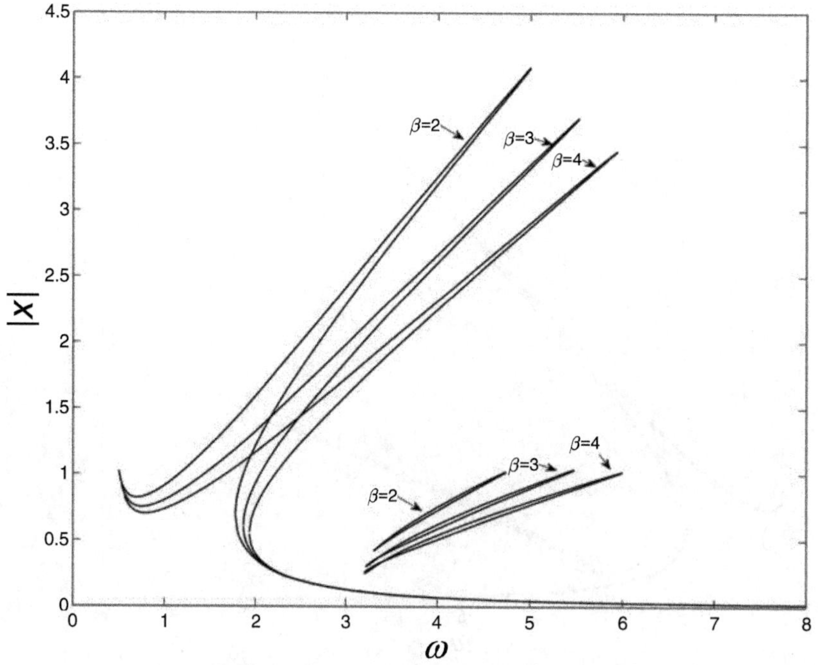

(A) harmonic and 1/3 subharmonic response for β=2, 3 and 4.

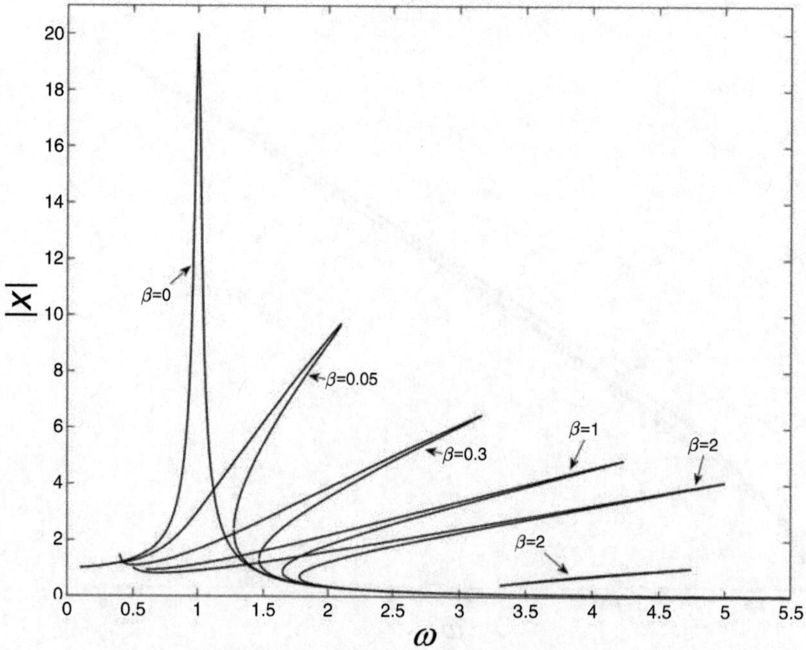

(B) harmonic and 1/3 subharmonic response for β=0, 0.05, 0.3, 1 and 2.

Figure 2.6 The amplitude vs frequency curves of the harmonic and 1/3 subharmonic solutions for the Duffing equation $\ddot{x} + 0.05\dot{x} + x + \beta x^3 = \cos\omega t$. Note that the 1/3 subharmonic response does not exist for $\beta = 0$, 0.05, 0.3 and 1.

We also see that the upper branch of the harmonic response, and the subharmonic response are bounded values, which is different from the undamped system. Also, the subharmonic response only exists in a finite frequency domain, which can be influenced by β. Fig. 2.6(A) indicates that smaller β narrows this domain. Fig. 2.6(B) shows that when β decreases to a certain value, the subharmonic response ceases to occur.

2.4.2.3 The effect of the amplitude F of the impressed force

The effect of the amplitude of the impressed force is shown in Fig. 2.7. It indicates qualitatively that F does not bend the response curve, which is similar to the damping ξ. Hence, all response curves have the same backbone. What is different from the effect of ξ is that F affects the response globally, while ξ only influences the tip area. As F increases, the peak amplitude of the harmonic response will increase, see Fig. 2.7(A). Fig. 2.7(B) shows that a larger F may enlarge the occurrence domain of the subharmonic solution. When F decreases to a certain value, the subharmonic solution ceases to occur. In the current case, when $F = 0.3$, $F = 0.5$ and $F = 1$, the subharmonic solution does not occur. When $F = 1.5$ and $F = 1.8$, it appears. It means a certain value between $1 \sim 1.5$ is the onset of the subharmonic solution.

Appendix A

$$\begin{cases} a_1\left[36 - 4\omega^2 + 27\beta\left(a_1^2 + a_1 a_2 + 2a_2^2 + b_1^2 + 2b_1 b_2 + 2b_2^2\right)\right] \\ = -3b_1(4\xi\omega - 9\beta a_2 b_1) \\ b_1\left[36 - 4\omega^2 + 27\beta\left(a_1^2 + 2a_2^2 - 2a_1 a_2 + b_1^2 - b_1 b_2 + 2b_2^2\right)\right] \\ = 3a_1(4\xi\omega - 9\beta a_1 b_2) \\ a_2\left[4 - 4\omega^2 + 3\beta\left(2a_1^2 + a_2^2 + 2b_1^2 + b_2^2\right)\right] = 4F - 4b_2\omega\xi - \beta a_1\left(a_1^2 - 3b_1^2\right) \\ b_2\left[4 - 4\omega^2 + 3\beta\left(2a_1^2 + a_2^2 + 2b_1^2 + b_2^2\right)\right] = \beta b_1\left(b_1^2 - 3a_1^2\right) + 4\omega\xi a_2 \end{cases} \tag{2.53}$$

Appendix B

$$\begin{cases} A_1\left[4 - 4\omega^2 + 3\beta\left(A_1^2 + A_1 A_2 + 2A_2^2 + B_1^2 + 2B_1 B_2 + 2B_2^2\right)\right] \\ = 4F + B_1(3\beta A_2 B_1 - 4\xi\omega) \\ B_1\left[4 - 4\omega^2 + 3\beta\left(A_1^2 - 2A_1 A_2 + 2A_2^2 + B_1^2 - B_1 B_2 + 2B_2^2\right)\right] \\ = 4\xi\omega A_1 - 3\beta A_1^2 B_2 \\ A_2\left[4 - 36\omega^2 + 3\beta\left(2A_1^2 + A_2^2 + 2B_1^2 + B_2^2\right)\right] = \beta A_1\left(3B_1^2 - A_1^2\right) - 12\xi\omega B_2 \\ B_2\left[4 - 36\omega^2 + 3\beta\left(2A_1^2 + A_2^2 + 2B_1^2 + B_2^2\right)\right] = 12\xi\omega A_2 + \beta B_1\left(B_1^2 - 3A_1^2\right) \end{cases} \tag{2.54}$$

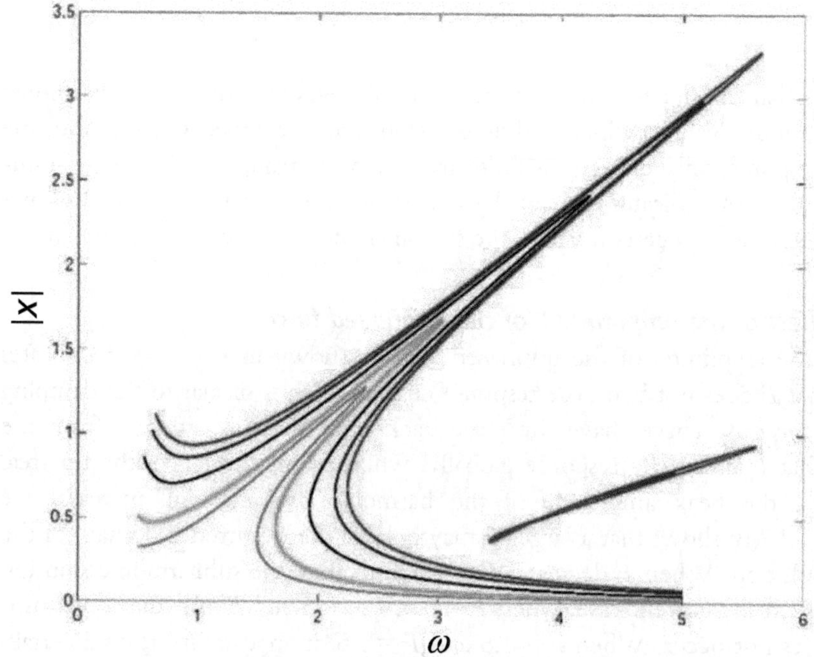

(A) harmonic and 1/3 subharmonic response for various F.

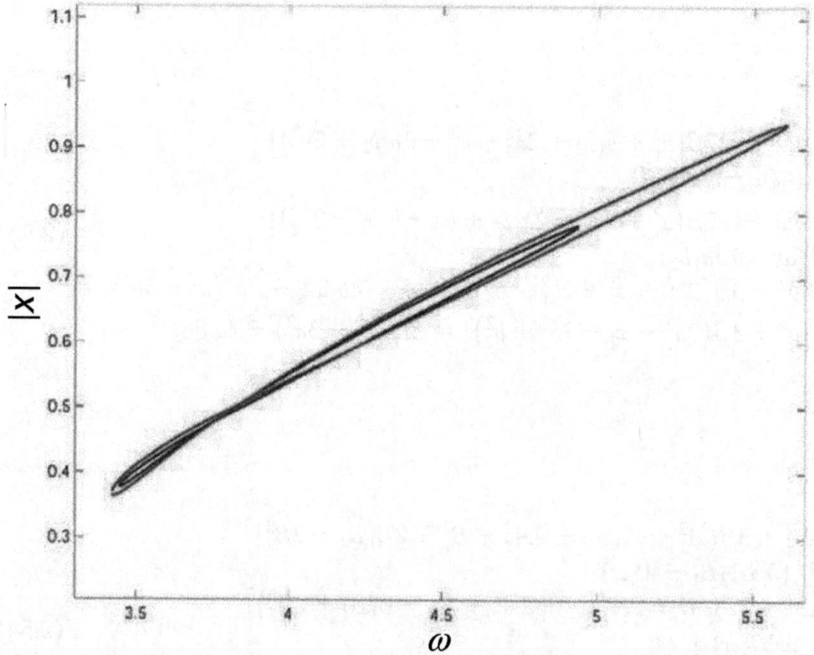

(B) 1/3 subharmonic response.

Figure 2.7 The amplitude vs frequency curves of the harmonic and 1/3 subharmonic solutions for the Duffing equation $\ddot{x} + 0.1\dot{x} + x + 4x^3 = F\cos\omega t$; the curves (left part of (A)) from bottom to top are $F = 0.3$, $F = 0.5$, $F = 1$, $F = 1.5$ and $F = 1.8$ sequentially. Note that subharmonic response does not occur for $F = 0.3$, $F = 0.5$, and $F = 1$. (A) Harmonic and 1/3 subharmonic response for various F. (B) 1/3 subharmonic response.

References

[1] S.N. Atluri, The meshless method (MLPG) for domain & BIE discretizations, volume 677, Tech Science Press, 2004.

[2] S.N. Atluri, Methods of computer modeling in engineering & the sciences, Tech Science Press, 2005.

[3] S.N. Atluri, S. Shen, The meshless local Petrov-Galerkin (MLPG) method, Crest, 2002.

[4] S.N. Atluri, T. Zhu, A new meshless local Petrov-Galerkin (MLPG) approach in computational mechanics, Computational Mechanics 22 (1998) 117–127.

[5] K.C. Hall, J.P. Thomas, W.S. Clark, Computation of unsteady nonlinear flows in cascades using a harmonic balance technique, AIAA Journal 40 (2002) 879–886.

[6] J.P. Thomas, E.H. Dowell, K.C. Hall, Nonlinear inviscid aerodynamic effects on transonic divergence, futter, and limit-cycle oscillations, AIAA Journal 40 (2002) 638–646.

[7] M.E. Levenson, Harmonic and subharmonic response for the Duffing equation: $\ddot{x} + \alpha x + \beta x^3 = F\cos\omega t(\alpha > 0)$, Journal of Applied Physics 20 (11) (1949) 1045–1051.

[8] H. Moriguchi, T. Nakamura, Forced oscillations of system with nonlinear restoring force, Journal of the Physical Society of Japan 52 (3) (1983) 732–743.

[9] P.A. Sturrock, Non-linear effects in electron plasmas, Proceedings of the Royal Society of London. Series A. Mathematical and Physical Sciences 242 (1230) (1957) 277–299.

[10] J.J. Stoker, Nonlinear Vibrations, Interscience, 1950.

[11] C. Hayashi, Forced oscillations with nonlinear restoring force, Journal of Applied Physics 24 (2) (1953) 198–207.

[12] C. Hayashi, Stability investigation of the nonlinear periodic oscillations, Journal of Applied Physics 24 (3) (1953) 344–348.

[13] C. Hayashi, Subharmonic oscillations in nonlinear systems, Journal of Applied Physics 24 (5) (1953) 521–529.

[14] W.Y. Tseng, J. Dugundji, Nonlinear vibrations of a beam under harmonic excitation, Journal of applied mechanics 37 (2) (1970) 292–297.

[15] W.Y. Tseng, J. Dugundji, Nonlinear vibrations of a buckled beam under harmonic excitation, Journal of applied mechanics 38 (2) (1971) 467–476.

[16] J.P. Thomas, K.C. Hall, E.H. Dowell, A harmonic balance approach for modeling nonlinear aeroelastic behavior of wings in transonic viscous flow, AIAA paper 1924 (2003) 2003.

[17] L. Liu, E.H. Dowell, J.P. Thomas, P. Attar, K.C. Hall, A comparison of classical and high dimensional harmonic balance approaches for a Duffing oscillator, Journal of Computational Physics 215 (2006) 298–320.

[18] M. Urabe, Numerical investigation of subharmonic solutions to Duffing's equation, vol. 5, Publ. RIMS Kyoto Univ, 1969, pp. 79–112.

[19] H.H. Dai, J.K. Paik, S.N. Atluri, The global nonlinear galerkin method for the analysis of elastic large deflections of plates under combined loads: A scalar homotopy method for the direct solution of nonlinear algebraic equations, Computers Materials and Continua 23 (1) (2011) 69.

[20] H.H. Dai, J.K. Paik, S.N. Atluri, The Global Nonlinear Galerkin Method for the Solution of von Karman Nonlinear Plate Equations: An Optimal & Faster Iterative Method for the Direct Solution of Nonlinear, 2011b.

[21] H.H. Dai, J.K. Paik, S.N. Atluri, The global nonlinear galerkin method for the analysis of elastic large deflections of plates under combined loads: a scalar homotopy method for the direct solution of nonlinear algebraic equations, Computers Materials and Continua 23 (1) (2011) 69.

[22] H.H. Dai, J.K. Paik, S.N. Atluri, The global nonlinear galerkin method for the solution of von karman nonlinear plate equations: an optimal & faster iterative method for the direct solution of nonlinear algebraic equations $\boldsymbol{F}(\boldsymbol{x}) = \boldsymbol{0}$, using $\boldsymbol{x} = \lambda[\alpha\boldsymbol{F} + (1-\alpha)\boldsymbol{B}^{\mathrm{T}}\boldsymbol{F}]$, Computers Materials and Continua 23 (2) (2011) 155.

[23] C.S. Liu, H.H. Dai, S.N. Atluri, Iterative solution of a system of nonlinear albegraic equations $\boldsymbol{F}(\boldsymbol{x}) = \boldsymbol{0}$, using $\dot{\boldsymbol{x}} = \lambda[\alpha\boldsymbol{R} + \beta\boldsymbol{P}]$ or $\lambda[\alpha\boldsymbol{F} + \beta\boldsymbol{P}^*]$, $\boldsymbol{R}$ is normal to a hyper-surface function of $\boldsymbol{F}$, $\boldsymbol{P}$ normal to $\boldsymbol{R}$, and $\boldsymbol{P}^*$ normal to $\boldsymbol{F}$, CMES: Computer Modeling in Engineering & Sciences 81 (4) (2011) 335–362.

[24] C.S. Liu, H.H. Dai, S.N. Atluri, A further study on using $\dot{\boldsymbol{x}} = \lambda[\alpha \boldsymbol{R} + \beta \boldsymbol{P}]$ $(\boldsymbol{P} = \boldsymbol{F} - \boldsymbol{R}(\boldsymbol{F} \cdot \boldsymbol{R})/\|\boldsymbol{R}\|^2)$ and $\dot{\boldsymbol{x}} = \lambda[\alpha \boldsymbol{F} + \beta \boldsymbol{P}^*]$ $(\boldsymbol{P}^* = \boldsymbol{R} - \boldsymbol{F}(\boldsymbol{F} \cdot \boldsymbol{R})/\|\boldsymbol{F}\|^2)$ in iteratively solving the nonlinear system of algebraic equations $\boldsymbol{F}(\boldsymbol{x}) = \boldsymbol{0}$, CMES: Computer Modeling in Engineering & Sciences 81 (2) (2011) 195–227.

[25] M. Urabe, Galerkin's procedure for nonlinear periodic systems, Archive for Rational Mechanics and Analysis 20 (1965) 120–152.

Chapter 3

Dealiasing for Harmonic Balance and Time Domain Collocation Methods

SUMMARY

The objective of this chapter is to investigate the effects of the number of harmonics and the frequency modulation on the accuracy of the HB method. The spectral analysis is applied to the periodic pitch motions to explore the distribution of the dominant frequencies/harmonics to offer an inference to the HB method to choose the appropriate number of harmonics. A frequency modulation phenomenon is observed, and its effect on the accuracy of the HB method is examined both numerically and analytically. Another objective is to reduce the large amount of computational costs arising from the application of the traditional high order HB method [1]. The explicit Jacobian matrix is derived to reduce the computational cost in each iteration, avoiding the normally used far more expensive numerical approximation of the Jacobian matrix. A comparative study of the effects of using the explicit and numerical Jacobian matrices on the accuracy and efficiency of the HB method is implemented.

In this chapter, the aeroelastic behavior of a 2 DOF airfoil is used as a demonstration. Aeroelasticity is considered to be a source of instability and vibration problems for aircraft wings; it is concerned with the physical phenomena which involve significant mutual interaction among inertial, elastic and aerodynamic forces [2]. The nonlinearities mostly encountered in aeroelastic system are classified as structural and aerodynamics ones [3]. The former mainly refers to nonlinear stiffness resulting from large amplitude vibration, freeplay and/or hysteresis nonlinearities. One prominent consequence of the structural nonlinearity is that, once the flutter speed is exceeded, the wing goes into a bounded oscillation rather than an exponentially increasing oscillation predicted by the linear aeroelastic model. A vast number of studies have been conducted to investigate the two DOF wing model with cubic [4], freeplay [5] and hysteresis [6] nonlinearities. The two DOF airfoil model with cubic nonlinearity in the pitch motion is our present concern.

The structure of this chapter is as follows. In Section 3.1 the equations of motion of the two DOF airfoil are introduced. The formulation of the HB algebraic system is carried out in Section 3.2. In addition, the numerical Jacobian matrix is calculated via a forward three-point difference technique and the explicit expression of the Jacobian matrix of the HB algebraic system is derived. In Section 3.3 we transform the original

Computational Methods for Nonlinear Dynamical Systems
DOI: https://doi.org/10.1016/B978-0-323-99113-1.00003-0

TDC system to a mathematically equivalent compact TDC system and derive an explicit Jacobian matrix for this compact TDC algebraic system. Besides, aliasing of the TDC method is introduced in Subsection 3.3.2. Finally, a number of illustrative examples of all these fast harmonic balance methods are provided in the last section.

3.1 Governing equations of the airfoil model

Shown in Fig. 3.1 is the physical model of a two DOF airfoil oscillating in pitch and plunge. The positive directions of the pitch angle α and the plunge displacement h are defined in Fig. 3.1. The equations of motion for this two DOF airfoil model with linear springs were established by Fung [7]. Taking into account the cubic structural nonlinearities, Fung's linear equations can be readily modified into the nonlinear form [31]:

$$S\alpha'' + mh'' + C_h h' + \overline{G}(h) = p(t), \tag{3.1a}$$

$$I_\alpha \alpha'' + Sh'' + C_\alpha \alpha' + \overline{M}(\alpha) = r(t), \tag{3.1b}$$

where the symbols m, S, I_α, C_α and C_h are defined in the nomenclature and $(\ )'$ represents derivative with respect to time t. $\overline{G}(h)$ and $\overline{M}(\alpha)$ are the nonlinear plunge and pitch stiffness terms, and $p(t)$ and $r(t)$ are external force and moment exerted on the airfoil, respectively. In the present study, no external force and moment are applied, thus $p(t) = r(t) = 0$.

In order to reduce the number of control parameters, the non-dimensionalization procedure is usually performed before further analysis. Upon introducing the following definitions:

$$\xi = \frac{h}{b},\ K_\xi = K_h,\ X_\alpha = \frac{S}{bm},\ \omega_\xi = \left(\frac{K_\xi}{m}\right)^{1/2},$$

$$\omega_\alpha = \left(\frac{K_\alpha}{I_\alpha}\right)^{1/2},\ r_\alpha = \left(\frac{I_\alpha}{mb^2}\right)^{1/2},\ \zeta_\xi = \frac{1}{2}C(mK_h)^{1/2},\ \zeta_\alpha = \frac{1}{2}C(I_\alpha K_\alpha)^{1/2}.$$

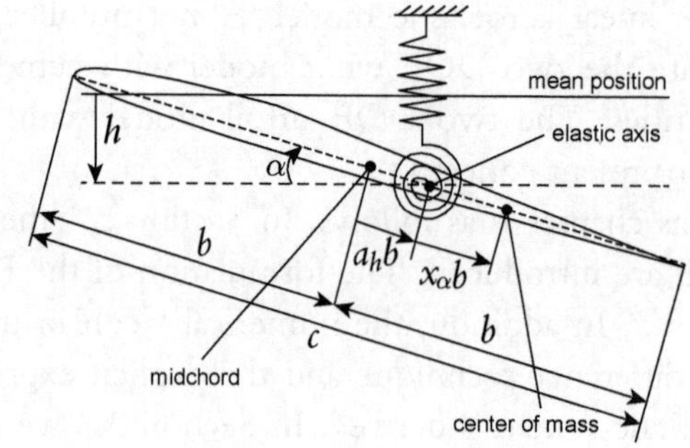

Figure 3.1 Sketch of a two DOF wing section model.

Eqs. (3.1) can be non-dimensionalized as follows

$$\ddot{\xi} + X_\alpha \ddot{\alpha} + 2\zeta_\xi \frac{\overline{\omega}}{U^*}\dot{\xi} + \left(\frac{\overline{\omega}}{U^*}\right)^2 G(\xi) = -\frac{1}{\pi\mu} C_L(\tau), \tag{3.2a}$$

$$\frac{X_\alpha}{r_\alpha^2}\ddot{\xi} + \ddot{\alpha} + 2\zeta_\alpha \frac{1}{U^*}\dot{\alpha} + \left(\frac{1}{U^*}\right)^2 M(\alpha) = \frac{2}{\pi\mu r_\alpha^2} C_M(\tau), \tag{3.2b}$$

where the related symbols are defined in Nomenclature. $(\cdot)$ represents $\mathrm{d}()/\mathrm{d}\tau$. The nonlinear pitch and plunge stiffness terms $M(\alpha)$ and $G(\xi)$ are $M(\alpha) = \alpha + \beta\alpha^3$, $G(\xi) = \xi + \gamma\xi^3$, and the linear aerodynamic force and moment $C_L(\tau)$ and $C_M(\tau)$ are [8]

$$\begin{aligned} C_L(\tau) &= \pi\left(\ddot{\xi} - a_h\ddot{\alpha} + \dot{\alpha}\right) + 2\pi\left\{\alpha(0) + \dot{\xi}(0) + \left(1/2 - a_h\right)\dot{\alpha}(0)\right\}\phi(\tau) \\ &\quad + 2\pi\int_0^\tau \phi(\tau - \sigma)\left[\dot{\alpha}(\sigma) + \ddot{\xi}(\sigma) + \left(1/2 - a_h\right)\ddot{\alpha}(\sigma)\right]\mathrm{d}\sigma, \\ C_M(\tau) &= \pi\left(1/2 + a_h\right)\left\{\alpha(0) + \dot{\xi}(0) + \left(1/2 - a_h\right)\dot{\alpha}(0)\right\}\phi(\tau) + \frac{\pi}{2}\left(\ddot{\xi} - a_h\ddot{\alpha}\right) - \frac{\pi}{16}\ddot{\alpha} \\ &\quad - \left(1/2 - a_h\right)\frac{\pi}{2}\dot{\alpha} + \pi\left(1/2 + a_h\right)\int_0^\tau \phi(\tau - \sigma)\left[\dot{\alpha}(\sigma) + \ddot{\xi}(\sigma) + \left(1/2 - a_h\right)\ddot{\alpha}(\sigma)\right]\mathrm{d}\sigma, \end{aligned} \tag{3.3}$$

respectively, in which the Wagner's function is $\phi(\tau) = 1 - \psi_1 e^{-\varepsilon_1\tau} - \psi_2 e^{-\varepsilon_2\tau}$ where $\psi_{1,2} = 0.165, 0.335$ and $\varepsilon_{1,2} = 0.0455, 0.3$. Since system (3.2) is essentially a system of integro-differential equations, it is difficult to solve this system directly due to the presence of the integral terms. Normally, the integro-differential equations should be transformed into pure differential equations, which can then be solved via either semi-analytical or numerical simulation methods.

The transformation from the integro-differential system to the differential system is by means of integral transformation methods. For the present two DOF airfoil problem, the following four transformations [9]

$$\begin{aligned} w_1 &= \int_0^\tau e^{-\varepsilon_1(\tau-\sigma)}\alpha(\sigma)\mathrm{d}\sigma, \; w_2 = \int_0^\tau e^{-\varepsilon_2(\tau-\sigma)}\alpha(\sigma)\mathrm{d}\sigma, \\ w_3 &= \int_0^\tau e^{-\varepsilon_1(\tau-\sigma)}\xi(\sigma)\mathrm{d}\sigma, \; w_4 = \int_0^\tau e^{-\varepsilon_2(\tau-\sigma)}\xi(\sigma)\mathrm{d}\sigma, \end{aligned} \tag{3.4}$$

are introduced to deal with the integral terms. Consequently, Eqs. (3.2) can be transformed into a system of pure differential equations:

$$\begin{cases} c_0\ddot{\xi}+c_1\ddot{\alpha}+c_2\dot{\xi}+c_3\dot{\alpha}+c_4\xi+c_5\alpha+c_6w_1+c_7w_2+c_8w_3+c_9w_4+c_{10}G(\xi)=f(\tau), \\ d_0\ddot{\xi}+d_1\ddot{\alpha}+d_2\dot{\xi}+d_3\dot{\alpha}+d_4\xi+d_5\alpha+d_6w_1+d_7w_2+d_8w_3+d_9w_4+d_{10}M(\alpha)=g(\tau), \\ \dot{w}_1=\alpha-\varepsilon_1 w_1, \\ \dot{w}_2=\alpha-\varepsilon_2 w_2, \\ \dot{w}_3=\xi-\varepsilon_1 w_3, \\ \dot{w}_4=\xi-\varepsilon_2 w_4. \end{cases} \tag{3.5}$$

The expressions for all coefficients in system (3.5) are given in the Appendix. Because only the periodic motions are considered in this chapter, both $f(\tau)$ and $g(\tau)$ can be set to zero [1].

Upon introducing $\boldsymbol{x}=[x_1,x_2,\ldots,x_8]^{\mathrm{T}}$, where $x_1=\alpha, x_2=\dot{\alpha}, x_3=\xi, x_4=\dot{\xi}$ and $x_{i+4}=\omega_i (i=1,2,3,4)$, system (3.5) can be written in the state-space form: $\dot{\boldsymbol{x}}=\boldsymbol{A}\boldsymbol{x}+\boldsymbol{N}(\boldsymbol{x})$, or explicitly as

$$\dot{\boldsymbol{x}}=\begin{bmatrix} 0 & 1 & 0 & 0 & 0 & 0 & 0 & 0 \\ a_{21}-g_{21} & a_{22} & a_{23}+g_{23} & a_{24} & a_{25} & a_{26} & a_{27} & a_{28} \\ 0 & 0 & 0 & 1 & 0 & 0 & 0 & 0 \\ a_{41}+g_{41} & a_{42} & a_{43}-g_{43} & a_{44} & a_{45} & a_{46} & a_{47} & a_{48} \\ 1 & 0 & 0 & 0 & -\varepsilon_1 & 0 & 0 & 0 \\ 1 & 0 & 0 & 0 & 0 & -\varepsilon_2 & 0 & 0 \\ 0 & 0 & 1 & 0 & 0 & 0 & -\varepsilon_1 & 0 \\ 0 & 0 & 1 & 0 & 0 & 0 & 0 & -\varepsilon_2 \end{bmatrix}\boldsymbol{x} + \begin{bmatrix} 0 \\ \gamma g_{23}x_3^3-\beta g_{21}x_1^3 \\ 0 \\ \beta g_{41}x_1^3-\gamma g_{43}x_3^3 \\ 0 \\ 0 \\ 0 \\ 0 \end{bmatrix}, \tag{3.6}$$

in which the relevant coefficients can be found in the Appendix. System (3.6) can be readily solved by numerical integration methods, e.g. the fourth-order Runge-Kutta (RK4) method, while system (3.5) is well suited for semianalytical methods.

3.2 Formulation of the HB method

Since we mainly focus on the periodic solutions, the HB method can be applied to solve the present system effectively. Liu et al. [1] employed the HB method with a moderately large number of harmonics to obtain an accurate prediction of the secondary bifurcation of the airfoil. This subject was revisited by Dai et al. [10], using a simple point-collocation method. In fact, both the HB and the collocation methods belong to the category of weighted residual methods, and generally speaking the HB method, i.e. time domain Galerkin method, is believed to be more accurate than the collocation method [11]. The accuracy of the weighted residual method depends on the number of harmonics included in the trial solutions as well as the type of the weighting functions. In the HB method, the trial functions and the weighting functions are the same, while in the method, the weighting functions are simple Dirac Delta functions. The collocation method, although simpler than the HB method, may generate unwanted solutions in addition to the physically meaningful ones. In Dai et al. [10], the relationships among the time domain collocation method, the well-known high dimensional harmonic balance (HDHB) method, and the classical harmonic balance method were explored, and the aliasing phenomenon arising from the HDHB method was clearly explained by means of explicit aliasing rules.

In this chapter, we only consider only the airfoil with nonlinearity in the pitch DOF, i.e. $\beta \neq 0$, $\gamma = 0$. In the followings, we first introduce the formulation of the HB method [1], and then derive the explicit Jacobian matrix for the purpose of reducing the computational cost. After that, the accuracy and computational efficiency of the present HB method are intensively studied.

It is known that solutions to dynamical systems that are smooth and periodic can be expressed accurately by a Fourier series with appropriate Fourier coefficients. In the HB method, the trial solutions to Eqs. (3.5) are assumed as truncated Fourier series expansions

$$
\begin{aligned}
&\alpha(\tau) = \alpha_0 + \sum_{n=1}^{N}[\alpha_{2n-1}\cos(n\omega\tau) + \alpha_{2n}\sin(n\omega\tau)],\\
&\xi(\tau) = \xi_0 + \sum_{n=1}^{N}\left[\xi_{2n-1}\cos(n\omega\tau) + \xi_{2n}\sin(n\omega\tau)\right],\\
&w_i(\tau) = w_0^i + \sum_{n=1}^{N}\left[w_{2n-1}^i\cos(n\omega\tau) + w_{2n}^i\sin(n\omega\tau)\right],\quad i = 1,2,3,4.
\end{aligned}
\tag{3.7}
$$

The remaining task is to determine the Fourier coefficients in the trial solutions (3.7). Theoretically, if the number of harmonics N in the trial solutions can be infinity, then the trial solutions (3.7) approach the true periodic solutions exactly. However, using an infinite number of harmonics is not feasible computationally. In practice, a finite number of harmonics are used. Substituting the trial solutions (3.7) into the governing Eqs. (3.5) and then balancing the Fourier coefficients of the first N harmonics yield

$$\begin{cases}\left(c_0\omega^2\boldsymbol{A}^2+c_2\omega\boldsymbol{A}+c_4\boldsymbol{I}+c_{10}\boldsymbol{I}\right)\mathbf{Q}_\xi+\left(c_1\omega^2\boldsymbol{A}^2+c_3\omega\boldsymbol{A}+c_5\boldsymbol{I}\right)\mathbf{Q}_\alpha+\sum_{i=1}^{4}c_{i+5}\mathbf{Q}_{w_i}=\mathbf{0},\\ \left(d_0\omega^2\boldsymbol{A}^2+d_2\omega\boldsymbol{A}+d_4\boldsymbol{I}\right)\mathbf{Q}_\xi+\left(d_1\omega^2\boldsymbol{A}^2+d_3\omega\boldsymbol{A}+d_5\boldsymbol{I}\right)\mathbf{Q}_\alpha+\sum_{i=1}^{4}d_{i+5}\mathbf{Q}_{w_i}+d_{10}\boldsymbol{M}_\alpha=\mathbf{0},\\ (\omega\boldsymbol{A}+\varepsilon_1\boldsymbol{I})\mathbf{Q}_{\omega_1}-\mathbf{Q}_\alpha=\mathbf{0},\\ (\omega\boldsymbol{A}+\varepsilon_2\boldsymbol{I})\mathbf{Q}_{\omega_2}-\mathbf{Q}_\alpha=\mathbf{0},\\ (\omega\boldsymbol{A}+\varepsilon_1\boldsymbol{I})\mathbf{Q}_{\omega_3}-\mathbf{Q}_\xi=\mathbf{0},\\ (\omega\boldsymbol{A}+\varepsilon_2\boldsymbol{I})\mathbf{Q}_{\omega_4}-\mathbf{Q}_\xi=\mathbf{0},\end{cases} \tag{3.8}$$

where

$$\mathbf{Q}_\alpha=\begin{bmatrix}\alpha_0\\ \alpha_1\\ \vdots\\ \alpha_{2N}\end{bmatrix},\ \mathbf{Q}_\xi=\begin{bmatrix}\xi_0\\ \xi_1\\ \vdots\\ \xi_{2N}\end{bmatrix},\ \mathbf{Q}_{w_i}=\begin{bmatrix}w_0^i\\ w_1^i\\ \vdots\\ w_{2N}^i\end{bmatrix},\ i=1,2,3,4, \tag{3.9}$$

and $\boldsymbol{I}$ is an identity matrix; $\boldsymbol{A}$ and $\boldsymbol{M}_\alpha$ are given by

$$\boldsymbol{A}=\begin{bmatrix}0&0&0&\cdots&0\\ 0&\boldsymbol{J}_1&0&\cdots&0\\ 0&0&\boldsymbol{J}_2&\cdots&0\\ \vdots&\vdots&\vdots&\ddots&\vdots\\ 0&0&0&\cdots&\boldsymbol{J}_N\end{bmatrix},\ \boldsymbol{J}_n=n\begin{bmatrix}0&1\\ -1&0\end{bmatrix},\ \boldsymbol{M}_\alpha=\begin{bmatrix}\hat{m}_0\\ \hat{m}_1\\ \vdots\\ \hat{m}_{2N}\end{bmatrix},$$

in which $\hat{m}_i(i=0,1,\ldots,2N)$ are the complicated Fourier coefficients for α^3.

Note that the last four equations in system (3.8) are linear, thus $\mathbf{Q}_{w_i}(i=1,2,3,4)$ can be expressed by $\mathbf{Q}_\alpha$ and $\mathbf{Q}_\xi$ easily. Then, upon substituting $\mathbf{Q}_{w_i}(i=1,2,3,4)$ into the first two equations in (3.8), we obtain

$$\begin{cases}\boldsymbol{A}_1\mathbf{Q}_\alpha+\boldsymbol{B}_1\mathbf{Q}_\xi=\mathbf{0}\\ \boldsymbol{A}_2\mathbf{Q}_\alpha+\boldsymbol{B}_2\mathbf{Q}_\xi+d_{10}\boldsymbol{M}_\alpha=\mathbf{0}\end{cases} \tag{3.10}$$

where

$$\begin{aligned}\boldsymbol{A}_1&=c_1\omega^2\boldsymbol{A}^2+c_3\omega\boldsymbol{A}+c_5\boldsymbol{I}+c_6(\omega\boldsymbol{A}+\varepsilon_1\boldsymbol{I})^{-1}+c_7(\omega\boldsymbol{A}+\varepsilon_2\boldsymbol{I})^{-1},\\ \boldsymbol{B}_1&=c_0\omega^2\boldsymbol{A}^2+c_2\omega\boldsymbol{A}+(c_4+c_{10})\boldsymbol{I}+c_8(\omega\boldsymbol{A}+\varepsilon_1\boldsymbol{I})^{-1}+c_9(\omega\boldsymbol{A}+\varepsilon_2\boldsymbol{I})^{-1},\\ \boldsymbol{A}_2&=d_1\omega^2\boldsymbol{A}^2+d_3\omega\boldsymbol{A}+d_5\boldsymbol{I}+d_6(\omega\boldsymbol{A}+\varepsilon_1\boldsymbol{I})^{-1}+d_7(\omega\boldsymbol{A}+\varepsilon_2\boldsymbol{I})^{-1},\\ \boldsymbol{B}_2&=d_0\omega^2\boldsymbol{A}^2+d_2\omega\boldsymbol{A}+d_4\boldsymbol{I}+d_8(\omega\boldsymbol{A}+\varepsilon_1\boldsymbol{I})^{-1}+d_9(\omega\boldsymbol{A}+\varepsilon_2\boldsymbol{I})^{-1}.\end{aligned}$$

Since the first equation of (3.10) is also linear, we derive $\mathbf{Q}_\xi = -\mathbf{B}_1^{-1}\mathbf{A}_1\mathbf{Q}_\alpha$. Substituting $\mathbf{Q}_\xi$ into the second equation of (3.10) yields the HB algebraic system as follows

$$\left(\mathbf{A}_2 - \mathbf{B}_2\mathbf{B}_1^{-1}\mathbf{A}_1\right)\mathbf{Q}_\alpha + d_{10}\mathbf{M}_\alpha = \mathbf{0}. \tag{3.11}$$

This system has $2N+2$ unknowns, that is, $\omega, \alpha_i (i = 0, 1, \ldots, 2N)$, for $2N+1$ algebraic equations, indicating an under-determined system. In order to obtain a well-determined system, we add an additional condition equation $\alpha_1 = 0$ (or alternatively, $\alpha_2 = 0$), since the phase of the first harmonic in pitch can be fixed for LCOs.

Generally speaking, the HB method is very accurate to obtain the periodic motions of the nonlinear dynamical systems [1,10,11], with just a few harmonics in the approximate solutions. However, when a system exhibits very complex dynamics, a relatively large number of harmonics should be considered. The present two DOF airfoil model has been revealed to show very complex dynamic properties, and at least nine harmonics are required to detect the subcritical Hopf bifurcation as revealed in [1].

Nevertheless, deriving the explicit Fourier coefficient vector $\mathbf{M}_\alpha$ involves intense algebraic operations. The more the harmonics are retained in the approximate solutions, the heavier the algebraic operations would be. The time consumed for the symbolic operations increases dramatically with the increase of the number of retained harmonics. In the present study, the nonlinear term of the HB algebraic system is handled with the help of Mathematica, which eases the pain arising from deriving the Fourier coefficient vector $\mathbf{M}_\alpha$ by hand.

3.2.1 Numerical approximation of Jacobian matrix

Normally, the Jacobian matrix of the HB algebraic system is obtained through numerical approximations. Here, the forward three-point difference formulas are employed to calculate the numerical Jacobian matrix in each iteration. The HB algebraic system (3.11) in conjunction with an additional condition equation can be written in a general form as follows:

$$F_i\left(x_j\right) = 0, \quad i, j = 1, 2, \ldots, n, \tag{3.12}$$

where $n = 2N+2$; $x_j = \alpha_{j-1} (j = 1, 2, \ldots, n-1)$ and $x_n = \omega$.

The analytical expression of the Jacobian matrix of Eqs. (3.12) is

$$\mathbf{J} = \begin{bmatrix} \frac{\partial F_1}{\partial x_1} & \frac{\partial F_1}{\partial x_2} & \cdots & \frac{\partial F_1}{\partial x_n} \\ \frac{\partial F_2}{\partial x_1} & \frac{\partial F_2}{\partial x_2} & \cdots & \frac{\partial F_2}{\partial x_n} \\ \vdots & \vdots & \ddots & \vdots \\ \frac{\partial F_n}{\partial x_1} & \frac{\partial F_n}{\partial x_2} & \cdots & \frac{\partial F_n}{\partial x_n} \end{bmatrix}, \tag{3.13}$$

which will be analytically derived in the next subsection. Here, the numerical differentiation is used to calculate its approximate numerical representation.

Let $\boldsymbol{F}=[F_1,F_2,\ldots,F_n]^{\mathrm{T}}$. Therefore, $\partial\boldsymbol{F}/\partial x_j$ represents the j-th column of $\boldsymbol{J}$. Employing the three-point difference formulas yields

$$\left(\frac{\partial \boldsymbol{F}}{\partial x_j}\right)_{\boldsymbol{x}=\boldsymbol{x}^k}=\frac{1}{2h}\left[-3\boldsymbol{F}\left(\boldsymbol{x}^k\right)+4\boldsymbol{F}\left(\boldsymbol{x}^k+h\boldsymbol{I}_j\right)-\boldsymbol{F}\left(\boldsymbol{x}^k+2h\boldsymbol{I}_j\right)\right],\ j=1,2,\ldots,n, \tag{3.14}$$

where $\boldsymbol{x}^k(k=1,2,\ldots)$ is the solution vector $\boldsymbol{x}$ at the k-th iteration step of an NAE solution algorithm; h is the step size in the three-point difference technique; vector $\boldsymbol{I}_j$ is the j-th column of the n-by-n identity matrix $\boldsymbol{I}$. Note that in the k-th step, the numerical Jacobian matrix is assembled as follows:

$$\boldsymbol{J}^k=\left[\frac{\partial \boldsymbol{F}}{\partial x_1},\frac{\partial \boldsymbol{F}}{\partial x_2},\ldots,\frac{\partial \boldsymbol{F}}{\partial x_n}\right]_{\boldsymbol{x}=\boldsymbol{x}^k}. \tag{3.15}$$

The iterative procedure will terminate when a prescribed convergence criterion of the NAE solver is satisfied. For brevity, the HB method with numerical Jacobian matrix is referred to as HBNJ. Similarly, the HB method with explicit Jacobian matrix is labeled as HBEJ.

3.2.2 Explicit Jacobian matrix of HB

Generally, the Jacobian matrix is required in solving an NAE system. Normally, the Jacobian matrix of an algebraic system is numerically calculated as discussed above. Nevertheless, it is always nice if the Jacobian matrix can be explicitly and succinctly derived. That is because an explicit form of the Jacobian matrix, if available, can significantly accelerate the computing rate of the algebraic solver. Next, we derive the Jacobian matrix of the HB algebraic system.

The HB algebraic system (3.11) is denoted by $\boldsymbol{g}=\boldsymbol{0}$, $\boldsymbol{g}\in\mathbb{R}^{2N+1}$, where

$$\boldsymbol{g}(\boldsymbol{\alpha},\omega)=\left(\boldsymbol{A}_2-\boldsymbol{B}_2\boldsymbol{B}_1^{-1}\boldsymbol{A}_1\right)\mathbf{Q}_\alpha+d_{10}\boldsymbol{M}_\alpha, \tag{3.16}$$

supplemented by an additional equation $g_{ic}=\alpha_1$, which is responsible for remedying the underdetermined system. Consequently, the Jacobian matrix of the algebraic system is expressed as

$$\boldsymbol{B}=\begin{bmatrix}\dfrac{\partial \boldsymbol{g}}{\partial \boldsymbol{\alpha}} & \dfrac{\partial \boldsymbol{g}}{\partial \omega}\\[2ex] \dfrac{\partial g_{ic}}{\partial \boldsymbol{\alpha}} & 0\end{bmatrix}, \tag{3.17}$$

where

$$\frac{\partial g_{ic}}{\partial \boldsymbol{\alpha}} = \left(1, \underbrace{0, \ldots, 0}_{2N}\right). \tag{3.18}$$

It is seen that the Jacobian $\boldsymbol{B}$ has two nontrivial components $\partial \boldsymbol{g}/\partial \boldsymbol{\alpha}$ and $\partial \boldsymbol{g}/\partial \omega$. For the first component,

$$\frac{\partial \boldsymbol{g}}{\partial \boldsymbol{\alpha}} \stackrel{def}{=} \left[\frac{\partial g_i}{\partial \alpha_j}\right] = \left(\boldsymbol{A}_2 - \boldsymbol{B}_2 \boldsymbol{B}_1^{-1} \boldsymbol{A}_1\right) + d_{10} \left[\frac{\partial \hat{m}_i}{\partial \alpha_j}\right], \tag{3.19}$$

in which the explicit expression of $\partial \hat{m}_i/\partial \alpha_j$ can be derived with the aid of Mathematica. $\boldsymbol{M}_\alpha$ is the Fourier coefficient vector of the cubic term $\alpha(\tau)^3$. Considering the first trial solution in Eq. (3.7), the cubic term can be expressed in terms of truncated Fourier series with $3N$ harmonics retained:

$$[\alpha(\tau)]^3 = \hat{m}_0 + \sum_{k=1}^{3N} [\hat{m}_{2k-1} \cos(k\omega\tau) + \hat{m}_{2k} \sin(k\omega\tau)], \tag{3.20}$$

where the Fourier coefficients $\hat{m}_0, \hat{m}_1, \ldots, \hat{m}_{6N}$ are given by the following formulas:

$$\hat{m}_0 = \frac{1}{2\pi} \int_0^{2\pi} \left\{ \alpha_0 + \sum_{n=1}^{N} [\alpha_{2n-1} \cos(n\theta) + \alpha_{2n} \sin(n\theta)] \right\}^3 d\theta, \tag{3.21a}$$

$$\hat{m}_{2k-1} = \frac{1}{\pi} \int_0^{2\pi} \left\{ \alpha_0 + \sum_{n=1}^{N} [\alpha_{2n-1} \cos(n\theta) + \alpha_{2n} \sin(n\theta)] \right\}^3 \cos(k\theta) d\theta, \tag{3.21b}$$

$$\hat{m}_{2k} = \frac{1}{\pi} \int_0^{2\pi} \left\{ \alpha_0 + \sum_{n=1}^{N} [\alpha_{2n-1} \cos(n\theta) + \alpha_{2n} \sin(n\theta)] \right\}^3 \sin(k\theta) d\theta, \tag{3.21c}$$

where $k = 1, 2, \ldots, 3N$, and $\theta \stackrel{def}{=} \omega\tau$, and n is a dummy index. In the HB method, one only needs to balance the coefficients associated with the constant term $\hat{m}_0$ and the first N harmonics $(\hat{m}_1, \ldots, \hat{m}_{2N})$ to obtain the NAEs. All the higher order harmonics, i.e., $k \geq N + 1$, are omitted.

In practice, however, the complex integral operations in Eqs. (3.21) are not necessary. Alternatively, one needs to perform the product-to-sum formula to the integrands of Eqs. (3.21) to obtain the Fourier coefficients $(\hat{m}_0, \hat{m}_1, \ldots, \hat{m}_{2N})$. In

Mathematica, the harmonic balance procedure can be immediately accomplished using *TrigReduce* and *Coefficient* functions. Furthermore, $\partial \hat{m}_i / \partial \alpha_j$ are given by

$$\left[\frac{\partial \hat{m}_i}{\partial \alpha_j}\right] = \begin{bmatrix} \frac{\partial \hat{m}_1}{\partial \alpha_1} & \frac{\partial \hat{m}_1}{\partial \alpha_2} & \cdots & \frac{\partial \hat{m}_1}{\partial \alpha_{2N+1}} \\ \frac{\partial \hat{m}_2}{\partial \alpha_1} & \frac{\partial \hat{m}_2}{\partial \alpha_2} & \cdots & \frac{\partial \hat{m}_2}{\partial \alpha_{2N+1}} \\ \vdots & \vdots & \ddots & \vdots \\ \frac{\partial \hat{m}_{2N+1}}{\partial \alpha_1} & \frac{\partial \hat{m}_{2N+1}}{\partial \alpha_2} & \cdots & \frac{\partial \hat{m}_{2N+1}}{\partial \alpha_{2N+1}} \end{bmatrix}. \tag{3.22}$$

Since we have obtained the Fourier coefficients $(\hat{m}_0, \hat{m}_1, \ldots, \hat{m}_{2N})$, the explicit expression of $\partial \hat{m}_i / \partial \alpha_j$ can be easily derived in Mathematica.

The second component $\partial \boldsymbol{g} / \partial \omega$ is

$$\frac{\partial \boldsymbol{g}}{\partial \omega} = \frac{\partial}{\partial \omega}\left(\boldsymbol{A}_2 - \boldsymbol{B}_2 \boldsymbol{B}_1^{-1} \boldsymbol{A}_1\right)\mathbf{Q}_\alpha. \tag{3.23}$$

Deriving Eq. (3.23) involves a considerable algebra. In Dai et al. [10], a time domain collocation (TDC) method was applied to the same two DOF airfoil model, and the TDC algebraic system was obtained therein. Moreover, the Jacobian matrix of the TDC algebraic system was explicitly derived. We realize that the component $\partial \boldsymbol{g} / \partial \omega$ appears in both the HB algebraic system's and the TDC algebraic system's Jacobian matrices. For the sake of brevity, the $\partial \boldsymbol{g} / \partial \omega$ is given bellow without derivation details.

Apparently, $\partial \boldsymbol{g} / \partial \omega$ consists of two terms

$$\frac{\partial}{\partial \omega}(\boldsymbol{A}_2 \mathbf{Q}_\alpha) \text{ and } -\frac{\partial}{\partial \omega}\left(\boldsymbol{B}_2 \boldsymbol{B}_1^{-1} \boldsymbol{A}_1 \mathbf{Q}_\alpha\right).$$

It is known that $\mathbf{Q}_\alpha$ is independent of ω, so we can derive

$$\frac{\partial}{\partial \omega}(\boldsymbol{A}_2) \text{ and } -\frac{\partial}{\partial \omega}\left(\boldsymbol{B}_2 \boldsymbol{B}_1^{-1} \boldsymbol{A}_1\right).$$

First, and then multiply both of them by $\mathbf{Q}_\alpha$.

The first term is

$$\frac{\partial \boldsymbol{A}_2}{\partial \omega} = 2 d_1 \omega \boldsymbol{A}^2 + d_3 \boldsymbol{A} + d_6 \boldsymbol{W}_{\varepsilon_1} + d_7 \boldsymbol{W}_{\varepsilon_2}, \tag{3.24}$$

where

$$\boldsymbol{W}_{\varepsilon_i}=\frac{\partial \boldsymbol{V}_{\varepsilon_i}}{\partial \omega}=\begin{bmatrix} 0 & & & & \\ & \dfrac{d\boldsymbol{v}_1^{\varepsilon_i}}{d\omega} & & & \\ & & \dfrac{d\boldsymbol{v}_2^{\varepsilon_i}}{d\omega} & & \\ & & & \ddots & \\ & & & & \dfrac{d\boldsymbol{v}_N^{\varepsilon_i}}{d\omega} \end{bmatrix},\quad i=1,2, \tag{3.25}$$

in which

$$\frac{d\boldsymbol{v}_n^{\varepsilon_i}}{d\omega}=-\frac{2n^2\omega}{\left[\varepsilon_i^2+(n\omega)^2\right]^2}\begin{bmatrix} \varepsilon_i & -n\omega \\ n\omega & \varepsilon_i \end{bmatrix}+\frac{1}{\varepsilon_i^2+(n\omega)^2}\begin{bmatrix} 0 & -n \\ n & 0 \end{bmatrix}.$$

The second term is

$$\frac{\partial}{\partial \omega}\left(\boldsymbol{B}_2\boldsymbol{B}_1^{-1}\boldsymbol{A}_1\right)=\boldsymbol{B}_2'\boldsymbol{B}_1^{-1}\boldsymbol{A}_1+\boldsymbol{B}_2\left(\boldsymbol{B}_1^{-1}\right)'\boldsymbol{A}_1+\boldsymbol{B}_2\boldsymbol{B}_1^{-1}\boldsymbol{A}_1', \tag{3.26}$$

where $(\ )'$ is defined as $d(\ )/d\omega$ in this section, and $\boldsymbol{A}_1'$ and $\boldsymbol{B}_2'$ are

$$\boldsymbol{A}_1'=2c_1\omega\boldsymbol{A}^2+c_3\boldsymbol{A}+c_6\boldsymbol{W}_{\varepsilon_1}+c_7\boldsymbol{W}_{\varepsilon_2}, \tag{3.27}$$

$$\boldsymbol{B}_2'=2d_0\omega\boldsymbol{A}^2+d_2A+d_8\boldsymbol{W}_{\varepsilon_1}+d_9\boldsymbol{W}_{\varepsilon_2}. \tag{3.28}$$

In addition, $\boldsymbol{B}_1^{-1}$ and $\left(\boldsymbol{B}_1^{-1}\right)'$ are derived as

$$\boldsymbol{B}_1^{-1}=\begin{bmatrix} \dfrac{1}{c_4+c_{10}+\dfrac{c_8}{\varepsilon_1}+\dfrac{c_9}{\varepsilon_2}} & & & & \\ & \boldsymbol{S}_1 & & & \\ & & \boldsymbol{S}_2 & & \\ & & & \ddots & \\ & & & & \boldsymbol{S}_N \end{bmatrix},$$
$$\left(\boldsymbol{B}_1^{-1}\right)'=\begin{bmatrix} 0 & & & & \\ & \ddots & & & \\ & & \delta_n & & \\ & & & \ddots & \\ & & & & \delta_N \end{bmatrix}, \tag{3.29}$$

where

$$\boldsymbol{S}_n = \boldsymbol{R}_n^{-1} = \frac{1}{r_1^2 + r_2^2}\begin{bmatrix} r_1 & -r_2 \\ r_2 & r_1 \end{bmatrix},$$

$$\delta_n = -\frac{2r_1r_1' + 2r_2r_2'}{\left(r_1^2 + r_2^2\right)^2}\begin{bmatrix} r_1 & -r_2 \\ r_2 & r_1 \end{bmatrix} + \frac{1}{r_1^2 + r_2^2}\begin{bmatrix} r_1' & -r_2' \\ r_2' & r_1' \end{bmatrix},$$

in which

$$r_1 = -c_0\omega^2n^2 + c_4 + c_{10} + \frac{c_8\varepsilon_1}{\varepsilon_1^2 + (n\omega)^2} + \frac{c_9\varepsilon_2}{\varepsilon_2^2 + (n\omega)^2},$$

$$r_2 = c_2\omega n - \frac{c_8 n\omega}{\varepsilon_1^2 + (n\omega)^2} - \frac{c_9 n\omega}{\varepsilon_2^2 + (n\omega)^2},$$

$$r_1' = -2c_0\omega n^2 - \frac{2c_8\varepsilon_1\omega n^2}{\left[\varepsilon_1^2 + (n\omega)^2\right]^2} + \frac{2c_9\varepsilon_2\omega n^2}{\left[\varepsilon_2^2 + (n\omega)^2\right]^2},$$

$$r_2' = c_2 n - \frac{c_8 n\left[\varepsilon_1^2 - (n\omega)^2\right]}{\varepsilon_1^2 + (n\omega)^2} - \frac{c_9 n\left[\varepsilon_2^2 - (n\omega)^2\right]}{\varepsilon_2^2 + (n\omega)^2}.$$

Having derived $\partial \boldsymbol{g}/\partial \omega$, the Jacobian matrix of the HB algebraic system can be assembled as Eq. (3.17). Then, the system of NAEs can be solved by an NAE solver. In numerical examples, a novel Jacobian-inversion-free method, named the optimal iterative algorithm (OIA) [12], which is more robust to initial conditions than the Newton-Raphson method, is applied to solve the NAEs: $\boldsymbol{F}(\boldsymbol{x}) = \boldsymbol{0}$. The procedures of the OIA algorithm are as follows:

(i) Choose a parameter $0 \leq \gamma < 1$, give an initial value of x_0 and compute $\boldsymbol{F}_0 = \boldsymbol{F}(x_0)$.

(ii) For $k = 0, 1, 2 \ldots$, repeat the following computations:

$$\boldsymbol{R}_k = \boldsymbol{B}_k T \boldsymbol{F}_k,$$

$$\boldsymbol{P}_k = \boldsymbol{R}_k - \frac{\left\|\boldsymbol{R}_k\right\|^2}{\boldsymbol{R}_k T \boldsymbol{C}_k \boldsymbol{R}_k}\boldsymbol{C}_k\boldsymbol{R}_k,$$

$$\boldsymbol{v}_1^k = \boldsymbol{B}_k\boldsymbol{R}_k,$$

$$\boldsymbol{v}_2^k = \boldsymbol{B}_k\boldsymbol{P}_k,$$

$$\alpha_k = \frac{1}{1 + \omega_k},$$

$$\beta_k = \frac{\omega_k}{1 + \omega_k},$$

$$\boldsymbol{u}_k = \alpha_k\boldsymbol{R}_k + \beta_k\boldsymbol{P}_k,$$

$$\boldsymbol{v}_k = \alpha_k\boldsymbol{v}_1^k + \beta_k\boldsymbol{v}_2^k,$$

$$\boldsymbol{x}_{k+1} = \boldsymbol{x}_k - (1 - \gamma)\frac{\boldsymbol{F}_k \cdot \boldsymbol{v}_k}{\left\|\boldsymbol{v}_k\right\|^2}\boldsymbol{u}_k.$$

If $\boldsymbol{x}_{k+1}$ satisfies a given stopping criterion $||\boldsymbol{F}_{k+1}|| < \varepsilon$, then stop; otherwise, go to step (ii). In the above, ω_k is computed from

$$\omega_k = -\frac{\left[\boldsymbol{v}_1^k, \boldsymbol{F}_k, \boldsymbol{v}_2^k\right]\cdot\boldsymbol{v}_1^k}{\left[\boldsymbol{v}_1^k, \boldsymbol{F}_k, \boldsymbol{v}_2^k\right]\cdot\boldsymbol{v}_2^k},$$

where [,] is a Jordan algebra formula defined by Liu [15] as $[\boldsymbol{a},\boldsymbol{b},\boldsymbol{c}] = (\boldsymbol{a}\cdot\boldsymbol{b})\boldsymbol{c} - (\boldsymbol{c}\cdot\boldsymbol{b})\boldsymbol{a}$. In the iterative procedures, $\boldsymbol{B}_k$ represents the Jacobian matrix of the k-th step. The parameter γ is 0.3 throughout the paper.

Since the explicit expression of the Jacobian matrix has been obtained, neither symbolic operations nor numerical calculations of the Jacobian are required in the iterative procedures any more, which saves a lot of computing time as will be demonstrated later. In a previous paper [16], the static large deflection of a simply supported von Kármán plate was studied. The governing PDEs with only spatial coordinates were transformed into a set of NAEs by a global Galerkin method. In a similar manner, the explicit Jacobian matrix of the complex Galerkin-resulting NAEs was derived for computational efficiency. This strategy would be applicable to other nonlinear problems, in which a set of NAEs finally emerges.

3.2.3 Mathematical aliasing of HB method

The HB algebraic system (3.11) can be written in a general form as

$$\boldsymbol{F}(\boldsymbol{x}) = \boldsymbol{0}, \tag{3.30}$$

where $\boldsymbol{F} \in \boldsymbol{R}^{2N+1}$, $\boldsymbol{x} \in \boldsymbol{R}^{2N+2}$, and $x_1 = \omega$, $x_{i+1} = \alpha_i, i = 1, 2, \ldots, 2N$. There are $2N + 1$ equations while $2N + 2$ unknowns in the nonlinear algebraic system. Suppose (α, ω) is a physical solution satisfying Eq. (3.30), α_1, α_2 corresponding to the first harmonic response should be dominant, and the harmonic amplitudes A_k should have a descending trend, where

$$A_0 = \alpha_0, \quad A_k = \sqrt{\alpha_{2k-1}^2 + \alpha_{2k}^2}, k = 1, 2, \ldots, N. \tag{3.31}$$

We can write a physical solution of Eq. (3.30) out in a vector form

$$\boldsymbol{x} = [\omega, \alpha_0, \alpha_1, \ldots, \alpha_{2N}]. \tag{3.32}$$

Recall that the assumed approximate solution is

$$\alpha(\tau) = \alpha_0 + \sum_{n=1}^{N}[\alpha_{2n-1}\cos(n\omega\tau) + \alpha_{2n}\sin(n\omega\tau)]. \tag{3.33}$$

Then alternatively, this periodic response having fundamental frequency ω can be represented through another form having frequency ω_a, where $\omega_a = \omega/m, (m = 2, 3, 4, \ldots)$. For example, when $m = 2$, the solution $\boldsymbol{x}$ to the NAEs turns out to be

$$\boldsymbol{x} = \left[\frac{1}{2}\omega, \alpha_0, 0, 0, \alpha_1, \alpha_2, 0, 0, \alpha_3, \alpha_4, \ldots\right]. \tag{3.34}$$

When $m = 3$, the solution is

$$\boldsymbol{x} = \left[\frac{1}{3}\omega, \alpha_0, 0, 0, 0, 0, \alpha_1, \alpha_2, 0, 0, 0, 0, \alpha_3, \alpha_4, \ldots\right], \tag{3.35}$$

and so on for other integer m.

We call the solution having $\omega_a = \omega/m$ a mathematically aliased solution. Because physically, the aliased solutions ($m = 2, 3, \ldots$) represent the same response as the physical solution ($m = 1$). In numerical examples, we find the mathematical aliasing phenomenon in the HB method.

The reason for the mathematical aliasing of the HB method lies in the variable frequency ω. Because ω is an unknown variable, the frequency and each harmonic amplitude can adjust their magnitudes in multiple ways ($m = 1, 2, \ldots$) to satisfy the governing equation (3.30). Among the multiple ways, however, only one is physically meaningful, viz. $m = 1$. The others are called the mathematically aliased solutions of the physical one. In general, this kind of aliasing of the HB method can arise from solving the self-excited dynamical systems where their frequency is treated as an unknown variable rather than a constant. Such examples include the Van der Pol's oscillator [16], a fluttering panel in flow.

3.3 Formulation of the TDC method

In Chapter 2, we proposed a TDC method for the periodic solutions to the Duffing equation and investigated the relationships among the HB method, the HDHB method and the TDC method. In the case of the Duffing oscillator, it is proved that the TDC method, whose number of collocation points is equal to the number of unknown coefficients, was equivalent to the HDHB method. Later, Dai et al. [18] extended the TDC method through increasing the number of collocation points to effectively reduce the effect of aliasing in the Duffing oscillator. Herein, we propose the TDC method for the two-dimensional airfoil problem.

In the TDC method, we first substitute the approximate solutions (3.7) into the governing equations (3.5). Since the solutions are not the exact ones, there must be residual errors. After rearrangement, we have

$$\begin{cases} R_1 = c_0\dddot{\xi} + c_1\ddot{\alpha} + c_2\ddot{\xi} + c_3\dot{\alpha} + (c_4 + c_{10})\xi + c_5\alpha + c_6w_1 + c_7w_2 + c_8w_3 + c_9w_4 \neq 0, \\ R_2 = d_0\ddot{\xi} + d_1\ddot{\alpha} + d_2\dot{\xi} + d_3\dot{\alpha} + d_4\xi + d_5\alpha + d_6w_1 + d_7w_2 + d_8w_3 + d_9w_4 + d_{10}\alpha^3 \neq 0, \\ R_3 = \dot{w}_1 + \varepsilon_1 w_1 - \alpha \neq 0, \\ R_4 = \dot{w}_2 + \varepsilon_2 w_2 - \alpha \neq 0, \\ R_5 = \dot{w}_3 + \varepsilon_1 w_3 - \xi \neq 0, \\ R_6 = \dot{w}_4 + \varepsilon_2 w_4 - \xi \neq 0, \end{cases} \tag{3.36}$$

where R_j represent $R_j\left(\alpha_0, \ldots, \alpha_{2N}, \xi_0, \ldots, \xi_{2N}, w_0^i, \ldots, w_{2N}^i, \omega, \tau\right), (i = 1, 2, 3, 4;\ j = 1, \ldots, 6)$. Now, we enforce the residual functions R_j to be zero at $2N+1$ equidistant time points τ_i over a period. Therefore, the TDC algebraic system (3.36) for the two-dimensional airfoil problem are obtained, and there are $6 \times (2N+1)$ equations and $6 \times (2N+1)+1$ unknowns.

Nevertheless, system (3.36) of the TDC method is not what we desired. Because in this system there are $2N+1$ harmonics which is the same as the HDHB and the HB method, however, the size of the system is six times larger. In this section, we will transform the original TDC system (3.36) to a mathematically equivalent compact TDC system.

We will process each collocation resulting term in the system (3.16). By collocating $\alpha(\tau)$ at $2N+1$ time points τ_i, we have

$$\alpha(\tau_i) = \alpha_0 + \sum_{n=1}^{N} [\alpha_{2n-1}\cos(n\omega\tau_i) + \alpha_{2n}\sin(n\omega\tau_i)]. \tag{3.37}$$

Let $\theta_i = \omega\tau_i$, Eq. (3.37) can be rewritten in a matrix form

$$\begin{bmatrix} \alpha(\tau_0) \\ \alpha(\tau_1) \\ \vdots \\ \alpha(\tau_{2N}) \end{bmatrix} = \begin{bmatrix} 1 & \cos\theta_0 & \sin\theta_0 & \cdots & \cos N\theta_0 & \sin N\theta_0 \\ 1 & \cos\theta_1 & \sin\theta_1 & \cdots & \cos N\theta_1 & \sin N\theta_1 \\ \vdots & \vdots & \vdots & \ddots & \vdots & \vdots \\ 1 & \cos\theta_{2N} & \sin\theta_{2N} & \cdots & \cos N\theta_{2N} & \sin N\theta_{2N} \end{bmatrix} \begin{bmatrix} \alpha_0 \\ \alpha_1 \\ \vdots \\ \alpha_{2N} \end{bmatrix}. \tag{3.38}$$

Therefore

$$\tilde{\mathbf{Q}}_\alpha \equiv \begin{bmatrix} \alpha(\tau_0) \\ \alpha(\tau_1) \\ \vdots \\ \alpha(\tau_{2N}) \end{bmatrix} = \boldsymbol{F}\mathbf{Q}_\alpha, \tag{3.39}$$

where $\boldsymbol{F}$ is in fact the discrete Fourier transform (DFT) matrix in Eq. (3.38). In the same manner, we have $\tilde{\mathbf{Q}}_{\xi} = \boldsymbol{F}\mathbf{Q}_{\xi}, \tilde{\mathbf{Q}}_{w_i} = \boldsymbol{F}\mathbf{Q}_{w_i}, (i = 1, 2, 3, 4)$. Similarly, collocating $\dot{\alpha}(\tau)$ at $2N + 1$ equidistant time points τ_i yields

$$\dot{\alpha}(\tau_i) = \sum_{n=1}^{N} [-n\omega\alpha_{2n-1}\sin n\omega\tau_i + n\omega\alpha_{2n}\cos n\omega\tau_i]. \tag{3.40}$$

The above equation can be written in a matrix form

$$\begin{bmatrix} \dot{\alpha}(\tau_0) \\ \dot{\alpha}(\tau_1) \\ \vdots \\ \dot{\alpha}(\tau_{2N}) \end{bmatrix} = \omega \begin{bmatrix} 0 & -\sin\theta_0 & \cos\theta_0 & \cdots & -N\sin N\theta_0 & N\cos N\theta_0 \\ 0 & -\sin\theta_1 & \cos\theta_1 & \cdots & -N\sin N\theta_1 & N\cos N\theta_1 \\ \vdots & \vdots & \vdots & \ddots & \vdots & \vdots \\ 0 & -\sin\theta_{2N} & \cos\theta_{2N} & \cdots & -N\sin N\theta_{2N} & N\cos N\theta_{2N} \end{bmatrix} \begin{bmatrix} \alpha_0 \\ \alpha_1 \\ \vdots \\ \alpha_{2N} \end{bmatrix}. \tag{3.41}$$

Interestingly, we find that the square matrix in the above equation is equal to $\boldsymbol{FA}$. Thus, we have

$$\tilde{\mathbf{Q}}_{\dot{\alpha}} = \omega\boldsymbol{FA}\mathbf{Q}_{\alpha}, \tag{3.42}$$

where

$$\tilde{\mathbf{Q}}_{\dot{\alpha}} \equiv \begin{bmatrix} \dot{\alpha}(\tau_0) \\ \dot{\alpha}(\tau_1) \\ \vdots \\ \dot{\alpha}(\tau_{2N}) \end{bmatrix}. \tag{3.43}$$

In the same manner, $\tilde{\mathbf{Q}}_{\dot{\xi}} = \omega\boldsymbol{FA}\mathbf{Q}_{\xi}, \tilde{\mathbf{Q}}_{\dot{w}_j} = \omega\boldsymbol{FA}\mathbf{Q}_{w_j}, (j = 1, 2, 3, 4)$, where $\tilde{\mathbf{Q}}_{\dot{\xi}}$ and $\tilde{\mathbf{Q}}_{\dot{w}_j}$ are defined similarly as $\tilde{\mathbf{Q}}_{\dot{\alpha}}$ in Eq. (3.43).

Furthermore, collocating $\ddot{\alpha}(\tau)$ at $2N + 1$ equidistant time points τ_i, we have

$$\ddot{\alpha}(\tau_i) = \sum_{n=1}^{N} \left[-n^2\omega^2\alpha_{2n-1}\cos n\omega\tau_i - n^2\omega^2\alpha_{2n}\sin n\omega\tau_i\right]. \tag{3.44}$$

Eq. (3.44) is written in a matrix form

$$\begin{bmatrix} \ddot{\alpha}(\tau_0) \\ \ddot{\alpha}(\tau_1) \\ \vdots \\ \ddot{\alpha}(\tau_{2N}) \end{bmatrix} = \omega^2 \begin{bmatrix} 0 & -\cos\theta_0 & -\sin\theta_0 & \cdots & -N^2\cos N\theta_0 & -N^2\sin N\theta_0 \\ 0 & -\cos\theta_1 & -\sin\theta_1 & \cdots & -N^2\cos N\theta_1 & -N^2\sin N\theta_1 \\ \vdots & \vdots & \vdots & \ddots & \vdots & \vdots \\ 0 & -\cos\theta_{2N} & -\sin\theta_{2N} & \cdots & -N^2\cos N\theta_{2N} & -N^2\sin N\theta_{2N} \end{bmatrix} \begin{bmatrix} \alpha_0 \\ \alpha_1 \\ \vdots \\ \alpha_{2N} \end{bmatrix}. \tag{3.45}$$

Note that the square matrix in the above equation is equal to $\boldsymbol{FA}^2$. Therefore,

$$\tilde{\mathbf{Q}}_{\ddot{\alpha}} \equiv \begin{bmatrix} \ddot{\alpha}(\tau_0) \\ \ddot{\alpha}(\tau_1) \\ \vdots \\ \ddot{\alpha}(\tau_{2N}) \end{bmatrix} = \omega^2 \boldsymbol{FA}^2 \mathbf{Q}_\alpha. \tag{3.46}$$

Similarly, $\tilde{\mathbf{Q}}_{\ddot{\xi}} = \omega^2 \boldsymbol{FA}^2 \mathbf{Q}_\xi$, where $\tilde{\mathbf{Q}}_{\ddot{\xi}}$ is defined in the same way as $\tilde{\mathbf{Q}}_{\ddot{\alpha}}$.

Upon transforming the TDC algebraic system (3.36) into the corresponding matrix form, we obtain:

$$\begin{cases} \left(c_0\omega^2\boldsymbol{D}^2 + c_2\omega\boldsymbol{D} + c_4\boldsymbol{I} + c_{10}\boldsymbol{I}\right)\tilde{\mathbf{Q}}_\xi + \left(c_1\omega^2\boldsymbol{D}^2 + c_3\omega\boldsymbol{D} + c_5\boldsymbol{I}\right)\tilde{\mathbf{Q}}_\alpha + \sum_{i=1}^{4} c_{i+5}\tilde{\mathbf{Q}}_{w_i} = \mathbf{0}, \\ \left(d_0\omega^2\boldsymbol{D}^2 + d_2\omega\boldsymbol{D} + d_4\boldsymbol{I}\right)\tilde{\mathbf{Q}}_\xi + \left(d_1\omega^2\boldsymbol{D}^2 + d_3\omega\boldsymbol{D} + d_5\boldsymbol{I}\right)\tilde{\mathbf{Q}}_\alpha + \sum_{i=1}^{4} d_{i+5}\tilde{\mathbf{Q}}_{w_i} + d_{10}\tilde{\boldsymbol{M}}_\alpha = \mathbf{0}, \\ (\omega\boldsymbol{D} + \varepsilon_1\boldsymbol{I})\tilde{\mathbf{Q}}_{\omega_1} - \tilde{\mathbf{Q}}_\alpha = \mathbf{0}, \\ (\omega\boldsymbol{D} + \varepsilon_2\boldsymbol{I})\tilde{\mathbf{Q}}_{\omega_2} - \tilde{\mathbf{Q}}_\alpha = \mathbf{0}, \\ (\omega\boldsymbol{D} + \varepsilon_1\boldsymbol{I})\tilde{\mathbf{Q}}_{\omega_3} - \tilde{\mathbf{Q}}_\xi = \mathbf{0}, \\ (\omega\boldsymbol{D} + \varepsilon_2\boldsymbol{I})\tilde{\mathbf{Q}}_{\omega_4} - \tilde{\mathbf{Q}}_\xi = \mathbf{0}, \end{cases} \tag{3.47}$$

where $\boldsymbol{D} = \boldsymbol{FAF}^{-1}$ and

$$\tilde{\boldsymbol{M}}_\alpha = \begin{bmatrix} \alpha^3(\tau_0) \\ \alpha^3(\tau_1) \\ \vdots \\ \alpha^3(\tau_{2N}) \end{bmatrix}. \tag{3.48}$$

It should be emphasized that no approximation is adopted during the entire derivation. Therefore, we have demonstrated the equivalence of the time domain collocation (TDC) method and the high dimensional harmonic balance (HDHB) method. Essentially, the HDHB method is a collocation method in disguise.

System (3.47) can be simplified to a compact form, because apart from the second equation, all other equations are linear. Thus, we can express $\tilde{\mathbf{Q}}_\xi$ and $\tilde{\mathbf{Q}}_{w_i}$ by $\tilde{\mathbf{Q}}_\alpha$ first, and then substitute them into the nonlinear equation, obtaining

$$\left(\boldsymbol{A}_{2\alpha} - \boldsymbol{B}_{2\xi}\boldsymbol{B}_{1\xi}^{-1}\boldsymbol{A}_{1\alpha}\right)\tilde{\mathbf{Q}}_\alpha + d_{10}\tilde{\boldsymbol{M}}_\alpha = \mathbf{0}, \tag{3.49}$$

in which $\boldsymbol{A}_{1\alpha}$, $\boldsymbol{B}_{1\xi}$, $\boldsymbol{A}_{2\alpha}$ and $\boldsymbol{B}_{2\xi}$ are as follows

$$\begin{aligned} \boldsymbol{A}_{1\alpha} &= c_1\omega^2\boldsymbol{D}^2 + c_3\omega\boldsymbol{D} + c_5\boldsymbol{I} + c_6(\omega\boldsymbol{D}+\varepsilon_1\boldsymbol{I})^{-1} + c_7(\omega\boldsymbol{D}+\varepsilon_2\boldsymbol{I})^{-1}, \\ \boldsymbol{B}_{1\xi} &= c_0\omega^2\boldsymbol{D}^2 + c_2\omega\boldsymbol{D} + (c_4 + c_{10})\boldsymbol{I} + c_8(\omega\boldsymbol{D}+\varepsilon_1\boldsymbol{I})^{-1} + c_9(\omega\boldsymbol{D}+\varepsilon_2\boldsymbol{I})^{-1}, \\ \boldsymbol{A}_{2\alpha} &= d_1\omega^2\boldsymbol{D}^2 + d_3\omega\boldsymbol{D} + d_5\boldsymbol{I} + d_6(\omega\boldsymbol{D}+\varepsilon_1\boldsymbol{I})^{-1} + d_7(\omega\boldsymbol{D}+\varepsilon_2\boldsymbol{I})^{-1}, \\ \boldsymbol{B}_{2\xi} &= d_0\omega^2\boldsymbol{D}^2 + d_2\omega\boldsymbol{D} + d_4\boldsymbol{I} + d_8(\omega\boldsymbol{D}+\varepsilon_1\boldsymbol{I})^{-1} + d_9(\omega\boldsymbol{D}+\varepsilon_2\boldsymbol{I})^{-1}. \end{aligned}$$

System (3.49) is the compact TDC algebraic system in terms of time domain variables and ω. Using $\tilde{\mathbf{Q}}_\alpha = \boldsymbol{F}\mathbf{Q}_\alpha$ and $\tilde{\boldsymbol{M}}_\alpha = (\boldsymbol{F}\mathbf{Q}_\alpha)^3$, system (3.49) can be transformed to

$$\left(\boldsymbol{A}_2 - \boldsymbol{B}_2\boldsymbol{B}_1^{-1}\boldsymbol{A}_1\right)\mathbf{Q}_\alpha + d_{10}\boldsymbol{F}^{-1}(\boldsymbol{F}\mathbf{Q}_\alpha)^3 = \mathbf{0}, \tag{3.50}$$

which is the compact TDC algebraic system with $2N + 2$ unknowns in terms of Fourier coefficient variables and ω. In this chapter, we use exclusively the compact TDC algebraic system (3.50) in the remainder of this chapter. It should be noted that the HB algebraic system (3.11) is distinct from the TDC algebraic system (3.50) by the last term, which is the source of the spurious solutions arising from the TDC system.

3.3.1 Explicit Jacobian matrix of TDC

In the course of solving a system of NAEs, the Jacobian matrix of the system is necessary. An explicit form Jacobian matrix, if available, may significantly accelerate the computing rate of the algebraic solver. Therefore, we derive explicitly the Jacobian matrix for the TDC algebraic system.

Let the TDC algebraic system be denoted by $\boldsymbol{g} \in \mathbb{R}^{2N+1}$ as

$$\boldsymbol{g}(\boldsymbol{\alpha}, \omega) = \left(\boldsymbol{A}_2 - \boldsymbol{B}_2\boldsymbol{B}_1^{-1}\boldsymbol{A}_1\right)\mathbf{Q}_\alpha + d_{10}\boldsymbol{E}(\boldsymbol{F}\mathbf{Q}_\alpha)^3, \tag{3.51}$$

where the vector $\alpha \in \mathbb{R}^{2N+1}$, the scalar $\omega \in \mathbb{R}$, and $\boldsymbol{E} = \boldsymbol{F}^{-1}$.

The supplemental equation corresponding to the imposed condition is denoted by g_{ic}. Consequently, the Jacobian matrix of the TDC algebraic system is

$$\boldsymbol{B} = \begin{bmatrix} \dfrac{\partial \boldsymbol{g}}{\partial \boldsymbol{\alpha}} & \dfrac{\partial \boldsymbol{g}}{\partial \omega} \\ \dfrac{\partial g_{ic}}{\partial \boldsymbol{\alpha}} & 0 \end{bmatrix}. \tag{3.52}$$

For the sake of brevity, let $\boldsymbol{L} = \boldsymbol{A}_2 - \boldsymbol{B}_2\boldsymbol{B}_1^{-1}\boldsymbol{A}_1$. Equation (3.51) can then be written as

$$\begin{aligned} g_i(\alpha_0, \alpha_1, \ldots \alpha_{2N}, \omega) &= L_{ij}(\omega)\alpha_j + d_{10}E_{ik}\left(F_{kj}\alpha_j\right)^3, \\ \frac{\partial g_i}{\partial \alpha_m} &= L_{ij}\delta_{jm} + d_{10}E_{ik} \times 3\left(F_{kj}\alpha_j\right)^2 \times \frac{\partial\left(F_{kj}\alpha_j\right)}{\partial \alpha_m} \\ &= L_{im} + d_{10}E_{ik} \times 3\left(F_{kj}\alpha_j\right)^2 \times F_{km}, \ (\text{sum over } j \text{ and } k), \end{aligned}$$

where δ_{jm} is the Kronecker delta. Deriving the component $\partial \boldsymbol{g}/\partial \omega$ would involve a considerable algebra, and the derivation details are provided in Appendix.

The imposed condition equation g_{ic} supplementing to the main algebraic equations is $\alpha_1 = 0$ in this problem. Thus, the component $\partial g_{ic}/\partial \boldsymbol{\alpha}$ in the Jacobian matrix can be immediately calculated. The imposed condition is applied for two purposes: (i) making the original under-determined system (3.50) into a well-determined system; (ii) fixing the phase of the response.

Having derived the explicit form Jacobian matrix, the TDC algebraic system can be readily solved by a NAE solver. With the explicit expression of the Jacobian matrix, the computational cost will be significantly reduced due to the avoidance of symbolic calculations required by the Jacobian matrix.

3.3.2 Mathematical aliasing of the TDC method

The HDHB method may produce spurious solutions in addition to the physically meaningful solutions. Previously, the HDHB method was misinterpreted as a kind of time domain harmonic balance method. Hence, the aliasing phenomenon belonging to the HDHB has not been explained clearly until now. In this section, based on the mathematical equivalence between the HDHB method and the TDC method, we will explicitly explain how the "aliasing" works in the TDC method (or equivalent HDHB method).

It is shown that the HB system (3.11) is distinct from the TDC system (3.50) by the last term. In general, the HB system does not accommodate non-physical spurious solutions (mathematically aliased solutions are physically meaningful) while the TDC/HDHB system does. Therefore, we ought to compare $\boldsymbol{M}_\alpha$ in the HB system and $\boldsymbol{E}(\boldsymbol{F}\boldsymbol{Q}_\alpha)^3$ of the TDC system carefully, since the difference between the two terms is obviously the source of spurious solutions.

It is straightforward to compare the two terms for a case with very few harmonics. Let

$$\boldsymbol{E}(\boldsymbol{F}\boldsymbol{Q}_\alpha)^3 = \boldsymbol{P}_\alpha = \begin{bmatrix} p_0 \\ p_1 \\ \vdots \\ p_{2N} \end{bmatrix}. \tag{3.53}$$

$\boldsymbol{M}_\alpha$ is as a previously defined coefficient vector of α^3. For $N=1$, the Fourier components for the cubic term of the HB1 system, that is i.e. HB system with one harmonic included, are

$$\boldsymbol{M}_\alpha^{(1)} = \left\{ \begin{array}{c} \left(A_0^2 + \dfrac{3}{2}A_1^2\right)\alpha_0 \\ \left(3A_0^2 + \dfrac{3}{4}A_1^2\right)\alpha_1 \\ \left(3A_0^2 + \dfrac{3}{4}A_1^2\right)\alpha_2 \end{array} \right\}. \tag{3.54}$$

In the TDC1 system, i.e. TDC system with one harmonic included, the Fourier components for the cubic term are

$$\boldsymbol{P}_{\alpha}^{(1)}=\left\{\begin{array}{l}\left(A_0^2+\dfrac{3}{2}A_1^2\right)\alpha_0+\dfrac{1}{4}\alpha_1^3-\dfrac{3}{4}\alpha_1\alpha_2^2\\\left(3A_0^2+\dfrac{3}{4}A_1^2\right)\alpha_1+\dfrac{3}{2}\left(\alpha_1^2-\alpha_2^2\right)\alpha_0\\\left(3A_0^2+\dfrac{3}{4}A_1^2\right)\alpha_2-3\alpha_0\alpha_1\alpha_2\end{array}\right\}. \tag{3.55}$$

Comparison between (3.54) and (3.55) shows that TDC1 system contains all the terms appearing in the HB1 system plus some additional terms.

Next, we examine HB2 and TDC2. The Fourier components of the HB2 system for the cubic term are

$$\boldsymbol{M}_{\alpha}^{(2)}\equiv\left\{\begin{array}{c}m_0^{(2)}\\m_1^{(2)}\\m_2^{(2)}\\m_3^{(2)}\\m_4^{(2)}\end{array}\right\}=\left\{\begin{array}{l}A_0^2\alpha_0+\dfrac{3}{2}A_1^2\alpha_0+\dfrac{3}{2}A_2^2\alpha_0+\dfrac{3}{4}\left(\alpha_1^2-\alpha_2^2\right)\alpha_3+\dfrac{3}{2}\alpha_1\alpha_2\alpha_4\\3A_0^2\alpha_1+\dfrac{3}{4}A_1^2\alpha_1+\dfrac{3}{2}A_2^2\alpha_1+3\alpha_0\alpha_1\alpha_3+3\alpha_0\alpha_2\alpha_4\\3A_0^2\alpha_2+\dfrac{3}{4}A_1^2\alpha_2+\dfrac{3}{2}A_2^2\alpha_2+3\alpha_0\alpha_1\alpha_4-3\alpha_0\alpha_2\alpha_3\\3A_0^2\alpha_3+\dfrac{3}{2}A_1^2\alpha_3+\dfrac{3}{4}A_2^2\alpha_3+\dfrac{3}{2}\left(\alpha_1^2-\alpha_2^2\right)\alpha_0\\3A_0^2\alpha_4+\dfrac{3}{2}A_1^2\alpha_4+\dfrac{3}{4}A_2^2\alpha_4+3\alpha_0\alpha_1\alpha_2\end{array}\right\}. \tag{3.56}$$

In the TDC2 system, the Fourier components for the cubic term are

$$\boldsymbol{P}_{\alpha}^{(2)}=\left\{\begin{array}{l}m_0+\dfrac{3}{4}\left(\alpha_3^2-\alpha_4^2\right)\alpha_1-\dfrac{3}{2}\alpha_2\alpha_3\alpha_4\\m_1+\dfrac{3}{4}\left(\alpha_1^2-\alpha_2^2\right)\alpha_3+\dfrac{3}{2}\left(\alpha_3^2-\alpha_4^2\right)\alpha_0+\dfrac{1}{4}\alpha_3^3-\dfrac{3}{4}\alpha_3\alpha_4^2-\dfrac{3}{2}\alpha_1\alpha_2\alpha_4\\m_2-\dfrac{3}{4}\left(\alpha_1^2-\alpha_2^2\right)\alpha_4-\dfrac{1}{4}\alpha_4^3+\dfrac{3}{4}\alpha_3^2\alpha_4-\dfrac{3}{2}\alpha_1\alpha_2\alpha_3-3\alpha_0\alpha_3\alpha_4\\m_3+\dfrac{1}{4}\alpha_1^3+\dfrac{3}{4}\left(\alpha_3^2-\alpha_4^2\right)\alpha_1-\dfrac{3}{4}\alpha_1\alpha_2^2+\dfrac{3}{2}\alpha_2\alpha_3\alpha_4+3\alpha_0\alpha_1\alpha_3-3\alpha_0\alpha_2\alpha_4\\m_4+\dfrac{1}{4}\alpha_2^3+\dfrac{3}{4}\left(\alpha_3^2-\alpha_4^2\right)\alpha_2-\dfrac{3}{4}\alpha_1^2\alpha_2-\dfrac{3}{2}\alpha_1\alpha_3\alpha_4-3\alpha_0\alpha_2\alpha_3-3\alpha_0\alpha_1\alpha_4\end{array}\right\}. \tag{3.57}$$

It reveals that the TDC2 system contains all terms in HB2 system, plus some additional terms. This conclusion has been drawn in Liu et al. [17] in comparing the HDHB and the HB methods, wherein they also proved that the extra terms in the HDHB system are in fact the higher order harmonic terms related to the nonlinear term in the HB system which are discarded in the course of balancing each individual harmonic in the HB method.

So far, we are partially clear about the aliasing phenomenon of the TDC/HDHB method. One may have a question in mind: are there some explicit aliasing rules to determine how the additional terms are plugged in? The exact deployment rules of the additional terms and the mechanism of their occurrence are of fundamental interests, and yet remain to be clarified. Next, we will explain the aliasing mechanism clearly.

To illustrate, we step back to the case with $N=1$. The assumed solution is

$$\alpha(\tau) = \alpha_0 + \alpha_1 \cos\omega\tau + \alpha_2 \sin\omega\tau. \tag{3.58}$$

Hence, the cubic term is

$$\begin{aligned}
\alpha^3(\tau) &= (\alpha_0 + \alpha_1 \cos\omega\tau + \alpha_2 \sin\omega\tau)^3 \\
&= \left(A_0^2 + \frac{3}{4}A_1^2\right)\alpha_0 + \left(3A_0^2 + \frac{3}{4}A_1^2\right)\alpha_1 \cos\omega\tau + \left(3A_0^2 + \frac{3}{4}A_1^2\right)\alpha_2 \sin\omega\tau \\
&\quad + \frac{3}{2}\left(\alpha_1^2 - \alpha_2^2\right)\alpha_0 \cos 2\omega\tau + 3\alpha_0\alpha_1\alpha_2 \sin 2\omega\tau + \frac{1}{4}\left(\alpha_1^2 - 3\alpha_2^2\right)\alpha_1 \cos 3\omega\tau \\
&\quad + \frac{1}{4}\left(3\alpha_1^2 - \alpha_2^2\right)\alpha_2 \sin 3\omega\tau.
\end{aligned} \tag{3.59}$$

In principle, the HB1 algebraic system retains the coefficients of the zeroth and the first harmonics during the harmonic balance procedure, while higher harmonics are not taken into account, see formula (3.54). However, expression (3.55) of the TDC1 system retains the coefficients of all harmonics in (3.59), but in a seemingly disordered way. In the present study, efforts have been made towards the discovery of a clear relationship between the TDC system and the HB system.

Firstly, according to observation, we reveal the aliasing rules of the TDC1 system to HB1 system as follows:

- $\cos 2\omega\tau$ is aliased to "$\cos\omega\tau$"
- $\sin 2\omega\tau$ is aliased to "$-\sin\omega\tau$"

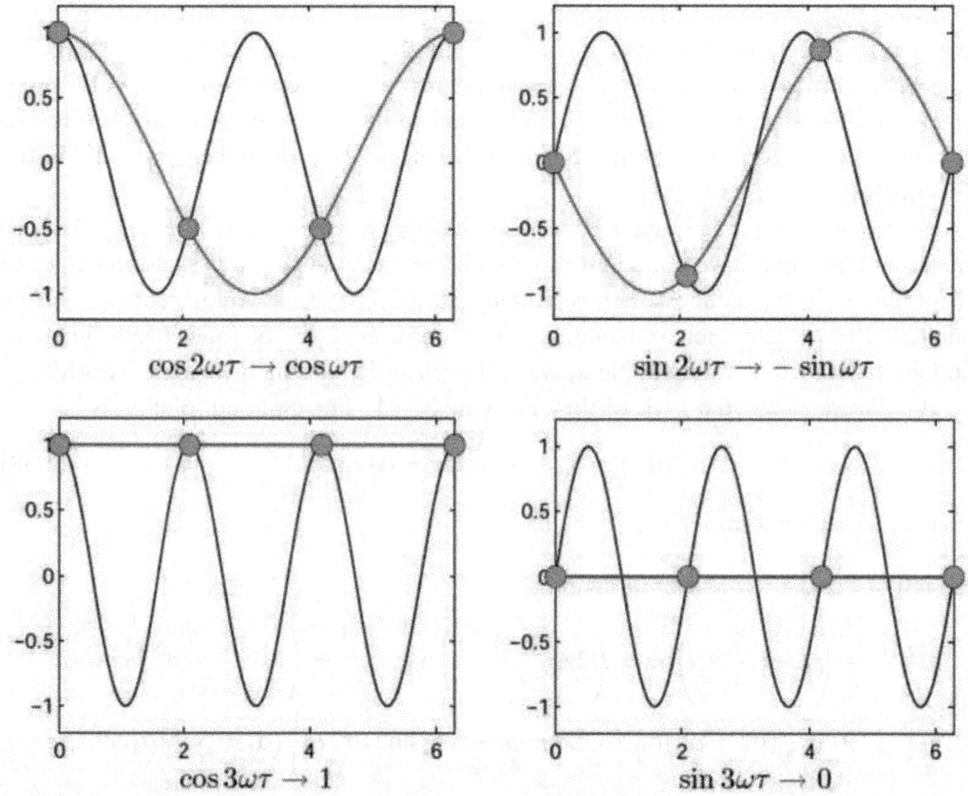

Figure 3.2 Schematic diagram: aliasing for the case with $N=1$.

- $\cos 3\omega\tau$ is aliased to "1"
- $\sin 3\omega\tau$ is aliased to "0"

The first rule means the coefficient of $\cos 2\omega\tau$ contributes to the coefficient of $\cos\omega\tau$, and so forth for the other three. For example, $\left(3A_0^2+\frac{3}{4}A_1^2\right)\alpha_1$ of $\cos\omega\tau$ plus $\frac{3}{2}\left(\alpha_1^2-\alpha_2^2\right)\alpha_0$ of $\cos 2\omega\tau$ of HB system compose the second component in (3.55) of the TDC system. The aliasing rules can be interpreted by Fig. 3.2.

Similarly, when $N=2$, the assumed solution is

$$\alpha(\tau)=\alpha_0+\alpha_1\cos\omega\tau+\alpha_2\sin\omega\tau+\alpha_3\cos 2\omega\tau+\alpha_4\sin 2\omega\tau, \tag{3.60}$$

and the cubic term is

$$
\begin{aligned}
\alpha^3(\tau) &= (\alpha_0+\alpha_1\cos\omega\tau+\alpha_2\sin\omega\tau+\alpha_3\cos2\omega\tau+\alpha_4\sin2\omega\tau)^3 \\
&= m_0^{(2)} + m_1^{(2)}\cos\omega\tau + m_2^{(2)}\sin\omega\tau + m_3^{(2)}\cos2\omega\tau + m_4^{(2)}\sin2\omega\tau \\
&\quad + \left(-\frac{3}{4}\alpha_1\alpha_2^2 + \frac{1}{4}\alpha_1^3 + \frac{3}{2}\alpha_2\alpha_3\alpha_4 - 3\alpha_0\alpha_2\alpha_4 + 3\alpha_0\alpha_1\alpha_2 + \frac{3}{4}(\alpha_3^2-\alpha_4^2)\alpha_1\right)\cos3\omega\tau \\
&\quad + \left(\frac{3}{4}\alpha_1^2\alpha_2 - \frac{1}{4}\alpha_2^3 + \frac{3}{2}\alpha_1\alpha_3\alpha_4 + 3\alpha_0\alpha_1\alpha_4 + 3\alpha_0\alpha_2\alpha_3 - \frac{3}{4}(\alpha_3^2-\alpha_4^2)\alpha_2\right)\sin3\omega\tau \\
&\quad + \left(-\frac{3}{2}\alpha_1\alpha_2\alpha_4 + \frac{3}{2}(\alpha_3^2-\alpha_4^2)\alpha_0 + \frac{3}{4}(\alpha_1^2-\alpha_2^2)\alpha_3\right)\cos4\omega\tau \\
&\quad + \left(\frac{3}{2}\alpha_1\alpha_2\alpha_3 + 3\alpha_0\alpha_3\alpha_4 + \frac{3}{4}(\alpha_1^2-\alpha_2^2)\alpha_4\right)\sin4\omega\tau \\
&\quad + \left(-\frac{3}{2}\alpha_2\alpha_3\alpha_4 + \frac{3}{4}(\alpha_3^2-\alpha_4^2)\alpha_1\right)\cos5\omega\tau + \left(\frac{3}{2}\alpha_1\alpha_3\alpha_4 + \frac{3}{4}(\alpha_3^2-\alpha_4^2)\alpha_2\right)\sin5\omega\tau \\
&\quad + \left(\frac{1}{4}\alpha_3^3 - \frac{3}{4}\alpha_3\alpha_4^2\right)\cos6\omega\tau + \left(-\frac{1}{4}\alpha_4^3 + \frac{3}{4}\alpha_3^2\alpha_4\right)\sin6\omega\tau.
\end{aligned}
\tag{3.61}
$$

The aliasing rules for the TDC2 system to the HB2 system are found to be

- $\cos3\omega\tau$ is aliased to "$\cos2\omega\tau$"
- $\sin3\omega\tau$ is aliased to "$-\sin2\omega\tau$"
- $\cos4\omega\tau$ is aliased to "$\cos\omega\tau$"
- $\sin4\omega\tau$ is aliased to "$-\sin\omega\tau$"
- $\cos5\omega\tau$ is aliased to "1"
- $\sin5\omega\tau$ is aliased to "0"
- $\cos6\omega\tau$ is aliased to "$\cos\omega\tau$"
- $\sin6\omega\tau$ is aliased to "$\sin\omega\tau$"

According to the aliasing rules, for example, the coefficient of $\sin\omega\tau$ of the TDC2 system is comprised of three members (see Fig. 3.3): the coefficient of $\sin\omega\tau$ of the HB2 system, $m_2^{(2)}$, the negative coefficient of $\sin4\omega\tau$, and the coefficient of $\sin6\omega\tau$. One can easily check the aliasing rules for the rest coefficients.

In the above, we have revealed some aliasing rules by observation. Next, a generalized aliasing rule will be provided rigorously. Hence, the aliasing phenomenon arising out of the TDC method will be interpreted clearly. It is important to realize that the TDC method is closely related to the pseudo-spectral method. As is known that the

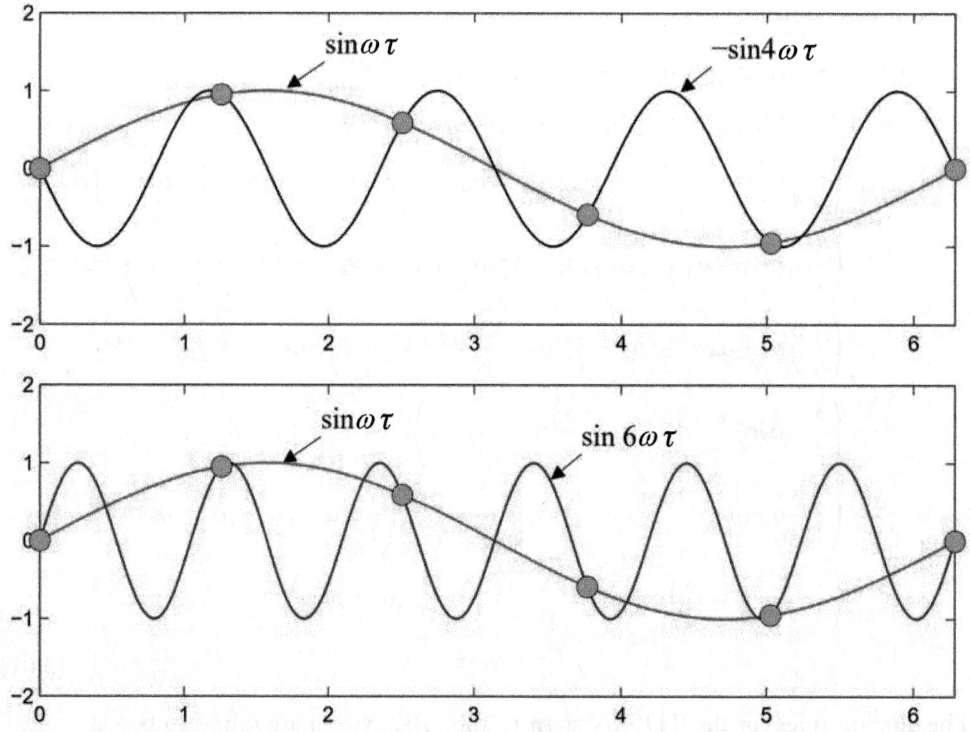

Figure 3.3 Illustration of aliasing.

pseudo-spectral method would give rise to the frequency aliasing problem in general. The aliasing of the pseudo-spectral method is mainly encountered in the community of fluid mechanics [19]. Herein based on the aliasing rules of the pseudo-spectral method, the aliasing of the TDC/HDHB can be explained by the following remark:

Remark

If the interval $\alpha \in [0, 2\pi]$ is discretized with a uniform collocation spacing h, then the wavenumbers in the trigonometric interpolant lie on the range $n \in [-L, L]$ where $L = \pi/h$, and is the so-called "aliasing limit" wavenumber. The Fourier functions are $\cos nx$, $\sin nx$ with n being an integer. Higher wavenumbers n such as $|n| > L$ will appear as if they were the wavenumbers

$$n_a = n \pm 2mL, \tag{3.62}$$

where $n_a \in [-L, L]$ and m is an integer. The wavenumbers beyond the range $[-L, L]$ are said to be aliased to wavenumbers within this range and n_a is the alias of n. The higher wavenumbers (higher harmonics) arise from the nonlinear term.

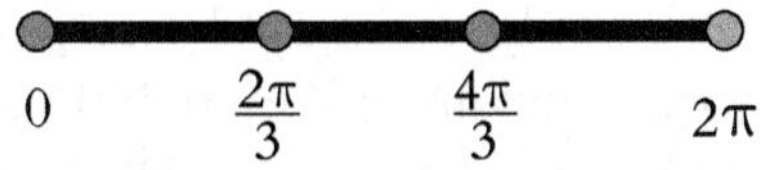

Figure 3.4 Three equally spaced collocation points in one period for $N = 1$.

Specifically, in the present two degree of freedom wing section model, there are three equally spaced collocation points in one period for the case with $N = 1$, see Fig. 3.4. Therefore, the spacing h is $2\pi/3$. So, the aliasing limit is $L = 1.5$. Due to the cubic nonlinearity, terms with wavenumbers 2 and 3 would occur. According to the Remark, any wavenumber n beyond the range $[-1.5, 1.5]$ is aliased to $n_a = n \pm 3m$, such that $n_a \in [-1.5, 1.5]$. A similar argument applies to the case with $N = 2$, where $n_a = n \pm 5m \in [-2.5, 2.5]$. This exactly verifies the previously proposed aliasing rules according to observation. Essentially, the aliasing arises from a misinterpretation of the high harmonic wave as its corresponding low harmonic wave due to the discrete collocation points.

The present aliasing rules are crucial in describing the relationship between the HB system and the TDC/HDHB system. For instance, if the explicit form of the HB system has been obtained, we can immediately write out its corresponding TDC system according to the present aliasing rules. In addition, the aliasing rules may be helpful to improve the dealiasing techniques [20]. Also, since the aliasing rules are extracted from the analysis of the nonlinear term, they are not confined to a specific model, and are applicable to other dynamical systems, such as the Duffing oscillator, the Van der Pol oscillator, et al.

3.4 Reconstruction harmonic balance method

An ingenious attempt to overcome the computational difficulty of HB is the alternating frequency—time (AFT) harmonic balance method [21]. It bypasses the complex symbolic calculations via computationally-cheap numerical calculations, thus finding wide applications. Based on the Shannon sampling theorem, the sampling rate of AFT is required to exceed twice the highest signal frequency to recover the nonlinear responses. Unexpectedly, numerical experiments reveal that the sampling rate could be far less than the prediction of Shannon theorem. Regarding that, a more concise and elegant high dimensional harmonic balance (HDHB) method was proposed using collocation, where the collocation number equals to the number of unknown Fourier coefficients, which is much less than the sampling rate of the AFT method. The HDHB method is computationally fast, and a series of modified versions, e.g., the Chebyshev-based Time-spectral Method (C-TSM), the Supplemental-frequency Harmonic Balance (SF-HB) method, etc., have been developed for specific problems. However, both the AFT method and the family of HDHB methods are impaired by aliasing when dealing with non-polynomial nonlinear problems. Worse yet, severe aliasing may cause non-physical solutions in the HDHB-like methods even for polynomial nonlinearity. Our study proved that the HDHB method is inherently not a variation of the HB method, but equivalently a time domain collocation method in disguise. Based on this finding, we extended the time domain collocation method by collocating at more nodes, and numerically suppressed the generation of non-physical solutions by using a least squares method [21]. However, the theoretical dealiasing rule is still unclear.

With the TDC method we proposed in this book, a collocation-based harmonic balance framework is developed to unify the HB, the AFT and the HDHB methods. Under this framework, the theoretical dealiasing rule can be revealed and plainly expressed by a novel aliasing matrix. It further leads to the discovery of an unprecedented conditional identity between the Fourier coefficients obtained from time domain analysis and frequency domain analysis. With these developments, we propose an advanced method for periodic analysis named as the reconstruction harmonic balance (RHB) method that can equivalently transform into the HB method, the AFT method, or the HDHB method by choosing corresponding collocation number. The RHB method completely addresses the computational difficulties of the HB method (computational burden due to symbolic calculations), the AFT method (aliasing in non-polynomial nonlinearity and redundant samplings) and the HDHB method (aliasing-induced non-physical solutions), so that very high order solutions can be obtained for strongly nonlinear dynamical systems with high accuracy and little computational effort.

3.5 Numerical examples

In the following examples, the system parameters of the two DOF airfoil model are fixed as in Table 3.1 unless otherwise stated. Both the HB method and the RK4 method are applied to the present problem. The results by the RK4 method with sufficiently small time step are used as the benchmark solutions.

3.5.1 RK4 results and spectral analysis

Inasmuch as the flow velocity under consideration is nondimensional, we should detect the linear flutter speed U_L^* first via the numerical integration method. The linear flutter speed is the velocity beyond which the linear system would diverge to an

Table 3.1 System parameters.

$\overline{\omega}$	x_α	β	γ	μ	r_α	a_h	ζ_α	ζ_ξ
0.2	0.25	80	0	100	0.5	−0.5	0	0

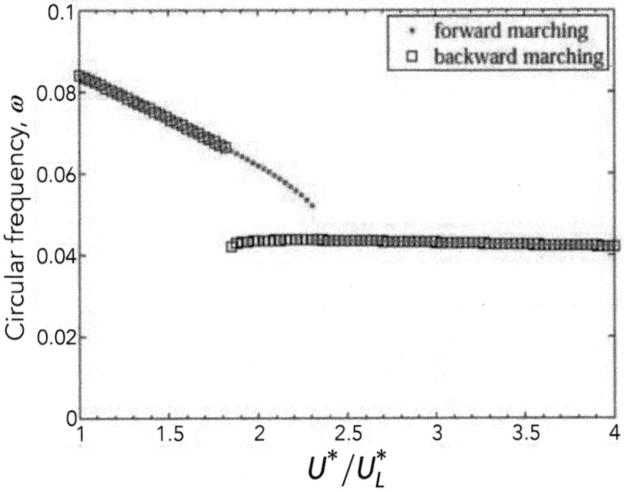

Figure 3.5 Bifurcation diagram of fundamental frequency vs. velocity, via the velocity marching procedure.

infinite amplitude. Thus, U_L^* can be obtained by removing the nonlinear term, and then increasing the flow speed U^* until divergent oscillations are observed. For the system with parameters in Table 3.1, the linear flutter speed is $U_L^* = 6.285$.

Shown in Fig. 3.5 is the bifurcation diagram of the frequency-vs.-velocity curve via a velocity marching procedure. The velocity marching procedure involves using the solution of the previous velocity step as the initial conditions of the next velocity step [4]. The velocity increment per step is $\Delta(U^*/U_L^*) = 0.01$. The circular frequency is obtained as the follows: (1) running a long time to damp out transients, (2) identifying two subsequent time instants t_1, t_2 which cause $\dot{\alpha}(t_1) = \dot{\alpha}(t_2) = 0^*$ and $\alpha(t_1) = \alpha(t_2)$, (3) then $\omega = 2\pi/(t_2 - t_1)$. It is seen that there is a hysteresis region [1.84, 2.35] for the velocity U^*/U_L^* in which two stable solutions coexist. It also shows that the frequency decreases continuously with the increase of the velocity until a bifurcation value 2.35 has been reached in the forward marching case. Beyond this critical value, the frequency jumps suddenly down to a small value, and keeps almost constant until the end. In contrast, the situation for the backward marching is opposite, with a jump-up at 1.84.

The FFT spectral analysis (with a rectangular window function) of the pitch motion is carried out to gain an insight into the distribution of the dominant frequencies, so that a prior knowledge on how to truncate the Fourier expansions in the HB method can be obtained.

We select four flow speed values, 1.2, 1.5, 1.8, and 2.0, on the upper branch, and four values, 2.0, 3.0, 3.5, and 4.0, on the lower branch. Unless otherwise noted, the lower and upper branches are referring to the curves in Fig. 3.5. Note that since

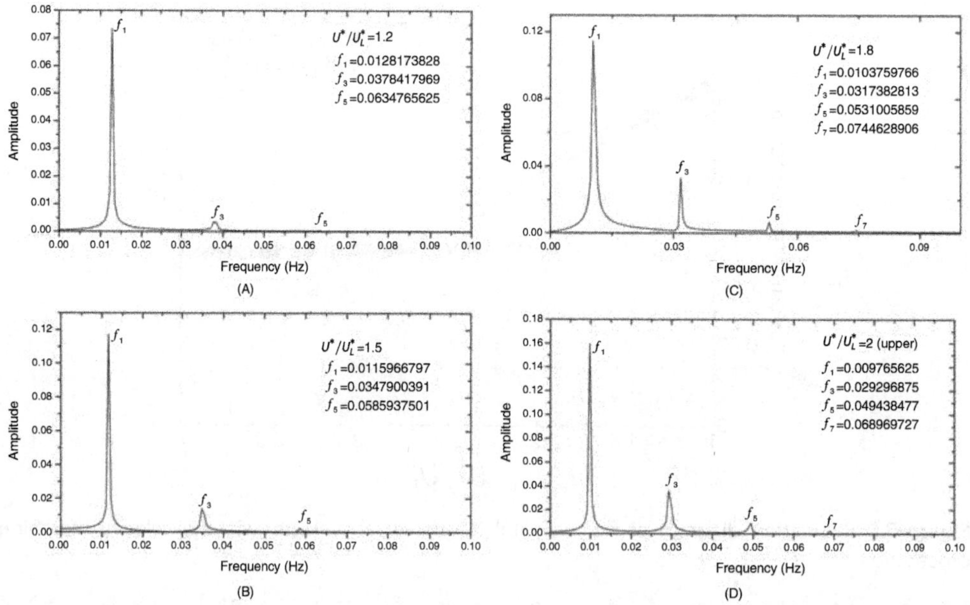

Figure 3.6 Amplitude spectra of the pitch motions of the numerical solutions for various values of U^*/U_L^* on the upper branch: (A) amplitude spectrum for $U^*/U_L^* = 1.2$, (B) amplitude spectrum for $U^*/U_L^* = 1.5$, (C) amplitude spectrum for $U^*/U_L^* = 1.8$, (D) amplitude spectrum for $U^*/U_L^* = 2.0$ at upper branch.

$U^*/U_L^* = 2.0$ is in the hysteresis region, there exist two stable solutions. The amplitude spectra of the FFT spectral analysis for all cases are given in Fig. 3.6 and Fig. 3.7, in which the dominant frequencies are labeled as $f_1, f_3, f_5, \ldots$ in sequence, because the second and the third dominant frequencies are approximately three and five times the fundamental frequency f_1. It is shown that for the cases on the upper branch, four dominant frequencies are sufficient to accurately approximate the pitch motion as implied in Fig. 3.6.

For $U^*/U_L^* = 1.2$, the fundamental circular frequency is $\omega_f = 2\pi f_1 \approx 0.080534$. Interestingly, upon checking the decimals of f_1, f_3, f_5 in Fig. 3.6(A), we find that

$$f_3 = \left(3 - \frac{1}{21}\right) f_1, f_5 = \left(5 - \frac{1}{21}\right) f_1.$$

It shows that the higher frequencies are not integer multiples of the fundamental one, but modulated by a proper fraction, $\Delta f = -1/21$. The similar frequency modulation phenomena are found for $U^*/U_L^* = 1.5, 1.8$, and 2.0. The relations between the fundamental frequencies and the high order frequencies are summarized in Table 3.2. It shows that the frequency modulation Δf is a small fraction which

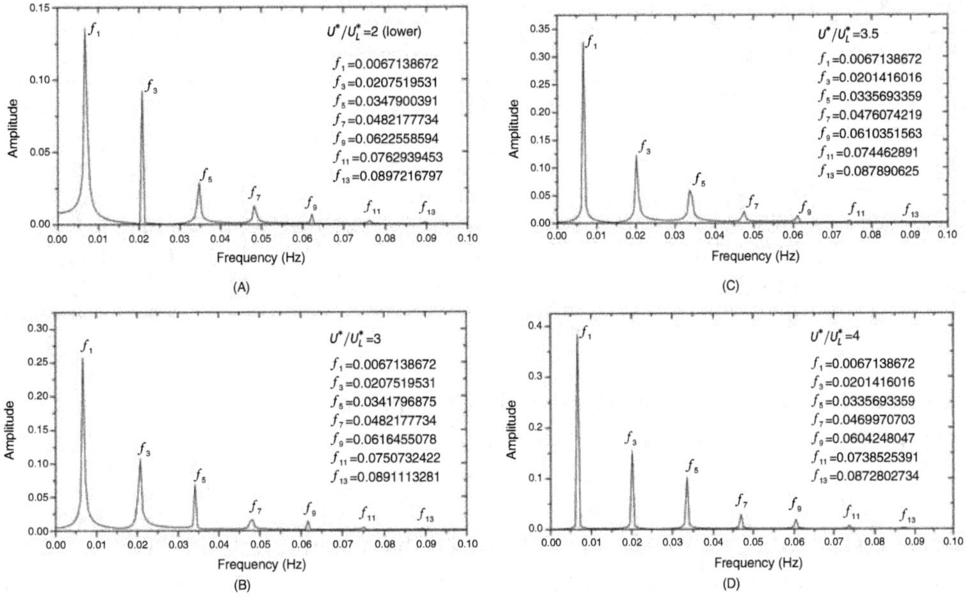

Figure 3.7 Amplitude spectra of the pitch motions of the numerical solutions for various values of U^*/U_L^* on the lower branch: (A) amplitude spectrum for $U^*/U_L^* = 2.0$ at lower branch, (B) amplitude spectrum for $U^*/U_L^* = 3.0$, (C) amplitude spectrum for $U^*/U_L^* = 3.5$, (D) amplitude spectrum for $U^*/U_L^* = 4.0$.

Table 3.2 Dominant harmonics for $U^*/U_L^* = 1.2, 1.5, 1.8$, and 2.0.

U^*/U_L^*	1.2	1.5	1.8	2.0 (Upper)
ω_f	0.080534	0.072864	0.065194	0.061359
ω_3	$\left(3-\frac{1}{21}\right)\omega_f$	$3\omega_f$	$\left(3+\frac{1}{17}\right)\omega_f$	$3\omega_f$
ω_5	$\left(5-\frac{1}{21}\right)\omega_f$	$\left(5+\frac{1}{19}\right)\omega_f$	$\left(5+\frac{2}{17}\right)\omega_f$	$\left(5+\frac{1}{16}\right)\omega_f$
ω_7	weak	weak	$\left(7+\frac{3}{17}\right)\omega_f$	$\left(7+\frac{1}{16}\right)\omega_f$

would vary case by case. Similar modulation phenomenon was reported in Lee et al. [13,21].

Provided in Table 3.3 are the circular frequencies for cases on the lower branch of the frequency-vs.-velocity curve. It is seen that a uniform modulation frequency $\Delta f = 1/11$ exists almost for all cases except for $U^*/U_L^* = 4.0$ where the higher harmonics are exactly integer multiples of the fundamental one. Also, it is noted that the fundamental frequencies on the lower branch are of the same value.

We revisit the case at $U^*/U_L^* = 4.0$, but with a stronger nonlinearity $\beta = 120$. It is found that the nonlinearity does not change the distribution of the dominant frequencies, which implies that Δf is independent of β. However, it is found that a stronger

Table 3.3 Dominant harmonics for $U^*/U_L^* = 2.0, 3.0, 3.5, 4.0$.

U^*/U_L^*	2.0 (Lower)	3.0	3.5	4.0
ω_f	0.042184	0.042184	0.042184	0.042184
ω_3	$\left(3+\frac{1}{11}\right)\omega_f$	$\left(3+\frac{1}{11}\right)\omega_f$	$3\omega_f$	$3\omega_f$
ω_5	$\left(5+\frac{2}{11}\right)\omega_f$	$\left(5+\frac{1}{11}\right)\omega_f$	$5\omega_f$	$5\omega_f$
ω_7	$\left(7+\frac{2}{11}\right)\omega_f$	$\left(7+\frac{2}{11}\right)\omega_f$	$\left(7+\frac{1}{11}\right)\omega_f$	$7\omega_f$
ω_9	$\left(9+\frac{3}{11}\right)\omega_f$	$\left(9+\frac{2}{11}\right)\omega_f$	$\left(9+\frac{1}{11}\right)\omega_f$	$9\omega_f$
ω_{11}	$\left(11+\frac{4}{11}\right)\omega_f$	$\left(11+\frac{2}{11}\right)\omega_f$	$\left(11+\frac{1}{11}\right)\omega_f$	$11\omega_f$
ω_{13}	$\left(13+\frac{4}{11}\right)\omega_f$	$\left(13+\frac{3}{11}\right)\omega_f$	$\left(13+\frac{1}{11}\right)\omega_f$	$13\omega_f$

nonlinearity diminishes the amplitude of the pitch motion, which is consistent with the concept that the nonlinearity added to a linear fluid-structure interaction system may restrain a divergent fluttering (of the linear system) to a bounded oscillation [7].

It is concluded from the spectral analysis that there are more dominant frequencies for the cases on the lower branch than those on the upper branch. In particular, seven dominant harmonics, $\omega_f, \omega_3, \ldots, \omega_{13}$, are required to accurately approximate the pitch motion on the lower branch, compared with four for the upper branch.

3.5.2 HBEJ vs. HBNJ

Comparisons of the accuracy and efficiency of the HB method with an explicit Jacobian matrix (HBEJ) and with a numerical Jacobian matrix (HBNJ) are carried out in this section. Firstly, we investigate the effect of the step size h on the accuracy of the HBNJ, since h is a crucial factor that affects the performance of the numerical differentiation.

The phase portraits of the plunge calculated by HBEJ as well as HBNJ are plotted in Fig. 3.8. As can be seen that, a small value of h is required to produce an accurate Jacobian matrix, thus an accurate result compared with the benchmark by HBEJ. Presently, $h=0.0001$ is suggested. A further smaller value of h, say $h=10^{-5}$ (not reported in Fig. 3.8), does not significantly improve the accuracy of the HBNJ.

Table 3.4 shows the comparison of the consumed computing times and iteration steps using the HBEJ and the HBNJ. In computations, the OIA (Optimal Iterative Algorithm) is employed, and the convergence criterion $\varepsilon=10^{-10}$, $h=10^{-4}$, and $U=3.5$. The Newton-Raphson (NR) method has also been tried, and it was found that the NR method was extremely sensitive to the initial guess. It normally cannot approach a solution, unless the initial guess is extremely near a solution. In general, the total computational time of the HBEJ is approximately two orders of magnitude less than that of the HBNJ. Fig. 3.9 shows the evolution histories of the residual errors of the NAEs by the HB9, HB11 and HB13 using both numerically calculated and analytical Jacobian matrices. It can be seen that the number of iterations using the HBEJ is

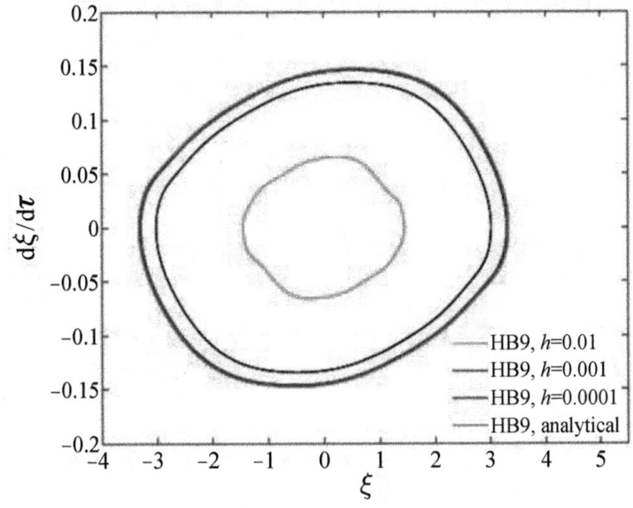

Figure 3.8 Phase portraits of the plunge motion at $U^*/U_L^* = 3.5$ via HBEJ, and HBNJ with different values of h.

Table 3.4 Comparison of computational efforts.

HB*n*	**Time (s)**			**Iterations**		**Time per iteration**		
	HBEJ	**HBNJ**	**Ratio**	**HBEJ**	**HBNJ**	**HBEJ**	**HBNJ**	**Ratio**
$n = 9$	3.1	170.2	1.82%	356	1173	0.0087	0.1451	6.00%
$n = 11$	11.2	258.8	4.43%	588	988	0.0190	0.2619	7.25%
$n = 13$	12.7	851.3	1.49%	614	1583	0.0207	0.5378	3.85%

roughly 2–3 times less than that of the HBNJ, which illustrates the HBEJ converges faster than the HBNJ.

In addition, the effect of the step size h on the convergence speed is revealed in Fig. 3.10. Given a prescribed criterion (10^{-7} in this case), using a smaller h can reduce the consumed number of iterations. Concretely, the consumed iterations for the HBNJ9 with $h = 10^{-3}$, 10^{-4}, 10^{-5}, and 10^{-6} are 596, 502, 439, and 425, respectively. Further decreasing h does not significantly enhance the convergence rate. On the other hand, the number of iterations for the HBEJ9 is 206, which is more than two times less than that for the HBNJ9. This can be explained by the fact that in the course of numerically calculating the Jacobian matrix (the Jacobian matrix is considered to provide an appropriate convergence direction), the HBNJ suffers from some numerical errors which can be avoided in the HBEJ. Note that if the step size h is relatively large, e.g. 10^{-2}, the numerical Jacobian matrix would be so coarse that the residual error cannot satisfy the prescribed criterion. Interestingly, the residual error history exhibits a fluctuation

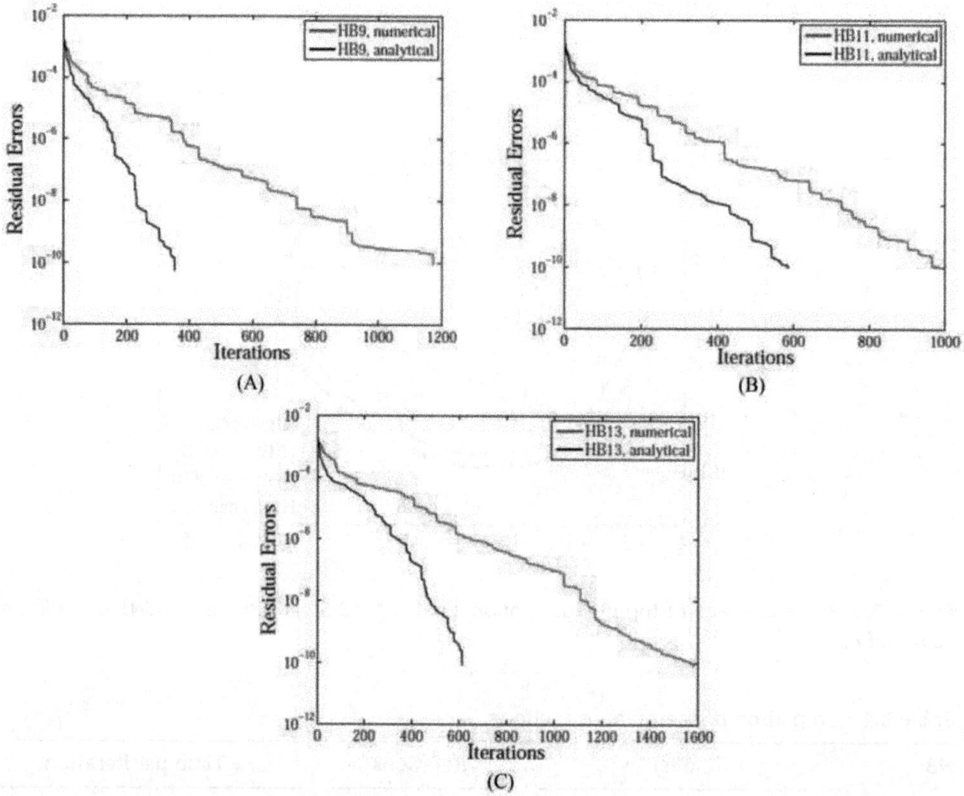

Figure 3.9 Comparisons of the residual error vs. iteration curves for (A) HB9, (B) HB11, and (C) HB13, both with and without an explicit Jacobian matrix.

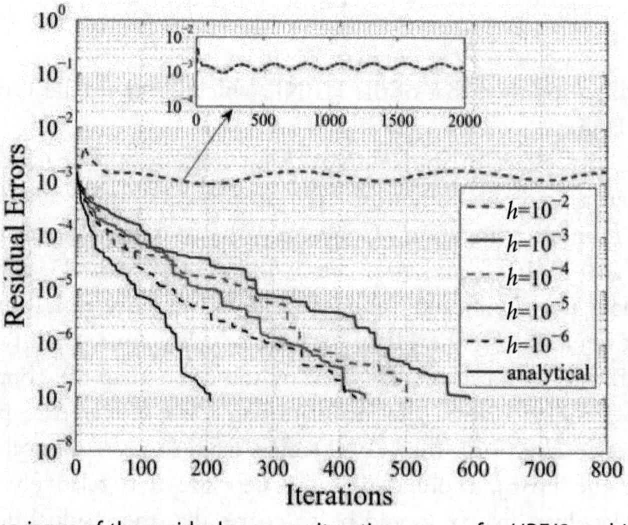

Figure 3.10 Comparisons of the residual error vs. iteration curves for HBEJ9 and HBNJ9 with different values of *h*.

around 10^{-3}. More importantly, the consumed time per iteration (TPI) of the HBEJ is less than 10% of the HBNJ (see Table 3.4), which is a reasonable measure of the absolute speed of the two methods. Overall, the HBEJ is superior to the HBNJ in two aspects: (i) fewer iterations and (ii) a shorter time per iteration (TPI).

3.5.3 Aliasing analysis of the HB and TDC methods

Liu et al. [21] demonstrated that the HDHB method may cause spurious solutions. We have shown that the TDC method is equivalent to the HDHB method. Therefore, the TDC method may produce spurious solutions, the aliasing phenomenon. Previous studies hold the opinion that the HB method only produces physical solutions. In this subsection, we will verify that a kind of mathematical aliasing may also come about in the HB method. Herein physical solutions consist of both stable and unstable steady state motions in contrast to the numerical integration method where only stable state solutions can be obtained. The RK4 is used to solve the case at $U^*/U_L^* = 2$. The initial conditions are chosen by letting $\alpha(0)$ be a specified value, and the rest conditions are zero. We carried out 1000 computations starting with 1000 randomly generated $\alpha(0) \in [-20^\circ, 20^\circ]$. The limit value 20° is imposed to ensure the aerodynamic operator given by Fung [7] to be valid. Monte Carlo simulation verified that there are two solutions for the present case. Similarly, we apply the HB and the TDC methods with many different initial guesses to check if they will produce spurious solutions.

3.5.3.1 Aliasing of the HB method

Three cases are used for the aliasing analysis of the HB method: (i) the HB3 with fully random initial conditions, (ii) the HB9 with partially random initial conditions, (iii) the HB9 with fully random initial conditions. In this section, "fully random" means that all unknown variables are randomly generated, while "partially random" means only α_1 and α_2 are randomly generated with others being 0.1.

HB3 with fully random initials

The Monte Carlo simulation is implemented for the analysis of the aliasing of the HB3. We carried out 1000 computations and the 1000 initial conditions are generated from within $[-0.5, 0.5]$. Fig. 3.11 shows that there are four possibilities, viz. "stagnant", $\omega_1 = 0.02538$, $\omega_2 = 0.03806$ and $\omega_3 = 0.06708$, for a given initial guess. Specifically, the probabilities for the occurrences of "stagnant", ω_1, ω_2 and ω_3 are 25%, 5%, 42% and 28%, respectively. For brevity, ω_1 is denoted by "POSS1", and so on.

However, we note that not all the obtained solutions are guaranteed to be real solutions. A further investigation is required to check on the motion of each solution. Fig. 3.12 shows the time histories and phase portraits for POSS1 and POSS2, which indicates that the two motions are identical. The two motions having different frequencies

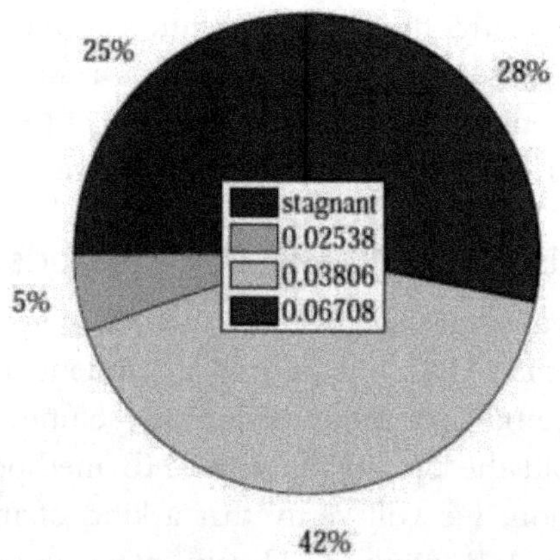

Figure 3.11 Monte Carlo simulation for HB3: probabilities for each possibility.

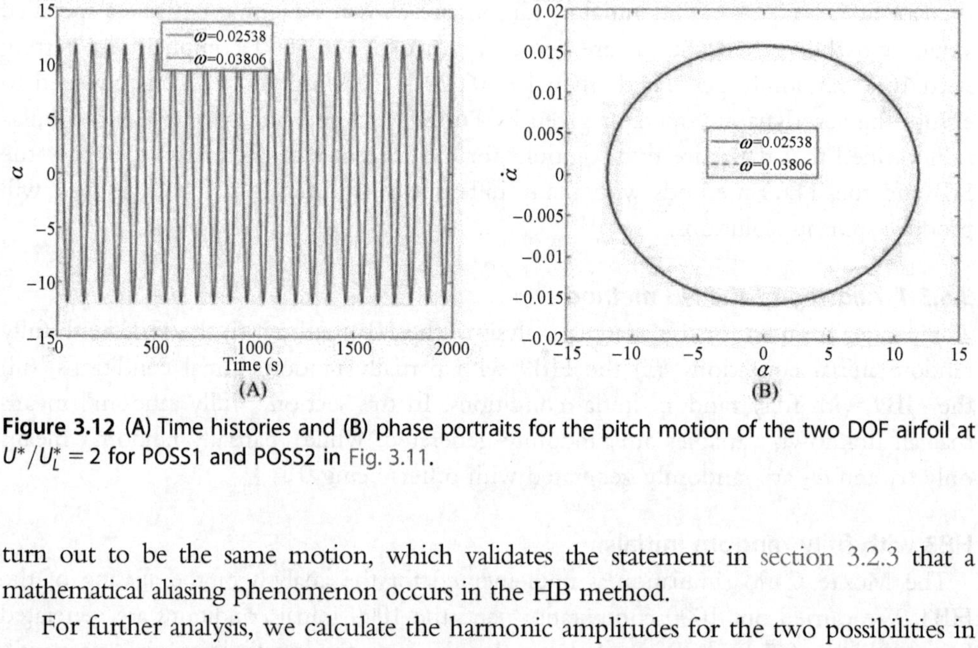

Figure 3.12 (A) Time histories and (B) phase portraits for the pitch motion of the two DOF airfoil at $U^*/U_L^* = 2$ for POSS1 and POSS2 in Fig. 3.11.

turn out to be the same motion, which validates the statement in section 3.2.3 that a mathematical aliasing phenomenon occurs in the HB method.

For further analysis, we calculate the harmonic amplitudes for the two possibilities in Table 3.5. It is shown that for neither POSS1 nor POSS2 the first harmonic is dominant, which suggests that the two possibilities are mathematically aliased solutions to a physical solution. For POSS1, the third harmonic dominates the motion, while the second harmonic dominates the motion in POSS2. The mechanism for the mathematical aliasing of the HB method is distinct from that of the physical aliasing of the TDC method. In the HB method, the mathematical aliasing takes place in such a way:

$$\cos 3\omega_{POSS2} = \cos 2\omega_{POSS3} = \cos\omega, \sin 3\omega_{POSS2} = \sin 2\omega_{POSS3} = \sin\omega. \tag{3.53}$$

Table 3.5 Harmonic amplitudes for POSS1–POSS3 via HB3.

	POSS1	POSS2	POSS3
1	0.00000000	0.00000000	0.00000000
A_1	0.00000000	0.00009948	0.17648060
A_2	0.00000017	0.20732346	0.00019259
A_3	0.20733870	0.00013174	0.04097729
ω	0.02525635	0.03788352	0.06701697

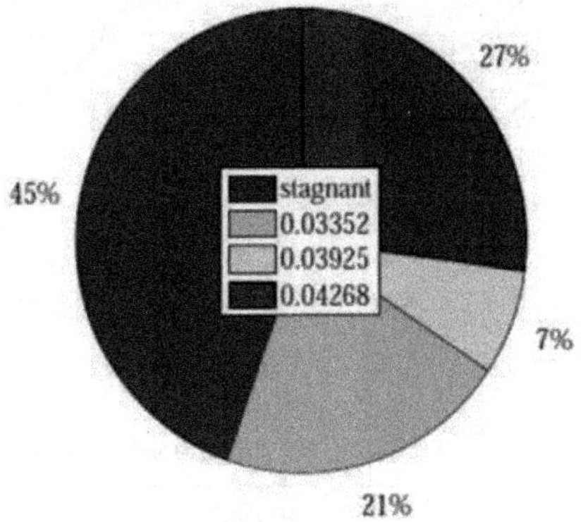

Figure 3.13 Monte Carlo simulation for HB9: probabilities for each possibility.

Therefore, we predict that a physically meaningful motion should be at $\omega = 3\omega_{POSS2} \approx 0.757$, where its first harmonic dominates the motion ($A_1 \approx 0.2073$). POSS3 has a dominant first harmonic. Therefore, it is a physical one, which may be stable or unstable.

HB9 with partially random initials

Shown in Fig. 3.13 is the Monte Carlo simulation result for the HB9 with partially random initial conditions. 1000 computations are carried out, with α_1 and α_2 generated from $[-0.5, 0.5]$. Four possibilities are detected for the HB9. In particular, the probabilities for the occurrences of "stagnant", POSS1, POSS2 and POSS3 are 45%, 21%, 7% and 27% respectively.

The corresponding phase portraits for POSS1–POSS3 are shown in Fig. 3.14. We see that the three motions are different so they are not mutually aliased. We need to check the harmonic amplitudes provided in Table 3.6 to determine whether a motion is physical or mathematically aliased. POSS1 has a dominant second harmonic and a significant sixth harmonic, which suggests that POSS1 is a mathematically aliased solution of a physical solution whose frequency should be approximately $2 \times \omega_{POSS1} \approx 0.067$, and whose first and third harmonic amplitudes should be approximately $(A_2)_{POSS1}$ and $(A_6)_{POSS1}$ respectively. As stated above, the frequency and

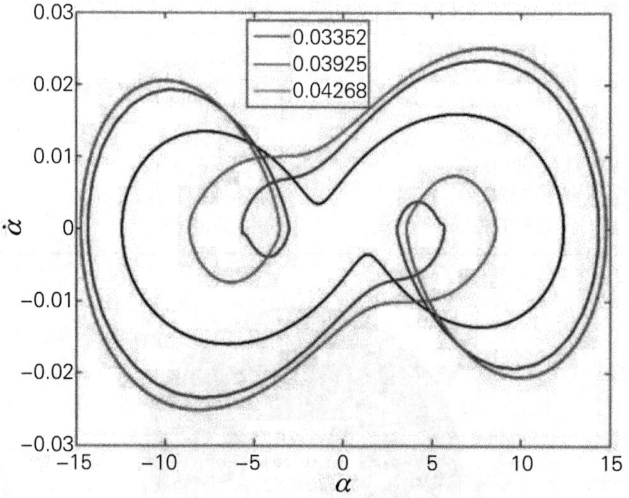

Figure 3.14 Phase portraits for the pitch motion of the two dof airfoil at $U^*/U_L^* = 2$ for the three possibilities in Fig. 3.13.

Table 3.6 Harmonic amplitudes for POSS1–POSS3 via HB9 with partially random initials.

	POSS1	POSS2	POSS3
1	0.00000000	0.00000000	0.00000000
A_1	0.00000364	0.13234863	0.15821813
A_2	0.17648212	0.00001101	0.00000588
A_3	0.00000623	0.08793958	0.09536424
A_4	0.00000000	0.00000921	0.00000327
A_5	0.00000612	0.03606737	0.03406593
A_6	0.04097708	0.00000606	0.00000266
A_7	0.00000318	0.01655326	0.01608910
A_8	0.00000000	0.00000352	0.00000134
A_9	0.00000217	0.00704681	0.00616772
ω	0.03350758	0.03924212	0.04264996

coefficient variables can adjust their magnitudes in several ways so as to satisfy the governing equations. Among them, only one is the physical one. The others are called mathematically aliased solutions of the physical one. This kind of aliasing of the HB method can be generated in the self-excited dynamical systems where the frequency is treated as a variable rather than a constant.

Both POSS2 and POSS3 have decaying harmonic amplitudes with their first harmonic being the dominant one. Thus, they are corresponding to the physical solutions. Specifically, POSS2 corresponds to an unstable solution. Alternatively, the stability of the two solutions can be determined by a perturbation technique [9].

HB9 with fully random initial guesses

In the present simulation, 10,000 initial guesses are randomly generated from within [− 0.2, 0.2]. In addition to the stagnant situation, there are eight possibilities in this case. The values of frequencies for all possibilities and their times of occurrence are given in HB9 with fully random initial conditions.

Table 3.8 and Table 3.9 show the harmonic amplitudes of each possibility. POSS1 has a dominant A_6, which suggests a corresponding physical frequency should be 6 $\times$ 0.01262659 $\approx$ 0.0757. POSS2−POSS5 do not have a dominant first harmonic, which suggests that they are mathematically aliased solutions. Conversely, POSS6−POSS8 have a dominant first harmonic response and a decaying trend for higher harmonics, so they are physical solutions.

By detection, we see that there are four different physical solutions in the eight possibilities. POSS4 and POSS5 are the aliased solutions of the physical (unstable) one with $\omega = 0.0667$. POSS6 corresponds to 0.0392 which might be an unstable steady

Table 3.7 Eight possibilities via HB9 with fully random initial conditions.

Frequency	Times
0.012628404080552	22
0.015151890123195	29
0.018942379174476	113
0.022341119181730	54
0.033517536097199	1144
0.039252396697425	886
0.042648003287537	17
0.061903613704298	28

Table 3.8 Harmonic amplitudes for POSS1−POSS4 via HB9 with fully random initial conditions.

	POSS1	POSS2	POSS3	POSS4
1	0.00000000	0.00000000	0.00000000	0.00000000
A_1	0.00000000	0.00000000	0.00000135	0.00000000
A_2	0.00000000	0.00000000	0.00000000	0.00000086
A_3	0.00000000	0.00000000	0.00000000	0.17648211
A_4	0.00000000	0.00000373	0.20730562	0.00000171
A_5	0.00000400	0.20730561	0.00000000	0.00000001
A_6	0.20730561	0.00000404	0.00000000	0.00000008
A_7	0.00000428	0.00000000	0.00000260	0.00000000
A_8	0.00000000	0.00000000	0.00000000	0.00000099
A_9	0.00000000	0.00000000	0.00000307	0.04097708
ω	0.01262659	0.01515186	0.01893981	0.02233841

Table 3.9 Harmonic amplitudes for POSS5−POSS8 via HB9 with fully random initial conditions.

	POSS5	POSS6	POSS7	POSS8
1	0.00000000	0.00000000	0.00000000	0.00000000
A_1	0.00000476	0.13234863	0.15821813	0.16499749
A_2	0.17648211	0.00001855	0.00000558	0.00000184
A_3	0.00000815	0.08793959	0.09536424	0.04845210
A_4	0.00000000	0.00001551	0.00000310	0.00000088
A_5	0.00000801	0.03606737	0.03406593	0.01149177
A_6	0.04097708	0.00001021	0.00000252	0.00000030
A_7	0.00000417	0.01655326	0.01608911	0.00267666
A_8	0.00000000	0.00000593	0.00000127	0.00000009
A_9	0.00000284	0.00704681	0.00616772	0.00061555
ω	0.03350761	0.03924209	0.04264998	0.06190453

state solution. POSS7 should be the stable solution on the lower branch of the frequency vs. flow speed curve. POSS8 corresponds to a stable upper branch solution.

In the framework of solving the Duffing equation, the HB method does not produce spurious solutions as demonstrated in References [19],[21]. However, it has been shown that a kind of mathematical aliasing of HB method arises in the present self-excited system. Therefore, we conclude that the HB method is aliasing-free when the system under consideration has a constant external frequency, and the HB method may produce mathematically aliased solutions if the frequency is variable.

3.5.3.2 Aliasing of the TDC method

Having investigated the aliasing phenomenon arising from the conventional HB method, we now shift our attention to the aliasing of TDC method. The TDC3 and TDC9 are taken as examples. Monte Carlo simulation with 1000 randomly generated initial guesses are applied. The initial conditions for the TDC3 are randomly generated for all eight variables from within $[-0.5, 0.5]$. The initial guesses for the TDC9 are given by the means that the two variables associated with the first harmonic are generated randomly from within $[-0.5, 0.5]$, the other 18 unknowns are set to 0.1. Monte Carlo histograms are plotted for the frequencies by both the TDC3 and TDC9 in Fig. 3.15.

In using the TDC3, we have detected 17 different solutions. However, upon checking the harmonic amplitudes of each individual solution, we find that none of them is the physically meaningful solution.

For the TDC9, up to 21 spurious solutions are obtained. Nevertheless, we cannot achieve a physical solution by means of the Monte Carlo simulation, which indicates that random initial conditions do not work for the TDC method to obtain a physical solution.

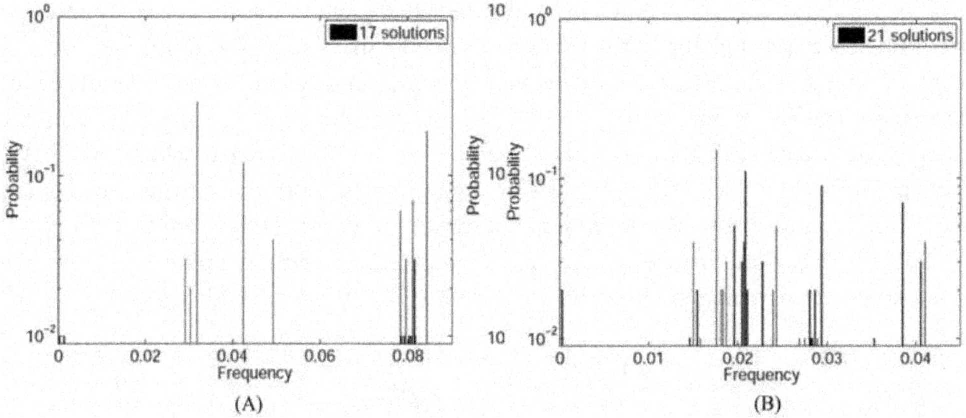

Figure 3.15 Monte Carlo histograms for the frequencies calculated by the TDC method at $U^*/U_L^* = 2$. (A) Results for TDC3. (B) Results for TDC9.

It is reasonable to conclude that the TDC method may produce many more solutions than the HB method, since the TDC method might possess two aliasing mechanisms: (i) the physical aliasing arising from the absorbed higher order harmonics in a disordered way, and (ii) (perhaps) the mathematical aliasing that may occur in the HB method due to the variation of frequency.

3.5.4 Dealiasing via a marching procedure

As shown in the previous section, the TDC method is very sensitive to the initial guess. We have a slim probability to obtain a physical solution if the initial guess is given randomly. To conquer this limitation, a marching procedure is introduced and used to supply an appropriate initial guess for the TDC method, as well as for the HB method. In this chapter, the amplitude vs. flow velocity curve is desired. It can be obtained by incrementally increasing or decreasing the velocity. At each step, the previous solution is employed as the initial condition for the next step. This procedure is named the velocity marching procedure.

Three things need to be noted in using the marching procedure. First, for the starting point or generating point [10], its initial guess should be chosen carefully. If the nonlinearity is weak, the solution of the linear system can be used as the initial guess at this point. If strong, one can perform a marching procedure along the parameter of the nonlinear term, i.e. β in this case, to attain an appropriate initial guess. One may also resort to a less sensitive method to generate a proper initial guess. Second, the marching increment should be small enough to ensure that the supplied initial guesses are within the radius of convergence of the solution. Third, the marching procedure is valid until a critical point is encountered, because a significant difference occurs at this point. Thus, using forward marching and backward marching, one can sweep out the upper and lower branches, once two generating points belonging to the two branches are found.

Fig. 3.16 and Fig. 3.17 show the fundamental frequency vs. flow speed curves by the TDC methods and the RK4 using the velocity marching procedure. For the forward marching case, the initial guess to the starting point is given by the solution of the corresponding linear system. Fig. 3.16(a) shows that for the forward marching the TDC9 agrees well with the RK4 until about $U^*/U_L^* = 2.2$, beyond which the TDC9 starts to differ from the RK4. The TDC9 results vary smoothly until the end. Hence, the TDC9 cannot predict the secondary bifurcation point for the forward marching.

For the backward marching, the initial guess at the starting point $U^*/U_L^* = 3$ is given through marching β. It is shown that the TDC9 agrees with the RK4 on the

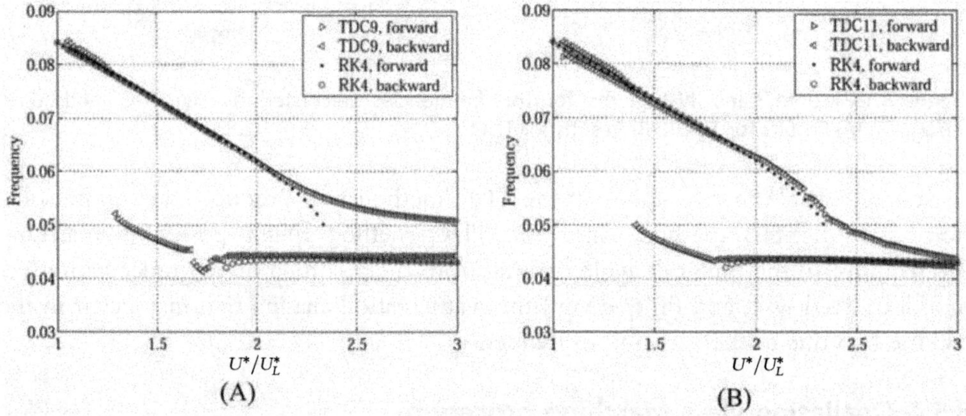

Figure 3.16 Fundamental frequency vs. flow speed by the TDC9 and TDC11 using velocity marching procedure. RK4 results are benchmarks for comparison. (A) Results for TDC9. (B) Results for TDC11.

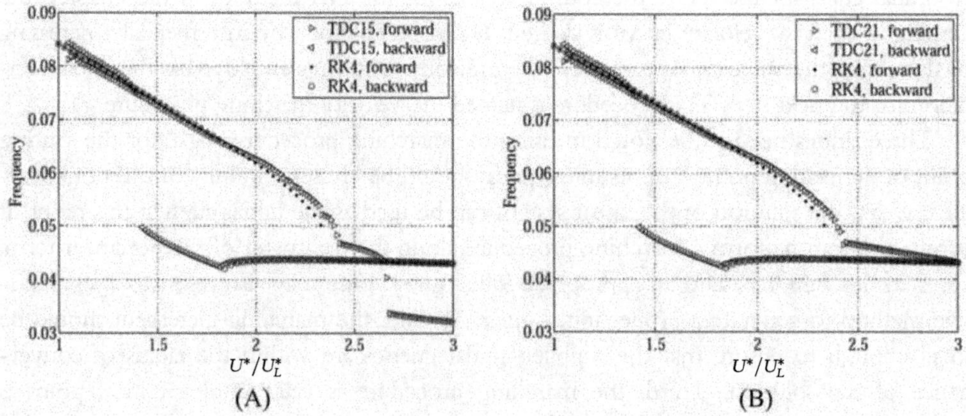

Figure 3.17 Fundamental frequency vs. flow speed by the TDC15 and TDC21 using velocity marching procedure. RK4 results are benchmarks for comparison. (A) Results for TDC15. (B) Results for TDC21.

entire lower branch. The bifurcation point predicted via RK4 is 1.84, beyond which the RK4 result jumps up to the upper branch. Prior to the bifurcation point, the TDC9 marches along the physical solution curve continuously. Beyond the bifurcation, the TDC9 zigzags along a nearby path instead of jumping to the upper branch. Because once a bifurcation point is reached, the motions before and after the bifurcation change dramatically. Thus, the initial guess supplied by marching procedure may be beyond the convergence radius of the solution, and does not work anymore.

In general, an initial guess has three destinations: (i) converging to a mathematically meaningful solution, (ii) stagnating at a stagnant point, (iii) diverging to infinity. Because the TDC system produces nonphysical solutions, mathematically there are many solutions for the initial guess to approach. Normally, an initial guess is prone to a nearby destination rather than a distant physical one, since mathematically the physical solution has no priority over the spurious solutions and stagnant points. The stopping criterion for the NAE solver is $\varepsilon = 10^{-6}$ or iteration = 20,000 whichever is satisfied first. The TDC9 sweeps out a nearby curve beyond the bifurcation point. It is shown that in [1.84, 1.68], the response curve is discontinuous, which implies stagnant points rather solutions are found. This portion of curve is physically meaningless. From 1.68 to 1.28, a smooth curve is swept out, which means mathematical solutions are obtained. Deceasing the flow speed below 1.28, a jump-up takes place. It is believed that when the nonlinearity is weak, the TDC system may accommodate few nonphysical solutions. The jump-up occurs because there are scarcely any nearby attractors (solutions and stagnant points) for the initial guess to approach.

Fig. 3.16(b) displays the frequency vs. flow speed curves by TDC11. For the forward marching, the response curve agrees well with the benchmark until 2.27, better than 1.8 by TDC9. Increasing the flow speed further, three segments are marched out which might correspond to mathematical solutions. The TDC11 cannot predict the forward marching bifurcation point as the TDC9. On the other hand, the backward marching curve via TDC11 sweeps out the entire lower branch curve, and the bifurcation point is predicted which is smaller than the benchmark 1.84. Beyond the bifurcation point, the TDC11 sweeps out a smooth curve which should be a mathematically meaningful solution. At $U^*/U_L^* = 1.38$, a jump-up occurs from the lower mathematically meaningful curve to the physically meaningful curve.

Fig. 3.17(a) shows that the TDC15 can predict the entire lower branch by the backward marching procedure accurately. Nevertheless, it cannot predict an accurate bifurcation value for the upper branch by the forward marching procedure. The results by the TDC21 are shown in Fig. 3.17(b). The forward marching marches out the entire upper branch. The bifurcation point of the forward marching is predicted to be 2.38, which is close to the benchmark 2.34. Beyond the critical point, a mathematical solution curve is marched out until the end. The backward marching curve is in excellent agreement with the benchmark. The entire lower branch curve has been detected precisely. The backward marching bifurcation point is accurately predicted by TDC21. Beyond the bifurcation point, a mathematical solution curve is marched out until $U^*/U_L^* = 1.42$. Then a jump-up from the mathematical solution curve to the upper physical solution curve takes place, and the follow-up portion agrees well with the benchmark.

In sum, it is concluded that the TDC method in conjunction with the marching procedure can be used to accurately solve the nonlinear dynamical problem. In general, the more harmonics are included, the more accurate the solution would be. In the present case, the TDC method with 21 harmonics can accurately march out the upper and lower physical solutions curves, as well as predicting the two bifurcation points.

Appendix

Coefficients in system (3.5):

$$c_0 = 1 + \frac{1}{\mu}, c_1 = x_\alpha - \frac{a_h}{\mu}, c_2 = \frac{2}{\mu}(1 - \psi_1 - \psi_2) + 2\zeta_\xi \frac{\overline{\omega}}{U^*},$$

$$c_3 = \frac{1}{\mu}\left[1 + (1 - 2a_h)(1 - \psi_1 - \psi_2)\right], c_4 = \frac{2}{\mu}(\varepsilon_1\psi_1 + \varepsilon_2\psi_2),$$

$$c_5 = \frac{2}{\mu}\left[1 - \psi_1 - \psi_2 + \left(\frac{1}{2} - a_h\right)(\varepsilon_1\psi_1 + \varepsilon_2\psi_2)\right],$$

$$c_6 = \frac{2}{\mu}\varepsilon_1\psi_1\left[1 - \varepsilon_1\left(\frac{1}{2} - a_h\right)\right], c_7 = \frac{2}{\mu}\varepsilon_2\psi_2\left[1 - \varepsilon_2\left(\frac{1}{2} - a_h\right)\right],$$

$$c_8 = -\frac{2}{\mu}\varepsilon_1^2\psi_1, c_9 = -\frac{2}{\mu}\varepsilon_2^2\psi_2, c_{10} = \left(\frac{\overline{\omega}}{U^*}\right)^2,$$

$$d_0 = \frac{x_\alpha}{r_\alpha^2} - \frac{a_h}{\mu r_\alpha^2}, d_1 = 1 + \frac{1 + 8a_h^2}{8\mu r_\alpha^2}, d_2 = -\frac{1 + 2a_h}{\mu r_\alpha^2}(1 - \psi_1 - \psi_2),$$

$$d_3 = \frac{1 - 2a_h}{2\mu r_\alpha^2} - \frac{(1 - 4a_h^2)(1 - \psi_1 - \psi_2)}{2\mu r_\alpha^2} + \frac{2\zeta_\alpha}{U^*}, d_4 = -\frac{1 + 2a_h}{\mu r_\alpha^2}(\varepsilon_1\psi_1 + \varepsilon_2\psi_2),$$

$$d_5 = -\frac{1 + 2a_h}{\mu r_\alpha^2}(1 - \psi_1 - \psi_2) - \frac{(1 - 4a_h^2)(\varepsilon_1\psi_1 + \varepsilon_2\psi_2)}{2\mu r_\alpha^2},$$

$$d_6 = -\frac{(1 + 2a_h)\varepsilon_1\psi_1}{\mu r_\alpha^2}\left[1 - \varepsilon_1\left(\frac{1}{2} - a_h\right)\right], d_7 = -\frac{(1 + 2a_h)\varepsilon_2\psi_2}{\mu r_\alpha^2}\left[1 - \varepsilon_2\left(\frac{1}{2} - a_h\right)\right],$$

$$d_8 = \frac{(1 + 2a_h)\varepsilon_1^2\psi_1}{\mu r_\alpha^2}, d_9 = \frac{(1 + 2a_h)\varepsilon_2^2\psi_2}{\mu r_\alpha^2}, d_{10} = \left(\frac{1}{U^*}\right)^2. \tag{A1}$$

Coefficients in system (3.6):

$$
\begin{aligned}
& c = 1/(c_0 d_1 - c_1 d_0),\ a_{21} = c(-d_5 c_0 + c_5 d_0),\ a_{22} = c(-d_3 c_0 + c_3 d_0), \\
& a_{23} = c(-d_4 c_0 + c_4 d_0),\ a_{24} = c(-d_2 c_0 + c_2 d_0),\ a_{25} = c(-d_6 c_0 + c_6 d_0), \\
& a_{26} = c(-d_7 c_0 + c_7 d_0),\ a_{27} = c(-d_8 c_0 + c_8 d_0),\ a_{28} = c(-d_9 c_0 + c_9 d_0), \\
& a_{41} = c(d_5 c_1 - c_5 d_1),\ a_{42} = c(d_3 c_1 - c_3 d_1),\ a_{43} = c(d_4 c_1 - c_4 d_1), \\
& a_{44} = c(d_2 c_1 - c_2 d_1),\ a_{45} = c(d_6 c_1 - c_6 d_1),\ a_{46} = c(d_7 c_1 - c_7 d_1), \\
& a_{47} = c(d_8 c_1 - c_8 d_1),\ a_{48} = c(d_9 c_1 - c_9 d_1), \\
& g_{21} = c c_0 d_{10},\ g_{23} = c d_0 c_{10},\ g_{41} = c c_1 d_{10},\ g_{43} = c d_1 c_{10}.
\end{aligned} \tag{A2}
$$

In the Jacobian matrix, we need to derive $\partial g/\partial \omega$. According to Eq. (3.2.3) we have

$$\frac{\partial g_i}{\partial \omega} = \frac{\partial L_{ij}(\omega)}{\partial \omega}\alpha_j, \tag{A3}$$

where

$$\boldsymbol{L} = \boldsymbol{A}_2 - \boldsymbol{B}_2 \boldsymbol{B}_1^{-1} \boldsymbol{A}_1. \tag{A4}$$

Explicit expressions of $\boldsymbol{A}_1$, $\boldsymbol{B}_1$, $\boldsymbol{A}_2$ and $\boldsymbol{B}_2$ are as follows:

$$
\begin{aligned}
\boldsymbol{A}_1 &= c_1\omega^2\boldsymbol{A}^2 + c_3\omega\boldsymbol{A} + c_5\boldsymbol{I} + c_6(\omega\boldsymbol{A}+\varepsilon_1\boldsymbol{I})^{-1} + c_7(\omega\boldsymbol{A}+\varepsilon_2\boldsymbol{I})^{-1}, \\
\boldsymbol{B}_1 &= c_0\omega^2\boldsymbol{A}^2 + c_2\omega\boldsymbol{A} + (c_4 + c_{10})\boldsymbol{I} + c_8(\omega\boldsymbol{A}+\varepsilon_1\boldsymbol{I})^{-1} + c_9(\omega\boldsymbol{A}+\varepsilon_2\boldsymbol{I})^{-1}, \\
\boldsymbol{A}_2 &= d_1\omega^2\boldsymbol{A}^2 + d_3\omega\boldsymbol{A} + d_5\boldsymbol{I} + d_6(\omega\boldsymbol{A}+\varepsilon_1\boldsymbol{I})^{-1} + d_7(\omega\boldsymbol{A}+\varepsilon_2\boldsymbol{I})^{-1}, \\
\boldsymbol{B}_2 &= d_0\omega^2\boldsymbol{A}^2 + d_2\omega\boldsymbol{A} + d_4\boldsymbol{I} + d_8(\omega\boldsymbol{A}+\varepsilon_1\boldsymbol{I})^{-1} + d_9(\omega\boldsymbol{A}+\varepsilon_2\boldsymbol{I})^{-1}.
\end{aligned} \tag{A5}
$$

For the first term in Eq. (3.54),

$$\frac{\partial \boldsymbol{A}_2}{\partial \omega} = 2\omega d_1\boldsymbol{A}^2 + d_3\boldsymbol{A} + \frac{\partial}{\partial \omega}\left[d_6(\omega\boldsymbol{A}+\varepsilon_1\boldsymbol{I})^{-1} + d_7(\omega\boldsymbol{A}+\varepsilon_2\boldsymbol{I})^{-1}\right]. \tag{A6}$$

In the above equation, we need to derive the differentiation of $(\omega\boldsymbol{A}+\varepsilon_i\boldsymbol{I})^{-1}, i = 1, 2$, with respect to ω. Let C_{ε_i} be

$$C_{\varepsilon_i} \equiv \omega\boldsymbol{A} + \varepsilon_i\boldsymbol{I} = \begin{bmatrix} \varepsilon_i & & & & \\ & C_1 & & & \\ & & C_2 & & \\ & & & \ddots & \\ & & & & C_N \end{bmatrix},$$

where

$$C_n = \begin{bmatrix} \varepsilon_i & n\omega \\ -n\omega & \varepsilon_i \end{bmatrix}.$$

Then the inverse of C_{ε_i} denoted by V_{ε_i} is

$$V_{\varepsilon_i} = C_{\varepsilon_i}^{-1} = \begin{bmatrix} \frac{1}{\varepsilon_i} & & & & \\ & v_1^{\varepsilon_i} & & & \\ & & v_2^{\varepsilon_i} & & \\ & & & \ddots & \\ & & & & v_N^{\varepsilon_i} \end{bmatrix}, \tag{A7}$$

where

$$v_n^{\varepsilon_i} = C_n^{-1} = \frac{1}{\varepsilon_i^2 + (n\omega)^2} \begin{bmatrix} \varepsilon_i & -n\omega \\ n\omega & \varepsilon_i \end{bmatrix}.$$

Then, we can calculate the derivative of V_{ε_i} with respect to ω

$$W_{\varepsilon_i} = \frac{\partial V_{\varepsilon_i}}{\partial \omega} = \begin{bmatrix} 0 & & & & \\ & \frac{dv_1^{\varepsilon_i}}{d\omega} & & & \\ & & \frac{dv_2^{\varepsilon_i}}{d\omega} & & \\ & & & \ddots & \\ & & & & \frac{dv_N^{\varepsilon_i}}{d\omega} \end{bmatrix}, \tag{A8}$$

where

$$\frac{dv_n^{\varepsilon_i}}{d\omega} = -\frac{2n^2\omega}{\left[\varepsilon_i^2 + (n\omega)^2\right]^2} \begin{bmatrix} \varepsilon_i & -n\omega \\ n\omega & \varepsilon_i \end{bmatrix} + \frac{1}{\varepsilon_i^2 + (n\omega)^2} \begin{bmatrix} 0 & -n \\ n & 0 \end{bmatrix}.$$

Therefore, the derivative of the first part of $\boldsymbol{L}$ with respect to ω is finally obtained as

$$\frac{\partial \boldsymbol{A}_2}{\partial \omega} = 2\omega d_1 \boldsymbol{A}^2 + d_3 \boldsymbol{A} + d_6 W_{\varepsilon_1} + d_7 W_{\varepsilon_2}. \tag{A9}$$

Now, we derive the derivative of the second part of $\boldsymbol{L}$ with respect to ω:

$$\frac{\partial}{\partial \omega}\left(\boldsymbol{B}_2 \boldsymbol{B}_1^{-1} \boldsymbol{A}_1\right) = \boldsymbol{B}_2' \boldsymbol{B}_1^{-1} \boldsymbol{A}_1 + \boldsymbol{B}_2 \left(\boldsymbol{B}_1^{-1}\right)' \boldsymbol{A}_1 + \boldsymbol{B}_2 \boldsymbol{B}_1^{-1} \boldsymbol{A}_1', \tag{A10}$$

where $(\ )'$ is defined as $\mathrm{d}()/\mathrm{d}\omega$. Since we have derived $\boldsymbol{A}_2'$, hence $\boldsymbol{A}_1'$ and $\boldsymbol{B}_2'$ can be immediately derived as follows

$$\begin{aligned} \boldsymbol{A}_1' &= 2\omega c_1 \boldsymbol{A}^2 + c_3 \boldsymbol{A} + c_6 W_{\varepsilon_1} + c_7 W_{\varepsilon_2}, \\ \boldsymbol{B}_2' &= 2\omega d_0 \boldsymbol{A}^2 + d_2 \boldsymbol{A} + d_8 W_{\varepsilon_1} + d_9 W_{\varepsilon_2}. \end{aligned} \tag{A11}$$

In expression (3.57), $\boldsymbol{B}_1^{-1}$ and $\left(\boldsymbol{B}_1^{-1}\right)'$ are also involved, which need a considerable algebra. Firstly, we write out the explicit expression for $\boldsymbol{B}_1$,

$$\boldsymbol{B}_1 = \begin{bmatrix} c_4 + c_{10} + \dfrac{c_8}{\varepsilon_1} + \dfrac{c_9}{\varepsilon_2} & & & & & \\ & R_1 & & & & \\ & & R_2 & & & \\ & & & \ddots & & \\ & & & & R_n & \\ & & & & & \ddots \\ & & & & & & R_N \end{bmatrix}, \tag{A12}$$

where

$$\begin{aligned} R_n &= c_0 \omega^2 J_n^2 + c_2 \omega J_n + (c_4 + c_{10}) I_2 + c_8 v_n^{\varepsilon_1} + c_9 v_n^{\varepsilon_2} \\ &= \begin{bmatrix} r_1 & r_2 \\ -r_2 & r_1 \end{bmatrix}, \end{aligned}$$

in which

$$\begin{aligned} r_1 &= -c_0 \omega^2 n^2 + c_4 + c_{10} + \frac{c_8 \varepsilon_1}{\varepsilon_1^2 + (n\omega)^2} + \frac{c_9 \varepsilon_2}{\varepsilon_2^2 + (n\omega)^2}, \\ r_2 &= c_2 \omega n - \frac{c_8 n\omega}{\varepsilon_1^2 + (n\omega)^2} - \frac{c_9 n\omega}{\varepsilon_2^2 + (n\omega)^2}. \end{aligned}$$

According to the property of the block diagonal matrix, $\boldsymbol{B}_1^{-1}$ can be derived immediately as

$$\boldsymbol{B}_1^{-1} = \begin{bmatrix} \dfrac{1}{c_4 + c_{10} + \dfrac{c_8}{\varepsilon_1} + \dfrac{c_9}{\varepsilon_2}} & & & & & \\ & S_1 & & & & \\ & & S_2 & & & \\ & & & \ddots & & \\ & & & & S_n & \\ & & & & & \ddots \\ & & & & & & S_N \end{bmatrix}, \tag{A13}$$

where

$$S_n = R_n^{-1} = \frac{1}{r_1^2 + r_2^2}\begin{bmatrix} r_1 & -r_2 \\ r_2 & r_1 \end{bmatrix}.$$

Further on, $\left(\boldsymbol{B}_1^{-1}\right)'$ can be derived as

$$\Delta = \left(\boldsymbol{B}_1^{-1}\right)' = \begin{bmatrix} 0 & & & & \\ & \ddots & & & \\ & & \delta_n & & \\ & & & \ddots & \\ & & & & \delta_N \end{bmatrix}, \tag{A14}$$

where

$$\delta_n = -\frac{2r_1r_1' + 2r_2r_2'}{\left(r_1^2 + r_2^2\right)^2}\begin{bmatrix} r_1 & -r_2 \\ r_2 & r_1 \end{bmatrix} + \frac{1}{r_1^2 + r_2^2}\begin{bmatrix} r_1' & -r_2' \\ r_2' & r_1' \end{bmatrix},$$

in which

$$r_1' = -2c_0\omega n^2 - \frac{2c_8\varepsilon_1\omega n^2}{\left[\varepsilon_1^2 + (n\omega)^2\right]^2} - \frac{2c_9\varepsilon_2\omega n^2}{\left[\varepsilon_2^2 + (n\omega)^2\right]^2},$$

$$r_2' = c_2 n - \frac{c_8 n\left[\varepsilon_1^2 - (n\omega)^2\right]}{\left[\varepsilon_1^2 + (n\omega)^2\right]^2} - \frac{c_9 n\left[\varepsilon_2^2 - (n\omega)^2\right]}{\left[\varepsilon_2^2 + (n\omega)^2\right]^2}.$$

So far, we have derived $\left(\boldsymbol{B}_1^{-1}\right)'$. Substituting $\boldsymbol{A}_1'$, $\boldsymbol{B}_2'$ and $\left(\boldsymbol{B}_1^{-1}\right)'$ into Eq. (3.57), we obtain

$$\begin{aligned} \frac{\partial}{\partial\omega}\left(\boldsymbol{B}_2\boldsymbol{B}_1^{-1}\boldsymbol{A}_1\right) &= \left(2\omega d_0\boldsymbol{A}^2 + d_2\boldsymbol{A} + d_8 W_{\varepsilon_1} + d_9 W_{\varepsilon_2}\right)\boldsymbol{B}_1^{-1}\boldsymbol{A}_1 \\ &\quad + \boldsymbol{B}_2\Delta\boldsymbol{A}_1 + \boldsymbol{B}_2\boldsymbol{B}_1^{-1}\left(2\omega c_1\boldsymbol{A}^2 + c_3\boldsymbol{A} + c_6 W_{\varepsilon_1} + c_7 W_{\varepsilon_2}\right). \end{aligned} \tag{A15}$$

Then, we obtain

$$\begin{aligned} \frac{\partial g}{\partial\omega} = \Big[&2\omega d_1\boldsymbol{A}^2 + d_3\boldsymbol{A} + d_6 W_{\varepsilon_1} + d_7 W_{\varepsilon_2} \\ &- \left(2\omega d_0\boldsymbol{A}^2 + d_2\boldsymbol{A} + d_8 W_{\varepsilon_1} + d_9 W_{\varepsilon_2}\right)\boldsymbol{B}_1^{-1}\boldsymbol{A}_1 \\ &- \boldsymbol{B}_2\Delta\boldsymbol{A}_1 - \boldsymbol{B}_2\boldsymbol{B}_1^{-1}\left(2\omega c_1\boldsymbol{A}^2 + c_3\boldsymbol{A} + c_6 W_{\varepsilon_1} + c_7 W_{\varepsilon_2}\right)\Big]Q_\alpha, \end{aligned} \tag{A16}$$

where W_{ε_i}, $(i = 1, 2)$, $\boldsymbol{B}_1$, $\boldsymbol{B}_1^{-1}$ and Δ are given in Eqs. (3.55), (3.58), (3.59) and (3.60) respectively. $\boldsymbol{B}_2$ is similar to $\boldsymbol{B}_1$.

References

[1] L. Liu, E.H. Dowell, The secondary bifurcation of an aeroelastic airfoil motion: effect of high harmonics, Nonlinear Dynamics 37 (1) (2004) 31–49.

[2] E.H. Dowell, E.F. Crawley, H. Curtiss Jr, D.A. Peters, R.H. Scanlan, F. Sisto, A Modern Course in Aeroelasticity, Kluwer Academic Publishers, Dodrecht, 1995.

[3] M.J. Patil, D.H. Hodges, On the importance of aerodynamic and structural geometrical nonlinearities in aeroelastic behavior of high-aspect-ratio wings, Journal of Fluids and Structures 19 (7) (2004) 905–915.

[4] B.H.K. Lee, L. Gong, Y.S. Wong, Analysis and computation of nonlinear dynamic response of a two-degree-of-freedom system and its application in aeroelasticity, Journal of Fluids and Structures 11 (3) (1997) 225–246.

[5] H. Alighanbari, S.J. Price, The post-hopf-bifurcation response of an airfoil in incompressible two-dimensional flow, Nonlinear Dynamics 10 (4) (1996) 381–400.

[6] K.W. Chung, Y.B. He, B.H.K. Lee, Bifurcation analysis of a two-degree-of-freedom aeroelastic system with hysteresis structural nonlinearity by a perturbation-incremental method, Journal of Sound and Vibration 320 (1) (2009) 163–183.

[7] Y.C. Fung, An Introduction to the Theory of Aeroelasticity, 1955.

[8] H.H. Dai, X.K. Yue, J.P. Yuan, S.N. Atluri, A time domain collocation method for studying the aeroelasticity of a two-dimensional airfoil with a structural nonlinearity, Journal of Computational Physics 270 (2014) 214–237.

[9] S.N. Atluri, Methods of Computer Modeling in Engineering & the Sciences, Tech Science Press, 2005.

[10] H.H. Dai, M. Schnoor, S.N. Atluri, A simple collocation scheme for obtaining the periodic solutions of the Duffing equation, and its equivalence to the high dimensional harmonic balance method: subharmonic oscillations, CMES: Computer Modeling in Engineering & Sciences 84 (5) (2012) 459–497.

[11] B.H.K. Lee, L. Liu, K.W. Chung, Airfoil motion in subsonic flow with strong cubic nonlinear restoring forces, Journal of Sound and Vibration 281 (3) (2005) 699–717.

[12] C.S. Liu, H.H. Dai, S.N. Atluri, A further study on using $\dot{x} = \lambda[\alpha R + \beta P](P = F\text{-}R(F \cdot R)/||R||^2)$ and $\dot{x} = \lambda[\alpha F + \beta P^*](P^* = R\text{-}F(F \cdot R)/||F||^2)$ in iteratively solving the nonlinear system of algebraic equations $\boldsymbol{F(x)} = 0$, CMES: Computer Modeling in Engineering & Sciences 81 (2) (2011) 195–227.

[13] C.S. Liu, Cone of non-linear dynamical system and group preserving schemes, International Journal of Non-Linear Mechanics 36 (2001) 1047–1068.

[14] H.H. Dai, X.K. Yue, S.N. Atluri, Solutions of the von K′arm′an plate equations by a Galerkin method, without inverting the tangent stiffness matrix, Journal of Mechanics of Materials and Structures 9 (2) (2014) 195–226.

[15] P.J. Attar, Cantilevered plate Rayleigh-Ritz trial function selection for von Kármán's plate equations, Journal of Aircraft 44 (2) (2007) 654–661.

[16] L. Liu, E.H. Dowell, K.C. Hall, A novel harmonic balance analysis for the van der pol oscillator, International Journal of Non-Linear Mechanics 42 (1) (2007) 2–12.

[17] J.P. Boyd, Chebyshev and Fourier Spectral Methods, Courier Dover Publications, 2001.

[18] A. LaBryer, P.J. Attar, High dimensional harmonic balance dealiasing techniques for a Duffing oscillator, Journal of Sound and Vibration 324 (3–5) (2009) 1016–1038.

[19] B.H.K. Lee, S.J. Price, Y.S. Wong, Nonlinear aeroelastic analysis of airfoils: bifurcation and chaos, Progress in Aerospace Sciences 35 (3) (1999) 205–334.

[20] H. Dai, Z. Yan, X. Wang, X. Yue, S.N. Atluri, Collocation-based harmonic balance framework for highly accurate periodic solution of nonlinear dynamical system. arXiv preprint arXiv: 2203.02990 (2022).

[21] L. Liu, J.P. Thomas, E.H. Dowell, P. Attar, K.C. Hall, A comparison of classical and high dimensional harmonic balance approaches for a Duffing oscillator, Journal of Computational Physics 215 (2006) 298–320.

CHAPTER 4

Application of Time Domain Collocation in Formation Flying of Satellites

SUMMARY

Relative motion of satellites in earth orbit has attracted a great deal of attentions in the past four decades. Various models accounting for the relative motions of two orbital satellites have been proposed, and applied in a variety of space missions. The first developed and perhaps also the most famous mathematical model for the relative motions is the Clohessy-Wiltshire (C-W) equations [1], which have been shown to be useful in solving the rendezvous problems. The C-W equations also provide an approach to determine the initial conditions for bounded or periodic relative motions for satellites in near-circular orbits. However, some of the assumptions that are used in the C-W equations introduce modeling errors that cannot be ignored. That is, for instance, the initial condition based on the C-W equations is valid only in some special cases with circular reference orbits, spherical earth, and linearized gravitational acceleration. Since the periodic relative orbits are important in relative orbit maintenance as well as in the tasks of spacecraft formation flying, more accurate models for relative motions and new approaches for obtaining periodic orbits are necessary.

In literature, many studies have been conducted to generalize the C-W equations. Tschauner and Hempel [2] incorporated the eccentricity for the reference orbit, based on which analytical expressions for relative motions as well as closed form solutions to the periodic relative motion were derived. Multifarious state transition matrices for relative motion near an orbit of arbitrary eccentricity were also developed, including the work of Melton [3] and Broucke [4], which used time as an independent variable, while the others use true anomaly instead. In the work of Inalhan [5] the initial conditions starting from which the Tschauner-Hempel (T-H) equations generate periodic solutions were obtained by using the homogenous solutions of the linearized relative equations of motion, which originated from the results by Lawden [6], Carter and Humi [7], and Carter [8]. This can be used to initialize to a closed-form aperture for a large fleet of vehicles with an eccentric reference orbit. However, considering nonlinear differential gravity, J_2 perturbation, and large eccentricity, the initial conditions derived by Inalhan do not work anymore.

To take into account the effects of nonlinear terms of gravity potential, eccentricity of reference orbit, and earth oblateness, many efforts have been done to establish more

Computational Methods for Nonlinear Dynamical Systems
DOI: https://doi.org/10.1016/B978-0-323-99113-1.00004-2

exact models of relative motion for the spacecraft in Earth orbit. The T-H equations perturbed by nonlinear differential gravity with no assumption on eccentricity were first presented by Euler and Shulman [9]. After that, many extended versions dealing with various nonlinearities and perturbations (e.g. the atmospheric drag and third body effects, etc.) were proposed. Unfortunately, there are no analytical solutions to these models due to the existence of nonlinearity, which nullifies the additivity and homogeneity pertaining to the linear systems.

Xu and Wang [10] presented a much more generalized dynamic model for satellite relative motions, which took J_2 perturbation, nonlinear terms of gravity potential, and eccentricity into account in their study. Since the dynamic equations are derived without approximation, it is an exact J_2 nonlinear relative model. In this chapter, we use the time domain collocation (TDC) method to search for periodic relative orbits in this model. In previous chapters, the TDC method has been applied to obtain the periodic solutions of the Duffing oscillator and the aeroelastic airfoil. Herein, the TDC method is extended to approximate the initial conditions for periodic relative orbits of spacecraft. The initial conditions for generating periodic relative motions of nearly circular orbits without J_2 perturbation can be obtained directly by C-W equations. However, in the presence of perturbations, the C-W initial conditions have to be refined to yield closed orbits. Using the TDC method, some adjustments of the initial conditions can be made to accommodate the perturbations.

The structure of this chapter is organized as follows. In Section 4.1 the TDC searching scheme for periodic relative orbits is proposed. In Section 4.2 the method of selecting initial values for the TDC method is introduced. In Section 4.3 the results of the TDC searching scheme are evaluated by simulating, the fuel consumption. In Section 4.4, some numerical results are presented to demonstrate the efficiency of this approach.

4.1 TDC searching scheme for periodic relative orbits

To implement the TDC method, a function in the form of a truncated Fourier expansion is considered as the approximation of the periodic solution, which can be expressed as follows:

$$f(t) = f_0 + \sum_{n=1}^{N} (f_{2n-1}\sin(n\omega_f t) + f_{2n}\cos(n\omega_f t)). \tag{4.1}$$

Here N is the number of harmonics included in the approximation, ω_f is the supposed frequency of the periodic motion, and f_i $(i = 0, 1, \ldots, 2N)$ are the harmonic coefficient variables. Obviously, the results of TDC method will be inherently periodic.

Collocate $f(t_j)$ $(j=1,2,\ldots,K)$ at K time points that are equally spaced in a period T of $f(t)$, then from the Eq. (4.1) we get

$$f(t_j)=f_0+\sum_{n=1}^{N}(f_{2n-1}\sin(n\omega_f t_j)+f_{2n}\cos(n\omega_f t_j)).$$

It can be seen that the $2N+1$ coefficients f_i are related to the collocation points $f(t_j)$ via the following transformation:

$$\begin{bmatrix} f(t_1) \\ f(t_2) \\ \vdots \\ f(t_K) \end{bmatrix}_{K\times 1} = \begin{bmatrix} 1 & \sin(\omega_f t_1) & \cos(\omega_f t_1) & \cdots & \cos(n\omega_f t_1) \\ 1 & \sin(\omega_f t_2) & \cos(\omega_f t_2) & \cdots & \cos(n\omega_f t_2) \\ \vdots & \vdots & \vdots & \ddots & \vdots \\ 1 & \sin(\omega_f t_K) & \cos(\omega_f t_K) & \cdots & \cos(n\omega_f t_K) \end{bmatrix}_{K\times(2N+1)} \begin{bmatrix} f_0 \\ f_1 \\ \vdots \\ f_{2N} \end{bmatrix}_{(2N+1)\times 1}. \quad (4.2)$$

For clear presentation we define the transformation matrix:

$$\boldsymbol{E}=\begin{bmatrix} 1 & \sin(\omega_f t_1) & \cos(\omega_f t_1) & \cdots & \cos(n\omega_f t_1) \\ 1 & \sin(\omega_f t_2) & \cos(\omega_f t_2) & \cdots & \cos(n\omega_f t_2) \\ \vdots & \vdots & \vdots & \ddots & \vdots \\ 1 & \sin(\omega_f t_K) & \cos(\omega_f t_K) & \cdots & \cos(n\omega_f t_K) \end{bmatrix}_{K\times(2N+1)}.$$

Thus if we can get the values of the K time points $f(t_j)$, the harmonic coefficients f_i can be determined by $[f_0,f_1,\ldots,f_{2N}]^{\mathrm{T}}=\boldsymbol{E}^{-1}[f(t_1),f(t_2),\ldots,f(t_K)]^{\mathrm{T}}$, where $\boldsymbol{E}^{-1}$ is the inverse $(K=2N+1)$ or pseudo inverse $(K\neq 2N+1)$ matrix of $\boldsymbol{E}$.

Using this expression, the first-order time derivative of $f(t)$ can be written as follows:

$$\dot{f}(t)=\frac{\mathrm{d}f(t)}{\mathrm{d}t}=\sum_{n=1}^{N}n\omega_f(f_{2n-1}\cos(n\omega_f t)-f_{2n}\sin(n\omega_f t)). \quad (4.3)$$

Collocating $\dot{f}(t)$ at t_j, we have

$$\dot{f}(t_j)=\sum_{n=1}^{N}n\omega_f(f_{2n-1}\cos(n\omega_f t_j)-f_{2n}\sin(n\omega_f t_j))$$

Referring to the derivation of Eq. (4.2), the relationship between the harmonic coefficients and the time derivatives $\dot{f}(t_j)$ $(j=1,2,\ldots,K)$ can be written in a matrix form

$$\begin{bmatrix} \dot{f}(t_1) \\ \dot{f}(t_2) \\ \vdots \\ \dot{f}(t_K) \end{bmatrix} = \omega_f \begin{bmatrix} 0 & \cos(\omega_f t_1) & -\sin(\omega_f t_1) & \cdots & -N\sin(N\omega_f t_1) \\ 0 & \cos(\omega_f t_2) & -\sin(\omega_f t_2) & \cdots & -N\sin(N\omega_f t_2) \\ \vdots & \vdots & \vdots & \ddots & \vdots \\ 0 & \cos(\omega_f t_K) & -\sin(\omega_f t_K) & \cdots & -N\sin(N\omega_f t_K) \end{bmatrix} \begin{bmatrix} f_0 \\ f_1 \\ \vdots \\ f_{2N} \end{bmatrix}. \quad (4.4)$$

It can also be expressed in a more subtle way:

$$\begin{bmatrix} \dot{f}(t_1) \\ \dot{f}(t_2) \\ \vdots \\ \dot{f}(t_K) \end{bmatrix} = \omega_f \boldsymbol{EA} \begin{bmatrix} f_0 \\ f_2 \\ \vdots \\ f_{2N} \end{bmatrix}, \tag{4.5}$$

where

$$\boldsymbol{A} = \begin{bmatrix} 0 & & & & \\ & A_1 & & & \\ & & A_2 & & \\ & & & \ddots & \\ & & & & A_N \end{bmatrix}_{(2N+1)\times(2N+1)}, \quad A_n = \begin{bmatrix} 0 & n \\ -n & 0 \end{bmatrix}.$$

Note that using Eq. (4.2), f_i can be expressed by $f(t_j)$. Substituting Eq. (4.2) into Eq. (4.5), we get

$$\begin{bmatrix} \dot{f}(t_1) \\ \dot{f}(t_2) \\ \vdots \\ \dot{f}(t_K) \end{bmatrix} = \omega_f \boldsymbol{EAE}^{-1} \begin{bmatrix} f(t_1) \\ f(t_2) \\ \vdots \\ f(t_K) \end{bmatrix}. \tag{4.6}$$

Thus the transformation from $f(t_j)$ to $\dot{f}(t_j)$ is established.

In the following parts, we will use the preceding equations and transformations to derive the TDC algebraic system for a nonlinear relative motion model, and thereby approximating the periodic solutions.

The exact J_2 nonlinear relative model between two elliptic satellite orbits derived by Xu and Wang is herein employed. Two Cartesian coordinate frames were used as shown in Fig. 4.1. The Earth centered inertial (ECI) coordinate frame is spanned by

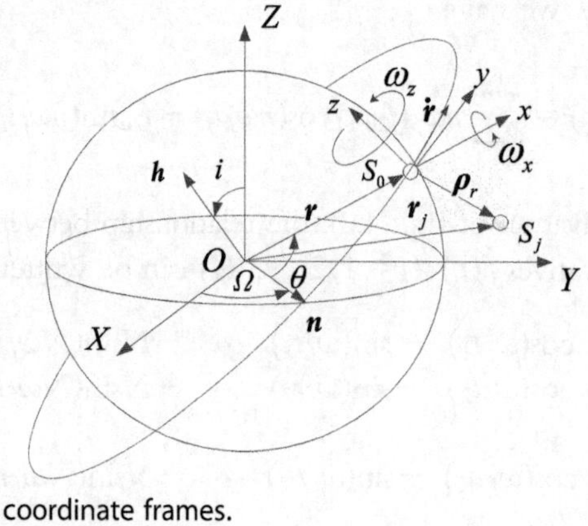

Figure 4.1 ECI and LVLH coordinate frames.

the unit vectors $(\boldsymbol{X}, \boldsymbol{Y}, \boldsymbol{Z})$. The local vertical local horizontal (LVLH) coordinate is attached on the reference satellite S_0.

With the J_2 gravity of the Earth incorporated, the relative motion of the satellite S_j in the LVLH frame is expressed by

$$\begin{aligned} \ddot{x}_r &= 2\dot{y}_r\omega_z - x_r(\eta_r{}^2 - \omega_z{}^2) + y_r\alpha_z - z_r\omega_x\omega_z - (\varsigma_r - \varsigma)s_is_\theta - r(\eta_r{}^2 - \eta^2) + F_x \\ \ddot{y}_r &= -2\dot{x}_r\omega_z + 2\dot{z}_r\omega_x - x_r\alpha_z - y_r(\eta_r{}^2 - \omega_z{}^2 - \omega_x{}^2) + z_r\alpha_x - (\varsigma_r - \varsigma)s_ic_\theta + F_y \\ \ddot{z}_r &= -2\dot{y}_r\omega_x - x_r\omega_x\omega_z - y_r\alpha_x - z_r(\eta_r{}^2 - \omega_x{}^2) - (\varsigma_r - \varsigma)c_i + F_z. \end{aligned} \tag{4.7}$$

If $v_x = \dot{x}_r$, $v_y = \dot{y}_r$, $v_z = \dot{z}_r$, then we get

$$\begin{aligned} \dot{x}_r &= v_x \\ \dot{y}_r &= v_y \\ \dot{z}_r &= v_z \\ \dot{v}_x &= 2v_y\omega_z - x_r(\eta_r{}^2 - \omega_z{}^2) + y_r\alpha_z - z_r\omega_x\omega_z - (\varsigma_r - \varsigma)s_is_\theta - r(\eta_r{}^2 - \eta^2) + F_x \\ \dot{v}_y &= -2v_x\omega_z + 2v_z\omega_x - x_r\alpha_z - y_r(\eta_r{}^2 - \omega_z{}^2 - \omega_x{}^2) + z_r\alpha_x - (\varsigma_r - \varsigma)s_ic_\theta + F_y \\ \dot{v}_z &= -2v_y\omega_x - x_r\omega_x\omega_z - y_r\alpha_x - z_r(\eta_r{}^2 - \omega_x{}^2) - (\varsigma_r - \varsigma)c_i + F_z, \end{aligned} \tag{4.8}$$

where ω_x, ω_z, α_x, α_z, r, ς, η, i, and θ are periodic time-varying parameters related to the chief orbit. They are described in the ECI coordinate frame by a set of differential equations which are independent of relative motion. For the reason that the equations of chief orbit are utilized directly in the searching procedure without treatment, they are listed in the chapter Appendix. F_x, F_y, and F_z are control forces of the satellite S_j in the LVLH coordinate. η_r and ς_r are functions of the relative positions x_r, y_r, and z_r and can be obtained from

$$\eta_r{}^2 = \frac{\mu}{r_r{}^3} + \frac{k_{J2}}{r_r{}^5} - \frac{5k_{J2}r_{rZ}{}^2}{r_r{}^7}, \quad \varsigma_r = \frac{2k_{J2}r_{rZ}}{r_r{}^5},$$

with

$$r_r = \sqrt{(r+x_r)^2 + y_r{}^2 + z_r{}^2}, \quad r_{rZ} = (r + x_r)s_is_\theta + y_rs_ic_\theta + z_rc_i, \quad k_{J2} = 3J_2\mu R_e^2/2.$$

Note that although Eq. (4.8) is equivalent to Eq. (4.7) mathematically, it suits the TDC method better. By increasing the number of equations in the system, the terms of second order time derivatives are removed. The advantage is that the approximate functions of relative velocities v_x, v_y, and v_z can be selected separately from those of relative positions x_r, y_r, and z_r with Eq. (4.8), that is, the information of both velocities and positions participate in the iteration procedure for solving this system and thus the error introduced by approximation shall be reduced.

For simplicity, we take $\dot{\boldsymbol{X}} = \mathbf{G}(\boldsymbol{X})$ on behalf of Eq. (4.8), with $\boldsymbol{X} = (x_ry_rz_rv_xv_yv_z)^{\mathrm{T}}$, which is the state vector of the relative motion. If there exists any periodic solution $\boldsymbol{X}(t)$ for the nonlinear relative model, the components can certainly be approximated by a truncated Fourier series expansions similar to $f(t)$.

Considering that the components of relative motion on three directions of LVLH frame can have different frequencies, we assume the following:

$$
\begin{aligned}
x_r(t) &= x_{r0} + \sum_{n=1}^{N} x_{r2n-1}\sin(n\omega_{xr}t) + x_{r2n}\cos(n\omega_{xr}t) \\
y_r(t) &= y_{r0} + \sum_{n=1}^{N} y_{r2n-1}\sin(n\omega_{yr}t) + y_{r2n}\cos(n\omega_{yr}t) \\
z_r(t) &= z_{r0} + \sum_{n=1}^{N} z_{r2n-1}\sin(n\omega_{zr}t) + z_{r2n}\cos(n\omega_{zr}t) \\
v_x(t) &= v_{x0} + \sum_{n=1}^{N} v_{x2n-1}\sin(n\omega_{xr}t) + v_{x2n}\cos(n\omega_{xr}t) \\
v_y(t) &= v_{y0} + \sum_{n=1}^{N} v_{y2n-1}\sin(n\omega_{yr}t) + v_{y2n}\cos(n\omega_{yr}t) \\
v_z(t) &= v_{z0} + \sum_{n=1}^{N} v_{z2n-1}\sin(n\omega_{zr}t) + v_{z2n}\cos(n\omega_{zr}t),
\end{aligned}
\tag{4.9}
$$

where x_r, y_r, z_r, v_x, v_y, v_z takes the same form of $f(t)$ except that the frequencies of relative motion around axis x, y, z are $\omega_{xr}, \omega_{yr}, \omega_{zr}$ separately. To determine the unknown coefficients and frequencies in Eq. (4.9), we have to derive the TDC algebraic system of Eq. (4.8) first.

If we collocate K points in a time period T of $\dot{\mathbf{X}} = \mathbf{G}(\mathbf{X})$, then we get

$$(\dot{\mathbf{X}}(t_1), \dot{\mathbf{X}}(t_2), \ldots \dot{\mathbf{X}}(t_K))^{\mathrm{T}} = (\mathbf{G}(\mathbf{X}(t_1)), \mathbf{G}(\mathbf{X}(t_2)), \ldots \mathbf{G}(\mathbf{X}(t_K)))^{\mathrm{T}}, \tag{4.10}$$

Recalling Eq. (4.6), the left-hand part of Eq. (4.10) can be converted to

$$
\begin{bmatrix} \dot{\boldsymbol{X}}_r \\ \dot{\boldsymbol{Y}}_r \\ \dot{\boldsymbol{Z}}_r \\ \dot{\boldsymbol{V}}_x \\ \dot{\boldsymbol{V}}_y \\ \dot{\boldsymbol{V}}_z \end{bmatrix}
=
\begin{bmatrix}
\omega_{xr}\boldsymbol{E}_x\boldsymbol{A}\boldsymbol{E}_x^{-1} & & & & & \\
& \omega_{yr}\boldsymbol{E}_y\boldsymbol{A}\boldsymbol{E}_y^{-1} & & & & \\
& & \omega_{zr}\boldsymbol{E}_z\boldsymbol{A}\boldsymbol{E}_z^{-1} & & & \\
& & & \omega_{xr}\boldsymbol{E}_x\boldsymbol{A}\boldsymbol{E}_x^{-1} & & \\
& & & & \omega_{yr}\boldsymbol{E}_y\boldsymbol{A}\boldsymbol{E}_y^{-1} & \\
& & & & & \omega_{zr}\boldsymbol{E}_z\boldsymbol{A}\boldsymbol{E}_z^{-1}
\end{bmatrix}
\begin{bmatrix} \boldsymbol{X}_r \\ \boldsymbol{Y}_r \\ \boldsymbol{Z}_r \\ \boldsymbol{V}_x \\ \boldsymbol{V}_y \\ \boldsymbol{V}_z \end{bmatrix}
$$

with the order of components in Eq. (4.10) being rearranged. For instance, $\boldsymbol{X}_r = (x_r(t_1), x_r(t_2), \ldots, x_r(t_K))^{\mathrm{T}}$, and so on. Furthermore, $\boldsymbol{E}_x$, $\boldsymbol{E}_y$ and $\boldsymbol{E}_z$ are transformation matrixes corresponding to ω_{xr}, ω_{yr}, and ω_{zr} separately. For brevity the block diagonal matrix on the right hand of the equation is denoted by $\tilde{\boldsymbol{E}}$ in the following parts.

By substituting the preceding equation into Eq. (4.10) and making some arrangements, we get the TDC algebraic system of Eq. (4.8)

$$\tilde{\mathbf{G}}(\mathbf{Q}) - \tilde{\boldsymbol{E}}\mathbf{Q} = \mathbf{0}, \tag{4.11}$$

where $\mathbf{Q} = (\boldsymbol{X}_r, \boldsymbol{Y}_r, \boldsymbol{Z}_r, \boldsymbol{V}_x, \boldsymbol{V}_y, \boldsymbol{V}_z)^{\mathrm{T}}$. $\tilde{\mathbf{G}}(\mathbf{Q})$ is the rearranged version of the vector $(\mathbf{G}(\boldsymbol{X}(t_1)), \mathbf{G}(\boldsymbol{X}(t_2)), \ldots \mathbf{G}(\boldsymbol{X}(t_K)))^{\mathrm{T}}$.

To solve this algebraic system, the Newton method can be very appropriate. Considering the complexity of Eq. (4.11), we derive the Jacobian matrix $\boldsymbol{J}$ in a numerical way. Denote $\boldsymbol{R} = \tilde{\mathbf{G}}(\mathbf{Q}) - \tilde{\boldsymbol{E}}\mathbf{Q}$, and assume that $\mathbf{Q}'_i = \mathbf{Q} + (0, \ldots \delta q_i \ldots 0)^{\mathrm{T}}$, δq_i is a small deviation from q_i, which is the i th component of $\mathbf{Q}$. Then a Jacobian matrix can be computed by $\boldsymbol{J}_i(\mathbf{Q}) = (\boldsymbol{R}(\mathbf{Q}'_i) - \boldsymbol{R}(\mathbf{Q}))/\delta q_i$, where $\boldsymbol{J}_i(\mathbf{Q})$ is the ith column of $\boldsymbol{J}$. Once $\mathbf{Q}$ is obtained, the relationship that is revealed in Eq. (4.2) can be used to give the approximate periodic solution $\boldsymbol{X}(t)$ of the relative model.

With $\boldsymbol{X}(t_j)$, which is the collocation point of $\boldsymbol{X}(t)$ at time $t_j (j = 1, 2, \ldots K)$, given by $\mathbf{Q}$, we can use it to initialize the relative orbits. The resulting orbits will be bounded if the approximation $\boldsymbol{X}(t)$ is close enough to the accurate periodic solution. In practical terms, the initialization procedure often takes place at a certain relative position. This can be realized by introducing additional constraints on the position of starting point, or in most cases, the first collocation point. The corresponding equations will be shown in Section 4.2. Here it is also worth noting that different selections of collocation points result in different approximation. Thus, a reasonable amount and spacement of collocation points is significant to reducing the approximation error.

4.2 Initial values for TDC method

To start an iterative process, one has to give some values for initialization. In this chapter, Newton method is applied to solve the resulting TDC algebraic system. In view of the inherent drawback of this method that initial values have to be close enough to the precise solution, some already presented relative dynamic models are used to provide proper relative state vectors for initialization. Under different conditions, the Clohessy-Wiltshire(C-W) equations and the Tschauner-Hempel(T-H) equations are employed separately.

Before we move on to the following parts, it is necessary to give a brief explanation on the significance of a chief orbit in relative motion. In the literature, a variety of chief orbits with different shapes and sizes were investigated. It was found that a circular or near circular chief orbit brings much convenience to the study of relative motion, while the elliptic orbit complicate the analysis. When the perturbations are incorporated, the relative motion theories could be more complex, because there is no circular orbit or elliptic orbit under the perturbations and thus the osculating orbital elements can hardly describe the satellite orbits conveniently. In the work of Schaub and Alfriend [11] the J_2 invariant relative orbits were investigated using mean orbit elements. With the constraints derived in the paper, it reached a conclusion that the size of relative orbit changes as the inclined angle of the chief orbit increase, and could be extremely large as it approaches a polar orbit. It was also revealed in the work of Inalhan [5] that the eccentricity of the chief orbit affects the shape of relative motion remarkably, not only on the in-plane motion, but also on the out-of-plane motion. Besides, the height of chief orbit affects the relative motion as well, in terms of the differential air drag. In this chapter, we focused on the effects of different eccentricities and inclined angles of chief orbit.

4.2.1 The C-W equations

As a customary analysis approach of rendezvous mechanics, the C-W equations of relative motion are linearized and can be readily solved to give periodic relative motion. As is known, the constraint on initial conditions $\dot{y}_r = -2\omega\dot{x}_r$ gives periodic relative motion, where $\omega = \sqrt{\mu/a^3}$, a being the constant radius of the chief orbit.

Applied in the exact J_2 relative model proposed by Xu and Wang [10], the preceding initial condition gives unbounded relative orbits that drift away. Suppose that the period of one orbit is time T, then the initial values $\mathbf{Q}_0$ for TDC algebraic system can be chosen by collocating points equally spaced in a period T on the unbounded orbits (Fig. 4.2).

Typically, the number of collocation points is chosen to be the same with that of harmonic coefficients in an approximate function. Note that the frequencies $\omega_{xr}, \omega_{yr}, \omega_{zr}$ are unknown in the above analysis. In order to solve Eq. (4.11), three more constraints are necessary. Herein, we impose a condition on the starting point, i.e., $x_r(t_0) = p_1$, $y_r(t_0) = p_2$, and $z_r(t_0) = p_3$, p_1, p_2, p_3 are predetermined. The additional condition can also be described as follows

$$x_{r0} + \sum_{n=1}^{N} x_{r2n-1}\sin(n\omega_{xr}t_0) + x_{r2n}\cos(n\omega_{xr}t_0) = p_1, \tag{4.12}$$

$$y_{r0} + \sum_{n=1}^{N} y_{r2n-1}\sin(n\omega_{yr}t_0) + y_{r2n}\cos(n\omega_{yr}t_0) = p_2, \tag{4.13}$$

$$z_{r0} + \sum_{n=1}^{N} z_{r2n-1}\sin(n\omega_{zr}t_0) + z_{r2n}\cos(n\omega_{zr}t_0) = p_3. \tag{4.14}$$

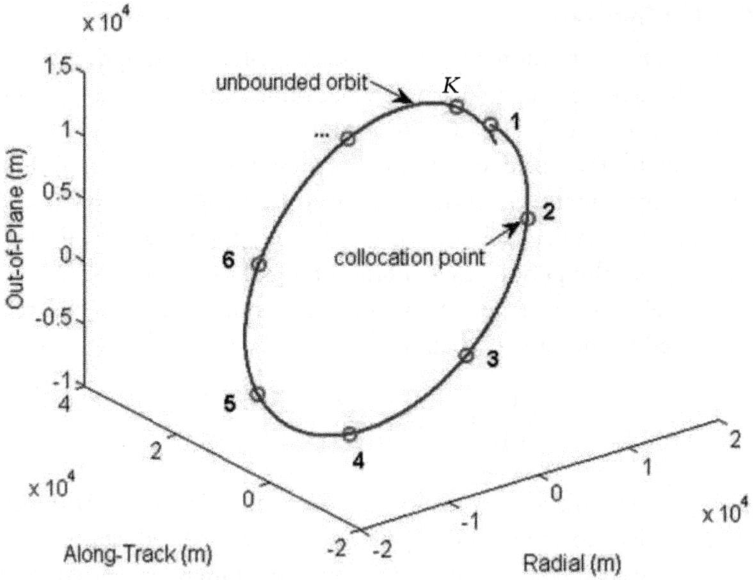

Figure 4.2 The selection of collocation points. Blue solid line: the unbounded orbit in a time period *T*; *red circles*: the collocation points.

Remind that Eq. (4.11) is about the positions and velocities of collocation points. The above equations are further transformed as follows:

$$\begin{bmatrix}1 & \sin(\omega_{xr}t_0) & \ldots & \cos(N\omega_{xr}t_0)\end{bmatrix}\boldsymbol{E}_x^{-1}\begin{bmatrix}x_r(t_1) & x_r(t_2) & \ldots & x_r(t_K)\end{bmatrix}^{\mathrm{T}} = p_1,$$

$$\begin{bmatrix}1 & \sin(\omega_{yr}t_0) & \ldots & \cos(N\omega_{yr}t_0)\end{bmatrix}\boldsymbol{E}_y^{-1}\begin{bmatrix}y_r(t_1) & y_r(t_2) & \ldots & y_r(t_K)\end{bmatrix}^{\mathrm{T}} = p_2,$$

$$\begin{bmatrix}1 & \sin(\omega_{zr}t_0) & \ldots & \cos(N\omega_{zr}t_0)\end{bmatrix}\boldsymbol{E}_z^{-1}\begin{bmatrix}z_r(t_1) & z_r(t_2) & \ldots & z_r(t_K)\end{bmatrix}^{\mathrm{T}} = p_3.$$

With the collocation points settled and the frequencies assumed, a closed orbit can be obtained through the transformations in Eq. (4.1) and Eq. (4.2). Unsurprisingly, this orbit does not satisfy Eq. (4.11) at first. However, with the Newton method or some other iteration methods, the position and velocity information of the collocation points as well as the values of frequencies can be corrected to minimize the residual error of Eq. (4.11). When the residual error caused by the collocation points is small enough, the corresponding orbit can be used to approximate the periodic relative motion.

Note that the C-W relative model is derived on with the assumptions of small relative distances, a circular reference orbit, and a central gravitation field. As the eccentricity of the chief orbit grows, the drift rate of the relative orbits increases, and thus the C-W equations can no longer provide appropriate initial values when the eccentricity is too large.

4.2.2 The T-H equations

The C-W equations of motion permit a simple analytic description of the relative model, yet it is limited to the circular reference orbit case. When an elliptical orbit is considered, we can use the T-H equations instead.

Many researchers have paid efforts on deriving the solution of T-H equations. Various state transition matrices, both those using true anomaly and those using time as the independent variable, are developed. Among them, the work presented by Inalhan et al. [5] gives the periodicity condition for T-H equations.

When the chief orbit is at perigee at the initial time, the bounded relative orbit condition expressed in the time domain is

$$\frac{\dot{y}(0)}{x(0)} = -\frac{n(2+e)}{(1+e)^{1/2}(1-e)^{2/3}}, \tag{4.15}$$

where $n = \sqrt{\mu/a^3}$, a being the semimajor radius of the chief orbit. As we can see, this expression is equivalent to the periodicity condition for C-W equation as $e \rightarrow 0$.

In the paper of Inalhan et al. [5], the Eq. (4.15) was further generalized to other values of θ; thus the initialization procedure can be carried out flexibly. In this chapter, however, only the special case when the initialization occurs at apogee is considered for simplicity.

4.3 Evaluation of TDC search scheme

In practical missions, the maintenance of the satellite formation asks that the deputy satellite flies along a reference orbit. Given that the use of the slowly drifting trajectory that is obtained with the TDC method as a reference is impractical in long-term missions, we project it to a periodic closed orbit (PCO). Then a closed loop control strategy based on the projected orbit is introduced to simulate the fuel consumption of the chief/deputy formation flying.

4.3.1 Projected closed orbit

Here, we introduce a straightforward approach to project the nearly bounded relative orbits that are found with the TDC searching scheme to a PCO.

Suppose the projected closed orbit can be expressed in a system like Eq. (4.9) and x_r, y_r, z_r, v_x, v_y, v_z are forced to have the same frequency ω_c. Then we collocate M points, normally spaced equally in a time domain, on the slowly drifting trajectory. If the frequency of the projected orbit ω_c is predetermined, it would be quite simple to derive the expression of the projected orbit. All we should do is to establish a formula similar to Eq. (4.2):

$$\mathbf{Q}_c = \widehat{\boldsymbol{E}}\hat{\mathbf{Q}}, \tag{4.16}$$

where $\widehat{\boldsymbol{E}}$ is a block diagonal matrix $\widehat{\boldsymbol{E}} = diag(\boldsymbol{E},\boldsymbol{E},\boldsymbol{E},\boldsymbol{E},\boldsymbol{E},\boldsymbol{E})$, $\hat{\mathbf{Q}}$ is the vector composed by the harmonic coefficients $\hat{\mathbf{Q}} = (x_0,\ldots,x_{2n},y_0,\ldots y_{2n},\ldots,v_{z0},\ldots,v_{z2n})^{\mathrm{T}}$, $\mathbf{Q}_c$ contains the state vectors of the M collocation points $\mathbf{Q}_c = (x(t_1),\ldots,x(t_M), y(t_1),\ldots,y(t_M),\ldots,v_z(t_M))^{\mathrm{T}}$.

Then multiply the equation with the pseudo-inverse of $\widehat{\boldsymbol{E}}$, we get the least squares solution $\hat{\mathbf{Q}} = \widehat{\boldsymbol{E}}^{-1}\mathbf{Q}_c$, due to the property of the pseudo inverse operation. Thus, we can derive the projected orbit as $\boldsymbol{F}(t) = \boldsymbol{E}(t)\hat{\mathbf{Q}}$, $\boldsymbol{E}(t) = diag(\boldsymbol{E}_1,\boldsymbol{E}_1,\boldsymbol{E}_1,\boldsymbol{E}_1,\boldsymbol{E}_1,\boldsymbol{E}_1)$, with $\boldsymbol{E}_1 = (1,\sin(\omega_c t),\cos(\omega_c t),\ldots,\cos(n\omega_c t))$.

However, if ω_c is unknown, the procedure to determine the projected closed orbit shall be laborious. Firstly we define a residual function $\boldsymbol{R} = \widehat{\boldsymbol{E}}\hat{\mathbf{Q}} - \mathbf{Q}_c$. Using the least squares method, we seek a solution $\omega_c{}^*$ and $\hat{\mathbf{Q}}^*$ that minimizes the squared error $\boldsymbol{R}^{\mathrm{T}}\boldsymbol{R}$. Thus the formula

$$\frac{\partial(\boldsymbol{R}^{\mathrm{T}}\boldsymbol{R})}{\partial\omega_c{}^*} = \mathbf{0}, \frac{\partial(\boldsymbol{R}^{\mathrm{T}}\boldsymbol{R})}{\partial\hat{\mathbf{Q}}^*} = \mathbf{0}. \tag{4.17}$$

has to be satisfied.

Substitute the residual function into Eq. (4.17), we get

$$\frac{\partial(\boldsymbol{R}^{\mathrm{T}}\boldsymbol{R})}{\partial\omega_c{}^*} = 2\hat{\mathbf{Q}}^{*\mathrm{T}}\widehat{\boldsymbol{A}}^{\mathrm{T}}\widehat{\boldsymbol{E}}^{*\mathrm{T}}\boldsymbol{T}^{\mathrm{T}}(\widehat{\boldsymbol{E}}^*\hat{\mathbf{Q}}^* - \mathbf{Q}_c) = \mathbf{0}$$

$$\frac{\partial(\boldsymbol{R}^{\mathrm{T}}\boldsymbol{R})}{\partial\hat{\mathbf{Q}}^*} = \widehat{\boldsymbol{E}}^*(\widehat{\boldsymbol{E}}^*\hat{\mathbf{Q}}^* - \mathbf{Q}_c) = \mathbf{0}. \tag{4.18}$$

The detailed information for matrix $\widehat{\boldsymbol{A}}$, $\boldsymbol{T}$ as well as the Jacobian matrix $\boldsymbol{J}$ of Eq. (4.18) is provided in the Appendix.

To solve this nonlinear algebraic equation, the globally optimal iterative algorithm (GOIA) is applied. The GOIA method is based on the concept of best decent vector u which takes the form of $u = \alpha_c\boldsymbol{F} + \boldsymbol{B}^{\mathrm{T}}\boldsymbol{F}$. It is proposed to solve a system of nonlinear algebraic equations (NAEs) without inverting the Jacobian matrix at each step. For more details of GOIA, refer to Liu and Atluri [12].

4.3.2 Closed loop control

The fuel consumption of the deputy satellite when it flies along the projected orbit is evaluated with a control technique called discrete linear quadratic regulator (DLQR). This technique is employed here due to its simplicity and ability to handle this sort of problems. Besides, the actuator is assumed to work impulsively, to approximate the way in which actual satellite thruster works.

Based on the assumption that the formation keeping is achieved by adjust the motion of the deputy satellite only, the linearized relative motion dynamics can be obtained as follows:

$$\dot{\boldsymbol{X}}_d = \overline{\boldsymbol{A}}\boldsymbol{X}_d + \overline{\boldsymbol{B}}\boldsymbol{u}, \tag{4.19}$$

where $\overline{\boldsymbol{A}} = \frac{\partial \mathbf{G}}{\partial \boldsymbol{X}_p}$, $\boldsymbol{X}_r$ is the state vector of the projected orbit, and $\boldsymbol{X}_d = \boldsymbol{X} - \boldsymbol{X}_p$ being the deviation of the actual trajectory from the projected orbit. $\overline{\boldsymbol{B}}$ is given as follows:

$$\overline{\boldsymbol{B}} = \begin{bmatrix} 0 & 0 & 0 \\ 0 & 0 & 0 \\ 0 & 0 & 0 \\ 1 & 0 & 0 \\ 0 & 1 & 0 \\ 0 & 0 & 1 \end{bmatrix}.$$

The detailed information about $\overline{\boldsymbol{A}}$ is provided in the chapter Appendix.

Then define the control vector $\boldsymbol{u}$ with a sampling interval T_s and a firing duration d as follows:

$$\boldsymbol{u}(t) = \begin{cases} \boldsymbol{u}_k/d & t_k < t < t_k + d \\ 0 & t_k + d < t < t_k + T_s \end{cases}, \tag{4.20}$$

where t_k is a sampling instant, $t_k = kT_s$ and $\boldsymbol{u}_k$ is the control signal provided by the discrete controller, $\boldsymbol{u}_k = \left[\Delta v_x, \Delta v_y, \Delta v_z\right]^{\mathrm{T}}$.

Thus we get the discrete time model expressed as

$$\boldsymbol{X}_{dk+1} = \tilde{\boldsymbol{A}}\boldsymbol{X}_{dk} + \tilde{\boldsymbol{B}}\boldsymbol{u}_k, \tag{4.21}$$

with $\tilde{\boldsymbol{A}} = e^{\overline{\boldsymbol{A}}T_s}$, $\tilde{\boldsymbol{B}} = e^{\overline{\boldsymbol{A}}T_s}\int_0^d e^{-\overline{\boldsymbol{A}}\tau} d\tau \overline{\boldsymbol{B}}/d$.

Using the DLQR control approach, the quadratic performance index below is minimized:

$$\boldsymbol{J} = \frac{1}{2}\sum_{k=0}^{\infty}\left[\boldsymbol{X}_{dk}{}^{\mathrm{T}}\boldsymbol{Q}\boldsymbol{X}_{dk} + \boldsymbol{u}_k{}^{\mathrm{T}}\boldsymbol{R}\boldsymbol{u}_k\right]. \tag{4.22}$$

And the optimal control law can be achieved simply by the well-known Ricatti solution [13]:

$$\boldsymbol{u}_k = -\boldsymbol{K}(\boldsymbol{X}(t_k) - \boldsymbol{X}_p(t_k)). \tag{4.23}$$

4.4 Numerical results

In the following parts, the numerical simulations demonstrate the effectiveness of the TDC approach on searching for periodic relative orbits, as well as the projected closed orbits and the corresponding fuel consumption.

It is assumed that the initialization procedure occurs at the apogee of the chief orbit where $r_0 = (1 + e)a$, $a = 8000$ km, $\dot{r}_0 = 0$ km/s, and the relative position of the deputy satellite is set to be $x_{r0} = 10$ km, $y_{r0} = 10$ km, $z_{r0} = 10$ km. For simplicity, the angular momentum of the chief orbit h_0 at the apogee is approximated with that of the Keplerian orbit. In the following simulations, the number of harmonics that are included in the approximate functions is $N = 4$, while the number of collocation points is $K = 9$. The resulting relative orbits are shown in the LVLH frame.

Fig. 4.3A shows the relative orbits obtained by using the TDC searching scheme when the eccentricity and the inclined angle of the chief orbit take the values $e = 0.005$ and $i = 0$, respectively. Fig. 4.3B shows the drifting orbits that are generated by the periodicity condition of C-W equations. The initial relative orbit is always shown as a bold solid line with a different color to the remaining orbits and the plots that are obtained with TDC approach always show the data of 10 orbits. In this case, the collocation points are selected in the time period $T = 7121$ s.

Nothing that both simulations above are conducted under the same assumptions, the improvement of the result is drastic. In Fig. 4.3A the relative orbit is nearly bounded and does not drift away after ten orbits, while in Fig. 4.3B, which is plotted for 20 orbits, the substantial drift is apparent.

The following simulations further investigate the performance of the TDC approach on searching for a periodic relative orbit. The values of eccentricity e and inclined angle i are changed, so that the TDC searching scheme can be tested in situations where the periodicity condition provided by C-W equations or T-H equations becomes completely inapplicable.

In the case $e = 0.02$, $i = pi/6$, the TDC searching scheme gives a fairly satisfying result. T is selected to be 7151 s.

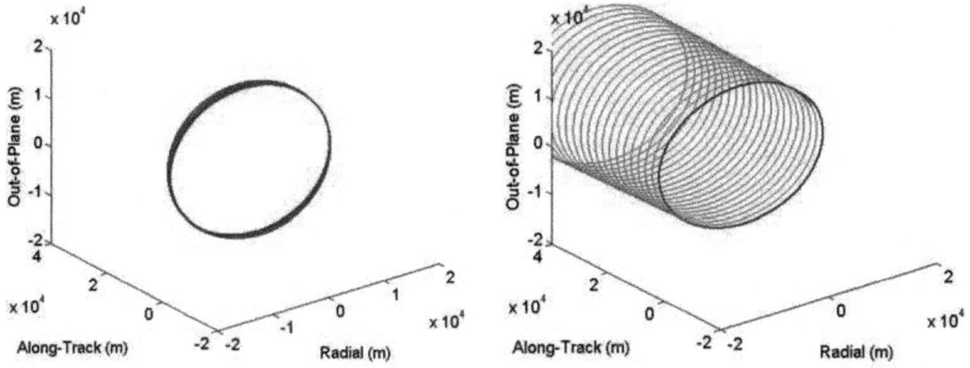

(A) (B)

Figure 4.3 Relative orbit with and without the correction of TDC searching scheme for the case $e = 0.005, i = 0$.

Fig. 4.4 makes a comparison between the orbits generated by the periodicity condition of the T-H equations and the TDC searching scheme. It is obvious that the drift of the orbits is relived greatly with the correction of the TDC method.

As the inclined angle increases up to $i = pi/3$, the resulting orbit is still nearly bounded and is demonstrated in Fig. 4.5, where $T = 7161$ s.

Fig. 4.6 illustrates the improvement in the result.

The orbit obtained by TDC approach is denoted by D1, which is shown with the solid red line, while the blue line marked with D2 is the orbit result from the T-H equations. Although D1 and D2 start from the same position $O(10\ \mathrm{km}, 10\ \mathrm{km}, 10\ \mathrm{km})$, the trajectories formed are quite different. In Fig. 4.6B clearly indicated that the drift rate of

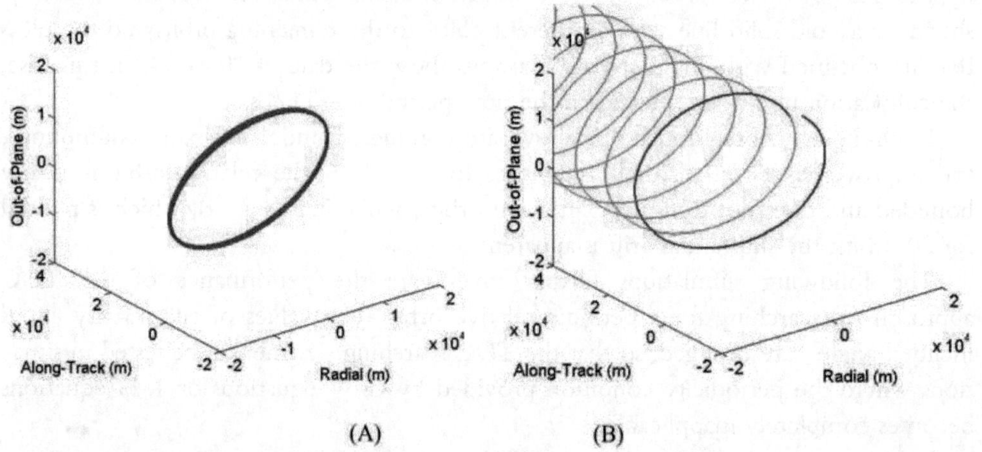

Figure 4.4 Relative orbit with and without the correction of TDC searching scheme for the case $e = 0.02, i = pi/6$.

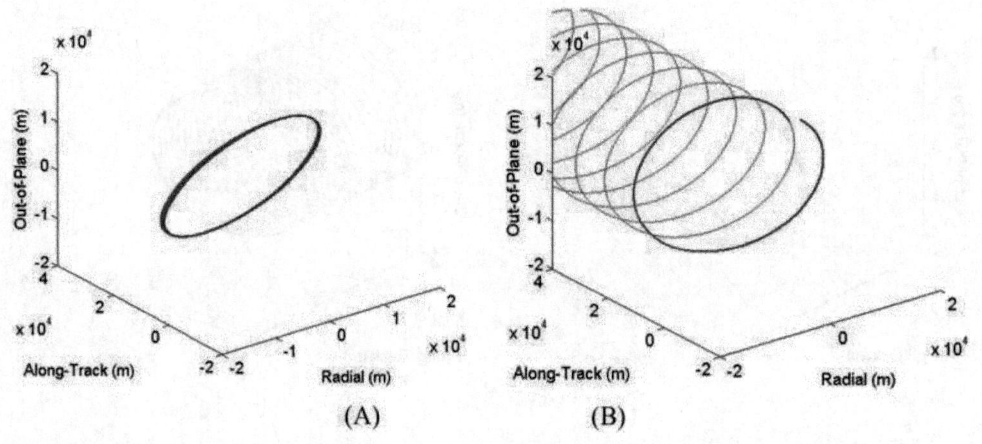

Figure 4.5 Relative orbit with and without the correction of TDC searching scheme for the case $e = 0.02, i = pi/3$.

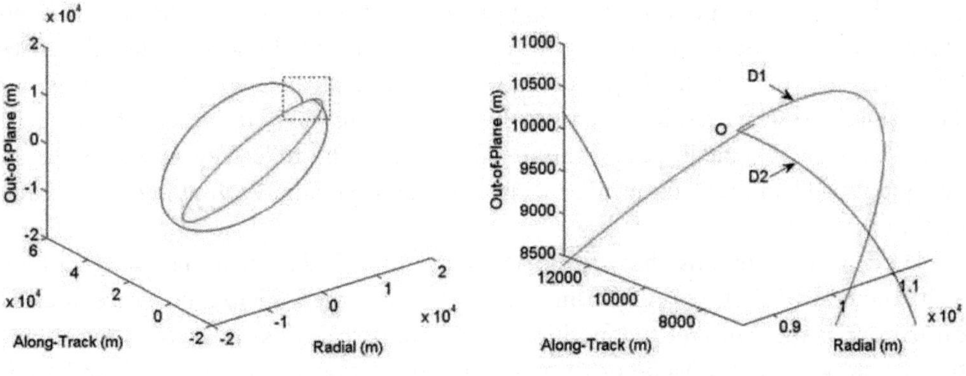

(A) (B)

Figure 4.6 Comparison of the initial relative orbits for the case $e = 0.02, i = pi/3$.

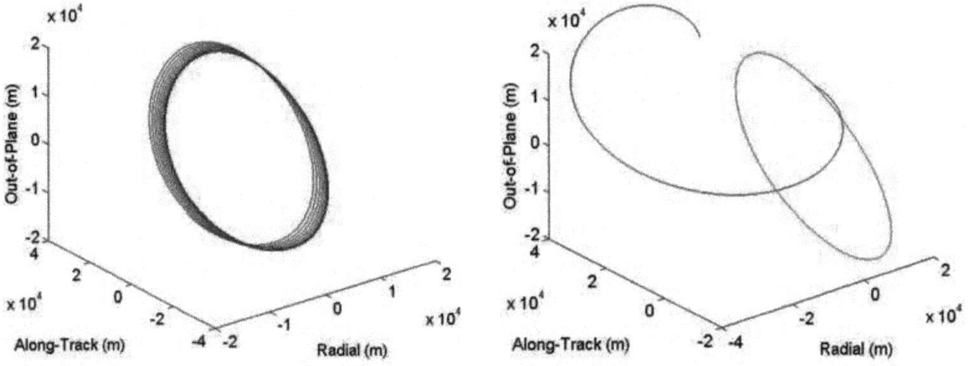

(A) (B)

Figure 4.7 Relative orbit with and without the correction of TDC searching scheme for the case $e = 0.1, i = pi/3$.

orbit D1 is much lower than D2. Besides, the shape of D1 is bounded tightly, making it possible to establish a fuel saving control strategy.

Here it is worth noting that D1 and D2 may appear to be retrograde because of the perspective employed in Fig. 4.6. Actually, the direction of D1 is consistent with that of D2 in the rough, although there indeed exists an acute angle between these two orbit planes. For further explanation, the intersection angle is caused by the slight differences between the initial velocities provided by the TDC searching scheme and the T-H equations, and the perspective is employed mainly for the purpose of demonstrating the drifting condition of D1 and D2.

In Fig. 4.7, the eccentricity of the reference orbit takes the value $e = 0.1$, with $i = pi/3$. The "period" T is 7191 s.

In this case, the periodicity condition arising from T-H equations is entirely impractical. As is shown in Fig. 4.7B, the drift rate on the along track direction is far from satisfying due to the J_2 perturbation and the inclined angle. Nevertheless, the TDC searching scheme still gives a slowly drifting orbit. Although the drift rate gains as the number of orbit increases, the initial orbit and the following several orbits are nearly bounded and can still provide a satisfying approximation to the periodic relative motion.

The initial conditions that are obtained by using T-H equations (IC1) and the TDC method (IC2) are presented in Table 4.1. All of the initialization procedures are carried out at the same relative position $O(10\text{ km}, 10\text{ km}, 10\text{ km})$. In IC1, the velocities v_{x0} and v_{z0} take the value of zero, for that there is no constraint on v_{x0} and v_{z0} in the periodicity condition developed by Inalhan. However, it is clear from the results that these two velocities are no longer negligible when the J_2 gravitational perturbation effect is included.

Table 4.2 shows the approximate frequencies of periodic solutions. As is shown, $\omega_x, \omega_y, \omega_z$ are slightly different; thus the approximate solutions are quasi periodic. Note that the knowledge of the perturbed frequencies due to J_2 aids in the design and tuning of the controllers for mitigating its adverse effects. The capability of approximating the in-plane and cross-track frequencies is an advantage of the TDC searching scheme. This table also includes the iteration steps for solving the algebraic equations obtained by the TDC method (i.e. the equations (4.11),(4.12),(4.13),(4.14)). It is shown that the iteration process is quickly convergent and hence costs little in computer resource.

The drift of the relative orbits on three directions is shown in Fig. 4.8 for the case $e = 0.02$ and $i = pi/3$. The figures plot the data of 20 orbits, corresponding to roughly 2 days of simulation time. Overall, the drift rates of the first several orbits are small and acceptable. However, as time accumulates, the orbits will eventually drift away. Thus, to generate a reliable projected closed orbit for control, the projection should be carried out in a time period in which the drift is unobvious.

By looking into the figures above with some attention, it can be easily found that the drift on the along-track direction is more severe than the others. Thus, the number of orbits used to generate the PCO is often determined by the drift rate on this direction. Normally less than 10 orbits should be up to the requirement.

Fig. 4.9 plots the projected orbits obtained under different situations, as well as the relative trajectories of the satellites which take the projected orbits as references and fly without control for 1 orbit. The tracking errors of different cases are listed in Table 4.3. As we can see, the tracking error grows as the eccentricity and the inclined angle increase. The reason for it is that the orbit of large eccentricity and inclined angle tends to drift away more quickly under the effect of the perturbations.

Table 4.1 Comparison of the initial conditions (IC) given by T-H equations and TDC method.

Case	$e = 0.005, i = 0$		$e = 0.02, i = \pi/6$		$e = 0.02, i = \pi/3$		$e = 0.1, i = \pi/3$	
	IC1	IC2	IC1	IC2	IC1	IC2	IC1	IC2
v_{x0} (m/s)	0	0.2990	0	−1.5751	0	−1.4593	0	4.3375
v_{y0} (m/s)	−17.6474	−17.5347	−17.6582	−17.1523	−17.6582	−17.1559	−17.9351	−15.3197
v_{z0} (m/s)	0	−0.4285	0	3.9841	0	5.5567	0	−8.9719

Table 4.2 Frequencies of periodic solutions and iteration steps of TDC searching scheme.

Case	$e=0.005, i=0$	$e=0.02, i=\pi/6$	$e=0.02, i=\pi/3$	$e=0.1, i=\pi/3$
ω_x (rad/s)	8.8323e − 4	8.8405e − 4	8.8553e − 4	8.8523e − 4
ω_y (rad/s)	8.8323e − 4	8.8658e − 4	8.9116e − 4	8.8669e − 4
ω_z (rad/s)	8.8507e − 4	8.8490e − 4	8.8450e − 4	8.8553e − 4
IS	14	6	8	16

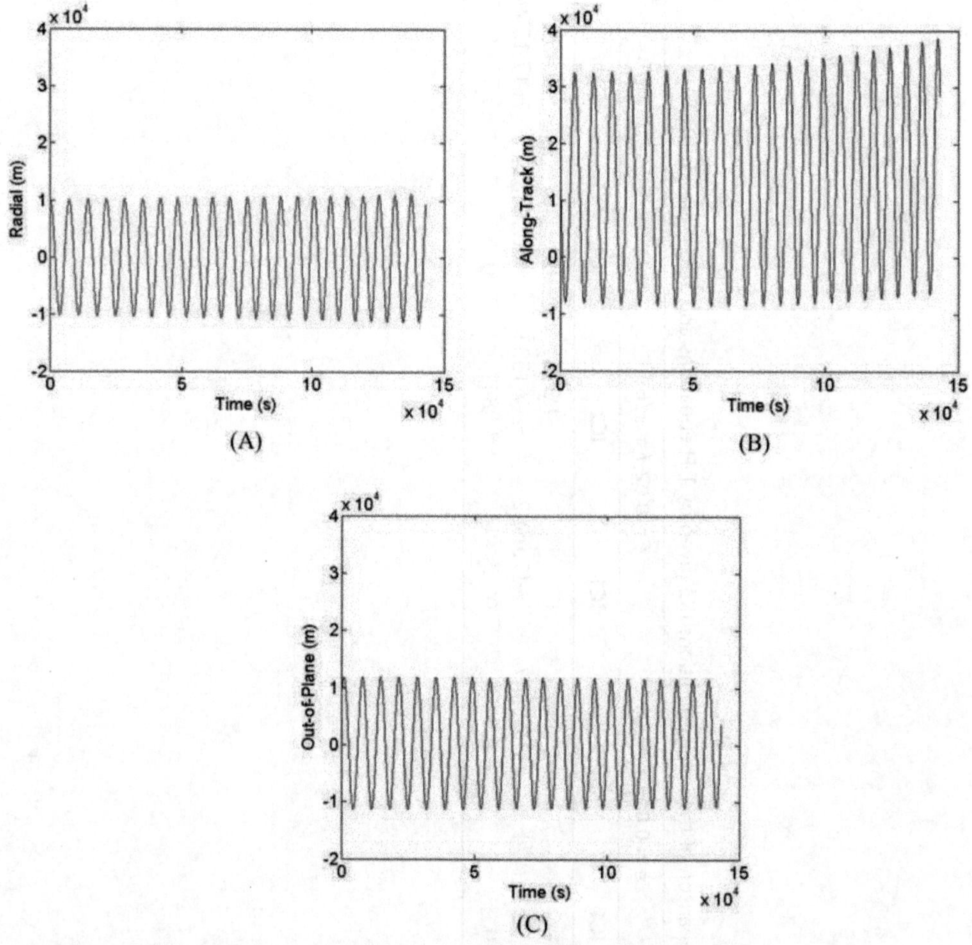

Figure 4.8 The drifts of the relative orbit in three directions, $e=0.02, i=pi/3$.

The number of points collocated to generate the PCOs is $M=50$. As is shown in Fig. 4.9, the PCOs (the red line) magnificently preserve the characteristics of the nearly bounded orbits presented above. The information of both the in-plane motion and the out-of-plane motion is embodied. The relative trajectories of the uncontrolled

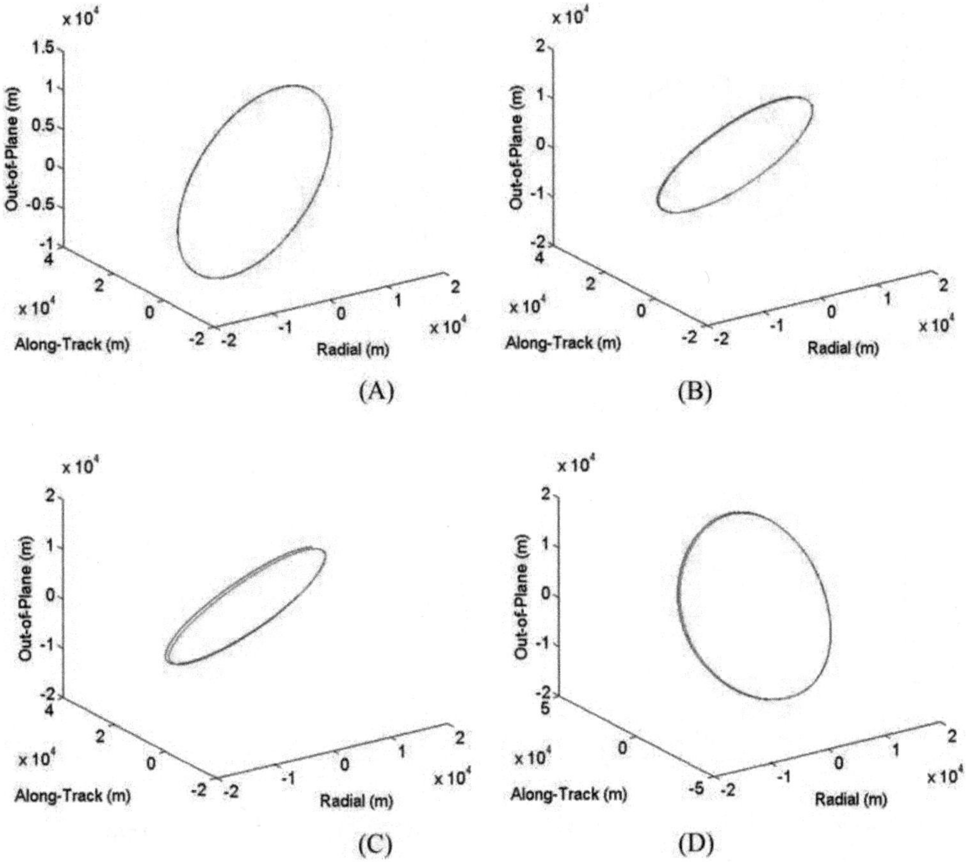

Figure 4.9 Relative trajectories without control and the PCOs. (A) $e=0.005, i=0$. (B) $e=0.02$, $i=pi/6$. (C) $e=0.02, i=pi/3$. (D) $e=0.1, i=pi/3$.

Table 4.3 Tracking errors of the relative trajectories without control after one orbit and the frequencies ω_c of the PCOs.

Case	Tracking error (m)	ω_c (rad/s)
$e=0.005$, $i=0$	57.5853	8.8345e − 4
$e=0.02$, $i=$ pi/6	423.0212	8.8628e − 4
$e=0.02$, $i=$ pi/3	1.2497e3	8.8870e − 4
$e=0.1$, $i=$ pi/3	888.1354	8.8572e − 4

satellites (blue line) indicate the efficiency of the PCOs. The tracking errors in some of the figures are not even visible on the scale that is shown. The frequencies of the PCOs are listed in Table 4.3 as well. They did not differ much from the original

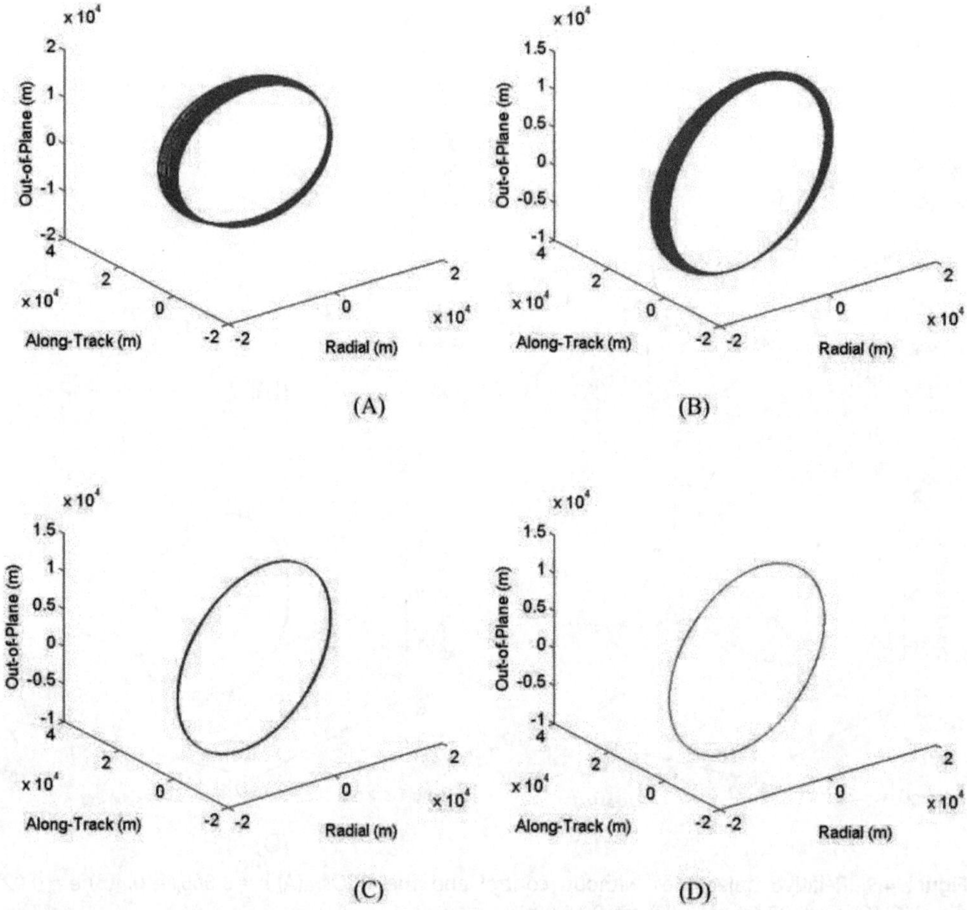

Figure 4.10 Relative trajectory under DLQR control. (A) Sampling time: 10,000 s. (B) Sampling time: 1000 s. (C) Sampling time: 100 s. (D) Sampling time: 10 s.

frequencies listed in Table 4.3. In the formation keeping process, they can help to design and tune the controller.

To further illustrate the efficiency of the projected closed orbit, we take the case where $e = 0.005$, $i = 0$. With the corresponding PCO as the reference orbit, it is simulated for 40 orbits, almost 4 days, to plot the relative trajectory under LQR control. The sampling time T_s is selected to be 10,000, 1000, 100, 10 s separately for each subfigure below. As is presented in Fig. 4.10, all the relative trajectories are bounded, tightly or loosely. In Fig. 4.10A the sampling time is obviously too long to keep the relative trajectory close to the PCO, even though, we haven't lost control of it. So it is indicated explicitly that the PCO is a reliable reference for periodic relative motion

Table 4.4 Delta V and Max. tracking error in the simulation using LQR control.

$e = 0.005$, $i = 0$, $x_r = 10$km, $y_r = 10$km, $z_r = 10$km, number of orbits $n = 40$		
Sampling (s)	Delta V (m/s)	Max. tracking error (m)
10,000	5.8320	6.4952e3
1000	11.0755	5.4374e3
100	15.8528	418.1591
10	9.0868	25.0111

control. The total simulated fuel consumption is denoted by Delta V. It is shown in Table 4.4 along with the maximal tracking error.

Here Delta V is calculated by the sum $\sum_{k=0}^{N_u} \|\boldsymbol{u}_k\|$, where the magnitude $\|\boldsymbol{u}_k\| = (\Delta v_x^2 + \Delta v_y^2 + \Delta v_z^2)^{\frac{1}{2}}$ is the norm of control signal $\boldsymbol{u}_k$ and $N_u = \text{floor}(40T/T_s)$ is the working times of the thruster during the 40-orbit period of relative motion. In this simulation, the firing duration d of the thrust pulses is supposed to be $d = 1$ s, and always occurs at the beginning of the sampling time.

As can be seen in Table 4.4, the sampling time is important to the efficiency of LQR control. The shorted the sampling interval is, the tighter the control could be. As $T_s = 10$ s, the maximal tracking error was reduced to 25.0111 meters, which is rarely satisfying reminding that the magnitude of the relative motion is 10^5 m. Correspondingly, the fuel consumption Delta V, however, did not change much. When $T_s = 10$ s, the Delta V needed to maintain the relative motion in a time period of almost 4 days was only 9.0868 m/s, roughly 0.227 m/s per orbit. It can also be easily found that the Delta V that is needed in the cases of $T_s = 100$ s and $T_s = 1000$ s is greater than that when the sampling time is 10 s. For the numerical simulations it was revealed that there exists a balance between the sampling time and the magnitude Delta V, thus careful consideration must be given to tuning the sampling interval.

Appendix

The motion of the chief satellite S_0 can be described by a set of equations as follows:

$$
\begin{aligned}
\dot{r} &= v_x \\
\dot{v}_x &= -\frac{\mu}{r^2} + \frac{h^2}{r^3} - \frac{k_{J2}}{r^4}(1 - 3s_i^2 s_\theta^2) \\
\dot{h} &= -\frac{k_{J2} s_i^2 s_{2\theta}}{r^3}
\end{aligned}
$$

$$\dot{\theta} = \frac{h}{r^2} + \frac{2k_{J2}c_i^2 s_\theta^2}{hr^3}$$

$$\dot{i} = -\frac{k_{J2}s_{2i}s_{2\theta}}{2hr^3}$$

$$\dot{\Omega} = -\frac{2k_{J2}c_i s_\theta^2}{hr^3}$$

To generate the projected orbits, matrix $\hat{\boldsymbol{A}}$, $\boldsymbol{T}$, and the Jacobian matrix $\boldsymbol{J}$ is necessary. They are expressed as follows:

$$\hat{\boldsymbol{A}} = \begin{bmatrix} \boldsymbol{A} & & & \\ & \boldsymbol{A} & & \\ & & \ddots & \\ & & & \boldsymbol{A} \end{bmatrix}_{6\times 6}.$$

$$\boldsymbol{T} = \begin{bmatrix} \boldsymbol{t} & & & \\ & \boldsymbol{t} & & \\ & & \ddots & \\ & & & \boldsymbol{t} \end{bmatrix}_{6\times 6} \text{ with } \boldsymbol{t} = \begin{bmatrix} t_1 & & & \\ & t_2 & & \\ & & \ddots & \\ & & & t_k \end{bmatrix}$$

$$\boldsymbol{J} = \begin{bmatrix} J_{11} & \boldsymbol{J}_{1n} \\ \boldsymbol{J}_{m1} & \boldsymbol{J}_{mn} \end{bmatrix} \text{ with}$$

$$J_{11} = \hat{\mathbf{Q}}^{*\mathrm{T}}\hat{\boldsymbol{A}}^{\mathrm{T}}\hat{\boldsymbol{A}}^{\mathrm{T}}\hat{\boldsymbol{E}}^{*\mathrm{T}}\boldsymbol{T}^{\mathrm{T}}\boldsymbol{T}^{\mathrm{T}}(\hat{\boldsymbol{E}}^{*}\hat{\mathbf{Q}}^{*} - \mathbf{Q}_c) + \hat{\mathbf{Q}}^{*\mathrm{T}}\hat{\boldsymbol{A}}^{\mathrm{T}}\hat{\boldsymbol{E}}^{*\mathrm{T}}\boldsymbol{T}^{\mathrm{T}}\boldsymbol{T}\hat{\boldsymbol{E}}^{*}\hat{\boldsymbol{A}}\hat{\mathbf{Q}}^{*},$$

$$\boldsymbol{J}_{1n} = \hat{\boldsymbol{A}}^{\mathrm{T}}\hat{\boldsymbol{E}}^{*\mathrm{T}}\boldsymbol{T}^{\mathrm{T}}(\hat{\boldsymbol{E}}^{*}\hat{\mathbf{Q}}^{*} - \mathbf{Q}_c) + \hat{\boldsymbol{E}}^{*\mathrm{T}}\boldsymbol{T}\hat{\boldsymbol{E}}^{*}\hat{\boldsymbol{A}}\hat{\mathbf{Q}}^{*},$$

$$\boldsymbol{J}_{m1} = \boldsymbol{J}_{1n}^{\mathrm{T}},$$

$$\boldsymbol{J}_{mn} = \hat{\boldsymbol{E}}^{*\mathrm{T}}\hat{\boldsymbol{E}}^{*}.$$

The matrix $\overline{\boldsymbol{A}}$ in Eq. (4.16) is expressed as $\overline{\boldsymbol{A}} = \overline{\boldsymbol{A}}_1 + \overline{\boldsymbol{A}}_2$ with

$$\overline{\boldsymbol{A}}_1 = \begin{bmatrix} 0 & 0 & 0 & 1 & 0 & 0 \\ 0 & 0 & 0 & 0 & 1 & 0 \\ 0 & 0 & 0 & 0 & 0 & 1 \\ -\eta_r^2 + \omega_z^2 & \alpha_z & -\omega_x\omega_z & 0 & 2\omega_z & 0 \\ -\alpha_z & -(\eta_r^2 - \omega_z^2 - \omega_x^2) & \alpha_x & -2\omega_z & 0 & 2\omega_x \\ -\omega_x\omega_z & -\alpha_x & -(\eta_r^2 - \omega_x^2) & 0 & -2\omega_x & 0 \end{bmatrix},$$

$$\overline{\boldsymbol{A}}_2 = \begin{bmatrix} 0 & 0 & 0 & 0 & 0 & 0 \\ 0 & 0 & 0 & 0 & 0 & 0 \\ 0 & 0 & 0 & 0 & 0 & 0 \\ -2k_{J2}r^{-5}(\sin i\sin\theta)^2 + 3ur^{-3} & -2k_{J2}r^{-5}\sin^2 i\cos\theta\sin\theta & -2k_{J2}r^{-5}\cos i\sin i\sin\theta & 0 & 0 & 0 \\ -2k_{J2}r^{-5}\sin^2 i\sin\theta\cos\theta & -2k_{J2}r^{-5}\sin^2 i\cos^2\theta & -2k_{J2}r^{-5}\cos i\sin i\cos\theta & 0 & 0 & 0 \\ -2k_{J2}r^{-5}\sin i\sin\theta\cos\theta & -2k_{J2}r^{-5}\cos i\sin i\cos\theta & -2k_{J2}r^{-5}\cos^2 i & 0 & 0 & 0 \end{bmatrix}.$$

Because $\overline{\boldsymbol{A}}$ is a function of time, it changes along the reference trajectory. However, in the numerical simulation it is reasonable to assume that $\overline{\boldsymbol{A}}$ is constant for simplicity. The simulation results demonstrate that the error that is introduced by this assumption can be ignored.

References

[1] W.H. Clohessy, R.S. Wiltshire, Terminal guidance system for satellite rendezvous, Journal of the Aerospace Sciences 27 (1960) 653–658. 674.

[2] J. Tschauner, P. Hempel, Optimale Beschleunigeungs programme fur das Rendezvous-Manover, Astronautica Acta 10 (5–6) (1964) 296–307.

[3] R.G. Melton, Time-explicit representation of relative motion between elliptical orbits, Journal of Guidance, Control, and Dynamics 23 (4) (2000) 604–610.

[4] R.A. Broucke, Solution of the elliptic rendezvous problem with the time as independent variable, Journal of Guidance, Control, and Dynamics 26 (4) (2003) 615–621.

[5] G. Inalhan, M. Tillerson, J.P. How, Relative dynamics and control of spacecraft formations in eccentric orbits, Journal of Guidance, Control, and Dynamics 25 (1) (2002) 48–59.

[6] D.F. Lawden, Optimal Trajectories for Space Navigation, Butterworths, London, 1963.

[7] T.E. Carter, M. Humi, Fuel-optimal rendezvous near a point in general keplerian orbit, Journal of Guidance, Control, and Dynamics 10 (6) (1987) 567–573.

[8] T.E. Carter, New form for the optimal rendezvous equations near a keplerian orbit, Journal of Guidance, Control, and Dynamics 13 (1) (1990) 183–186.

[9] E.A. Euler, Y. Shulman, Second-order solution to the elliptic rendezvous problem, AIAA Journal 5 (5) (1967) 1033–1035.

[10] G. Xu, D. Wang, Nonlinear dynamic equations of satellite relative motion around an oblate earth, Journal of Guidance, Control, and Dynamics 31 (5) (2008) 1521–1524.

[11] H. Schaub, K.T. Alfriend, J_2 Invariant relative orbits for spacecraft formations, Celestial Mechanics and Dynamical Astronomy 79 (2001) 77–95.

[12] C.S. Liu, S.N. Atluri, A globally optimal iterative algorithm using the best descent vector $\dot{\mathbf{x}} = \lambda\left[\alpha_c \mathbf{F} + \mathbf{B}^{\mathrm{T}}\mathbf{F}\right]$, with the critical value α_c, for solving a system of nonlinear algebraic equations $\mathbf{F}(\mathbf{x}) = 0$, Computer Modeling in Engineering & Sciences 84 (6) (2012) 575–601.

[13] F.L. Lewis, Optimal Control, Wiley, New York, 1986.

CHAPTER 5

Local Variational Iteration Method

SUMMARY

Among the various asymptotic methods the variational iteration method (VIM) [1,2], the Adomian decomposition method (ADM) [3], and the Picard iteration method (PIM) [4] (often combined with other approximation techniques, such as the modified Chebyshev-Picard iteration (MCPI) method [5]) have received wide attention. Unlike in classical perturbation methods, these three methods do not depend on the existence of a "small parameter" in the nonlinear problem, and they are not limited to cases of weak nonlinearities. The applications of the three methods are relatively straightforward in that there is no need to determine the so-called small parameter or make linearization transformations to the original equation. Further, the results that are obtained with these methods are valid for the whole nonlinear parameter domain. Thus, they are regularly utilized in areas where the traditional techniques are not applicable. In literature, some comparisons have been made between them on the efficiency and scope of application. The advantages and disadvantages of each method are illustrated [2,6].

The Variational Iteration Method is capable of solving a large class of nonlinear problems, including both the systems expressed as either ordinary or partial differential equations. It is an analytical asymptotic approach [7] where the initial guess function is corrected step by step, and finally reaches the true solution. The correctional iterative formula of VIM is similar to that of the Newton-Raphson method, except that the former method is for functions while the latter is for fixed point of nonlinear algebraic equations (NAEs). From the view of Inokuti, Sekine and Mura [1], we can see that the VIM is somehow an extension of the Newton method to the problem in function space. And the extension can also be made to problems expressed in algebraic, differential, integral, or finite-difference equations, or the combination of them. This idea is illustrated in detail and referred to as the general use of Lagrange multipliers in Inokuti, Sekine and Mura [1]. However, for the case of general nonlinear terms, the derivation of exact generalized Lagrange multiplier is normally impossible. Considering that the VIM of He [2] omits the nonlinearity in the adjoint equation satisfied by the generalized Lagrange multipliers. Other than that, the VIM of He [2] incorporates an artificial parameter in the approximation of the periodic solution. By eliminating the secular terms including the parameter, the accuracy of the result can be improved significantly. But this trick does not always work. As is illustrated

Computational Methods for Nonlinear Dynamical Systems
DOI: https://doi.org/10.1016/B978-0-323-99113-1.00005-4

in Section 5.3, it seems that only in some special cases can the approach show its efficiency.

The ADM is a decomposition method. It treats the solution as the sum of a series of functions, and decomposes the nonlinear terms into the rearranged Taylor series expansion in terms of Adomian polynomials. With an initial guess being given, the solution is corrected in each iteration step by adding an integral of an Adomian polynomial. As the iteration proceeds, the nonlinear equation is solved gradually. This method provides great convenience for the calculation of the correctional function, but the construction of Adomian polynomials is somewhat complex, especially for the higher order terms. The conditions for the convergence of this method have been discussed in literature. It should be noted that the method is just locally convergent because it is based on a Taylor series expansion. Some modifications were made to it by recursively applying it over divided time segments, so that the global convergence can be achieved [8].

Compared to VIM and ADM, Picard's method takes a very simple form. However, its application is very limited, since it needs to integrate the nonlinear terms in each iteration step, which could be very difficult to implement. Compared to Adomian's method, the Picard iteration lacks the ease of computation and also the ability to solve a wide class of equations. But by combining with other computational techniques, it is still possible to make advances to PIM. For example, the MCPI method [5] combines the Chebyshev polynomials with the Picard's method and was applied to the two body gravitational integration problem. It is shown that the integration process of the Picard's method becomes very simple and the method achieves high accuracy and efficiency.

Despite all the apparent differences in VIM, ADM and PIM, it is the objective of this Chapter to show that they are all not unrelated. As will be illustrated in Section 5.2, the iteration formula of the Picard's method and the ADM can be completely deduced from the iterative formula of the VIM. This indicates that a common mathematical principle, which is the general use of Lagrange multipliers, underlies these methods, and a unification of the concepts underlying the three methods can be elucidated. It will help to explain the similarity of the three methods, and how we can combine the ideas in VIM, ADM and PIM to achieve possibly better analytical asymptotic methods.

After that, a local variational iteration method (LVIM) is proposed in Section 5.3. Firstly, the differential transform (DT) method [9] is introduced to help identify the generalized Lagrange multipliers. Then a piecewise solution is obtained by dividing the whole time domain into a finite number of small segments. The forcing term is approximated by a truncated Taylor series expansion in the LVIM, so that the calculation of integrals can be simplified. The advantages of the LVIM are as follows,

1. The initial guess function can take a very simple form. Unlike the conventional VIM, there is no need to pay much attention to the construction of an initial guess. A linear function could be good enough for the LVIM, as long as it meets the boundary condition of the problem.
2. Instead of restricting the nonlinear term from variation artificially, the nonlinear term can be included in the derivation of generalized Lagrange multipliers in the LVIM. So the correctional formula is more effective.
3. The introduction of the Differential Transform method provides the approximation of generalized Lagrange multipliers in a Taylor series expansion. It makes the derivation of generalized Lagrange multipliers much simpler, and also provides convenience to the integration of the correctional formula.
4. The LVIM has applicability to long term predictions of many kinds of complex nonlinear problems, including chaotic motions.
5. The accuracy and computational speed of LVIM is relatively very high, even compared to the numerical methods such as the fourth-order RK4.

To convey the ideas in a simple and clear way, a system of first order differential equations is taken for illustration in this paper. However, the conclusions made in the following sections are also applicable to the higher order differential equations. It is well known that a higher order differential equation can always be transformed into a system of first order differential equations. For example, the equation

$$\frac{\mathrm{d}^n x}{\mathrm{d}\tau^n} = f\left(\frac{\mathrm{d}^{n-1}x}{\mathrm{d}\tau^{n-1}}, \ldots \frac{\mathrm{d}x}{\mathrm{d}\tau}, x, \tau\right) \tag{5.1}$$

can be rewritten as a system of equations like

$$x = x_1, \quad \frac{\mathrm{d}x_1}{\mathrm{d}\tau} = x_2, \ldots \frac{\mathrm{d}x_n}{\mathrm{d}\tau} = f(xn, \ldots x_1, x_0, \tau) \tag{5.2}$$

by introducing the variables $x_1, \ldots x_n$. Therefore, the nonlinear ordinary differential equations can be expressed in a general form as

$$\frac{\mathrm{d}\boldsymbol{x}}{\mathrm{d}\tau} = \boldsymbol{F}(\boldsymbol{x}, \tau),$$

where

$$\boldsymbol{x} = (x_1, x_2, \ldots)^{\mathrm{T}}, \ \boldsymbol{F} = (f_1, f_2, \ldots)^{\mathrm{T}}, \ \tau \in [t_0, t]. \tag{5.3}$$

where, here onwards, a bold symbol indicates a vector or a matrix, and for brevity, the differential operator $\mathrm{d}/\mathrm{d}t$ is denoted as $\boldsymbol{L}$ in the following sections. In the preceding equation, $\boldsymbol{F}(\boldsymbol{x}, \tau)$ is a nonlinear function of the state vector $\boldsymbol{x}$ and the independent variable τ.

5.1 VIM and its relationship with PIM and ADM

The three methods, i.e., VIM, ADM and PIM, were proposed independently by different researchers at different times. ADM and PIM have been well known for a long time, while VIM has a history of no more than two decades. Although they are regarded as being unrelated to one another at the very first, the similarity of these three methods on handling nonlinear problems implies that there could be some hidden relationships inside and can possibly lead to unification with the perspective provided by VIM. In the following, the VIM is described and some transformations are made to elucidate the relationships between them.

5.1.1 VIM

Consider the following general nonlinear system:

$$\boldsymbol{L}\boldsymbol{x} = \boldsymbol{F}(\boldsymbol{x}, \tau), \quad \tau \in [t_0, t], \tag{5.4}$$

where $\boldsymbol{L}$ is the first order differential operator and $\boldsymbol{F}$ is a nonlinear operator. The solution of this system can be approximated with an initial approximation $\boldsymbol{x}_0(t)$ and the correctional formula as

$$\boldsymbol{x}_{n+1}(t) = \boldsymbol{x}_n(t) + \int_{t_0}^{t} \boldsymbol{\lambda}(\tau)\{\boldsymbol{L}\boldsymbol{x}_n(\tau) - \boldsymbol{F}[\boldsymbol{x}_n(\tau), \tau]\}\mathrm{d}\tau, \tag{5.5}$$

where $\boldsymbol{\lambda}(\tau)$ is a matrix of Lagrange multipliers which are yet to be determined. Suppose $\boldsymbol{\Pi}[\boldsymbol{x}(\tau), \boldsymbol{\lambda}(\tau)]$ is a vector function of $\boldsymbol{x}(\tau)$ and $\boldsymbol{\lambda}(\tau)$, where $t_0 \le \tau \le t$.

$$\boldsymbol{\Pi}[\boldsymbol{x}(t), \boldsymbol{\lambda}(t)] = \boldsymbol{x}(\tau)|_{\tau=t} + \int_{t_0}^{t} \boldsymbol{\lambda}(\tau)\{\boldsymbol{L}\boldsymbol{x}(\tau) - \boldsymbol{F}[\boldsymbol{x}(\tau), \tau]\}\mathrm{d}\tau. \tag{5.6}$$

Let $\hat{\boldsymbol{x}}(\tau)$ be the exact solution of $\boldsymbol{L}\boldsymbol{x}(\tau) = \boldsymbol{F}[\boldsymbol{x}(\tau), \tau]$. Naturally, it satisfies the expression:

$$\boldsymbol{\Pi}[\hat{\boldsymbol{x}}(t), \boldsymbol{\lambda}(t)] = \hat{\boldsymbol{x}}(\tau)|_{\tau=t} + \int_{t_0}^{t} \boldsymbol{\lambda}(\tau)\{\boldsymbol{L}\hat{\boldsymbol{x}}(\tau) - \boldsymbol{F}[\hat{\boldsymbol{x}}(\tau), \tau]\}\mathrm{d}\tau = \hat{\boldsymbol{x}}(\tau)|_{\tau=t}. \tag{5.7}$$

Now we want to make the function $\boldsymbol{\Pi}[\boldsymbol{x}(t), \boldsymbol{\lambda}(t)]$ stationary about $\boldsymbol{x}$ at $\boldsymbol{x}(t) = \hat{\boldsymbol{x}}(t)$. Firstly, the variation of $\boldsymbol{\Pi}[\boldsymbol{x}(t), \boldsymbol{\lambda}(t)]$ is derived as

$$\begin{aligned}
\delta\boldsymbol{\Pi}[\boldsymbol{x}(t), \boldsymbol{\lambda}(t)] &= \delta\boldsymbol{x}(\tau)|_{\tau=t} + \delta\int_{t_0}^{t} \boldsymbol{\lambda}(\tau)\{\boldsymbol{L}\boldsymbol{x}(\tau) - \boldsymbol{F}[\boldsymbol{x}(\tau), \tau]\}\mathrm{d}\tau \\
&= \delta\boldsymbol{x}(\tau)|_{\tau=t} + \int_{t_0}^{t} \delta\boldsymbol{\lambda}(\tau)\{\boldsymbol{L}\boldsymbol{x}(\tau) - \boldsymbol{F}[\boldsymbol{x}(\tau), \tau]\}\mathrm{d}\tau + \int_{t_0}^{t} \boldsymbol{\lambda}(\tau)\delta\{\boldsymbol{L}\boldsymbol{x}(\tau) - \boldsymbol{F}[\boldsymbol{x}(\tau), \tau]\}\mathrm{d}\tau \\
&= \int_{t_0}^{t} \delta\boldsymbol{\lambda}(\tau)\{\boldsymbol{L}\boldsymbol{x}(\tau) - \boldsymbol{F}[\boldsymbol{x}(\tau), \tau]\}\mathrm{d}\tau + \delta\boldsymbol{x}(\tau)|_{\tau=t} + \boldsymbol{\lambda}(\tau)\delta\boldsymbol{x}(\tau)|_{\tau=t_0}^{\tau=t} \\
&\quad - \int_{t_0}^{t} \left[\boldsymbol{L}\boldsymbol{\lambda}(\tau) + \boldsymbol{\lambda}(\tau)\frac{\partial \boldsymbol{F}(\boldsymbol{x}, \tau)}{\partial \boldsymbol{x}}\right]\delta\boldsymbol{x}(\tau)\mathrm{d}\tau - \int_{t_0}^{t} \boldsymbol{\lambda}(\tau)\frac{\partial \boldsymbol{F}(\boldsymbol{x}, \tau)}{\partial \tau}\delta\tau\mathrm{d}\tau.
\end{aligned} \tag{5.8}$$

If $\boldsymbol{F}$ is not an explicit function of τ, the term $[\partial \boldsymbol{F}(\boldsymbol{x}, \tau)/\partial\tau]\delta\tau$ can be omitted in the preceding formula.

Then we collect the terms including $\delta\boldsymbol{x}(\tau)|_{\tau=t}$ and $\delta\boldsymbol{x}(\tau)$:

$$\delta\boldsymbol{x}(\tau)|_{\tau=t} + \boldsymbol{\lambda}(\tau)\delta\boldsymbol{x}(\tau)|_{\tau=t}, \quad \int_{t_0}^{t} \left[\boldsymbol{L}\boldsymbol{\lambda}(\tau) + \boldsymbol{\lambda}(\tau)\frac{\partial \boldsymbol{F}(\boldsymbol{x}, \tau)}{\partial \boldsymbol{x}}\right]\delta\boldsymbol{x}(\tau)\mathrm{d}\tau. \tag{5.9}$$

Note that the boundary value of $\boldsymbol{x}(\tau)$ at $\tau = t_0$ is prescribed, that is to say, $\delta\boldsymbol{x}(\tau)|_{\tau=t_0} = \boldsymbol{0}$. Thus the stationary condition for $\boldsymbol{\Pi}[\boldsymbol{x}(\tau), \boldsymbol{\lambda}(\tau)]$ is obtained as

$$\begin{cases}
\delta\boldsymbol{x}(\tau)|_{\tau=t}\text{:} diag[1, 1, ...] + \boldsymbol{\lambda}(\tau)|_{\tau=t} = \boldsymbol{0} \\
\delta\boldsymbol{x}(\tau)\text{:} \boldsymbol{L}\boldsymbol{\lambda}(\tau) + \boldsymbol{\lambda}(\tau)\dfrac{\partial \boldsymbol{F}(\boldsymbol{x}, \tau)}{\partial \boldsymbol{x}} = \boldsymbol{0} \\
\delta\boldsymbol{\lambda}(\tau)\text{:} \boldsymbol{L}\boldsymbol{x}(\tau) = \boldsymbol{F}[\boldsymbol{x}(\tau), \tau].
\end{cases} \tag{5.10}$$

Noting that the exact solution $\hat{\boldsymbol{x}}$ is unknown; therefore the truly optimal $\boldsymbol{\lambda}(\tau)$ is not available herein. As an alternative, $\boldsymbol{\lambda}(\tau)$ is approximated by replacing $\hat{\boldsymbol{x}}$ with $\boldsymbol{x}_n$. If $\boldsymbol{x}_n$ is a neighbored function of $\hat{\boldsymbol{x}}$, i.e., $\hat{\boldsymbol{x}} - \boldsymbol{x}_n = \delta\hat{\boldsymbol{x}}$, the error caused will not exceed $O^2(\delta\hat{\boldsymbol{x}})$.

In the VIM of He [2], sometimes the nonlinear term is considered as being restricted from variation, in order to simplify the calculation of $\boldsymbol{\lambda}$. As an asymptotic method for solving nonlinear problems, VIM provides plenty of freedom to the user. It does not restrict the selection of initial guess functions and even allows the existence of unknown parameters in it. Further, depending on the selection of a restricted variation, the matrix of generalized Lagrange multipliers $\boldsymbol{\lambda}$ can be determined in various ways. The less restricted the variational terms are, the more accurate will the Lagrange multipliers be.

5.1.2 Comparison of VIM with PIM and ADM

Instead of PIM, the VIM of He [2] is utilized to solve the IVP governed by Eq. (5.4), we will get the corresponding correctional formula with $\boldsymbol{\lambda}$:

$$\boldsymbol{x}_{n+1}(t)=\boldsymbol{x}_n(t)+\int_{t_0}^{t}\boldsymbol{\lambda}[\boldsymbol{L}\boldsymbol{x}_n-\boldsymbol{F}(\boldsymbol{x}_n,\tau)]\mathrm{d}\tau=\boldsymbol{0}, \tag{5.11}$$

where $\boldsymbol{F}(\boldsymbol{x}_n,\tau)$ is considered as being restricted from variation. The matrix of Lagrange multipliers $\boldsymbol{\lambda}$ are determined from the following restricted stationary conditions:

$$\begin{cases}\boldsymbol{\lambda}(\tau)|_{\tau=t}+diag[1,1,...]=\boldsymbol{0}\\ \boldsymbol{L}\boldsymbol{\lambda}(\tau)=\boldsymbol{0}\end{cases}. \tag{5.12}$$

The Lagrange multipliers, therefore, can be identified as $\boldsymbol{\lambda}(\tau)=diag[-1,-1,...]$. As a result, we obtain the following iteration formula:

$$\boldsymbol{x}_{n+1}(t)=\boldsymbol{x}_n(t)-\int_{t_0}^{t}\{\boldsymbol{L}\boldsymbol{x}_n(\tau)-\boldsymbol{F}[\boldsymbol{x}_n(\tau),\tau]\}\mathrm{d}\tau=\boldsymbol{x}_n(t_0)+\int_{t_0}^{t}\boldsymbol{F}[\boldsymbol{x}_n(\tau),\tau]\mathrm{d}\tau. \tag{5.13}$$

Noting that $\boldsymbol{x}_n(t_0)$ is equal to the initial value $\boldsymbol{x}(t_0)$, so it is exactly the Picard iteration.

In the paper of He [2], a first-order differential equation is considered and the ADM is proved to be a specific version of VIM. This work is reviewed in the following.

Consider the equation $Lx+Rx+Nx=g(\tau)$, in which L is a first-order differential operator, while R is a linear operator, and N is nonlinear operator. For this equation, the correctional formula of VIM is

$$x_{n+1}=x_n+L^{-1}\{\lambda[Lx_n+Rx_n+Nx_n-g(\tau)]\}. \tag{5.14}$$

Considering Rx_n+Nx_n as restricted from variation, λ can be easily obtained as -1 in this case. Suppose $x_0=\overline{x}_0=x(t_0)-L^{-1}g(\tau)$, substituting it into the formula gives

$$x_1=\overline{x}_0-L^{-1}\{L\overline{x}_0+R\overline{x}_0+N\overline{x}_0\}=\overline{x}_0+\overline{x}_1. \tag{5.15}$$

Note that $L^{-1}L\overline{x}_0=0$. Therefore we have

$$\overline{x}_1=-L^{-1}\{L\overline{x}_0+R\overline{x}_0+N\overline{x}_0\}=-L^{-1}R\overline{x}_0-L^{-1}A_0. \tag{5.16}$$

Similarly the iteration process gives

$$x_2=(\overline{x}_0+\overline{x}_1)-L^{-1}\{L(\overline{x}_0+\overline{x}_1)+R(\overline{x}_0+\overline{x}_1)+N(\overline{x}_0+\overline{x}_1)\}=\overline{x}_0+\overline{x}_1+\overline{x}_2. \tag{5.17}$$

If $\overline{x}_1$ is regarded as being relatively small, expanding the nonlinear functional about $\overline{x}_0$ and ignoring the higher-level small functions leads to $N(\overline{x}_0+\overline{x}_1)=N(\overline{x}_0)+\overline{x}_1N'(\overline{x}_0)$. Then we have

$$\begin{aligned}\overline{x}_2 &= -L^{-1}\{L(\overline{x}_0+\overline{x}_1)+R(\overline{x}_0+\overline{x}_1)+N(\overline{x}_0+\overline{x}_1)\}\\&= -L^{-1}\{L\overline{x}_1+R(\overline{x}_0+\overline{x}_1)+N(\overline{x}_0)+\overline{x}_1N'(\overline{x}_0)\}\\&= -L^{-1}\{L(-L^{-1}R\overline{x}_0-L^{-1}A_0)+R(\overline{x}_0+\overline{x}_1)+A_0+\overline{x}_1N'(\overline{x}_0)\}\\&= -L^{-1}R\overline{x}_1-L^{-1}A_1.\end{aligned} \tag{5.18}$$

Further, in the nth step of iteration, regard $\overline{x}_{n-1}$ as the relatively small function and omit the terms $O(\overline{x}_{m1}\overline{x}_{m2}\ldots\overline{x}_{mk})$, where $m1+m2+\cdots+mk\geq n$. It leads to $\overline{x}_n=-L^{-1}R\overline{x}_{n-1}-L^{-1}A_{n-1}$, which is the same as ADM.

It is worth to note that the above comparison can also be made in the case of a multi-dimensional system. Herein, a general form of the nonlinear ordinary differential equations is taken for illustration.

$$\boldsymbol{L}\boldsymbol{x}=\boldsymbol{F}(\boldsymbol{x},\tau). \tag{5.19}$$

The ADM gives an iteration formula as

$$\overline{\boldsymbol{x}}_{n+1}=\boldsymbol{L}^{-1}\boldsymbol{A}_n,\ \boldsymbol{x}_n=\sum_{i=0}^{n}\overline{\boldsymbol{x}}_i, \tag{5.20}$$

while the correctional formula of VIM is

$$\boldsymbol{x}_{n+1}(t)=\boldsymbol{x}_n(t)+\int_{t_0}^{t}\boldsymbol{\lambda}[\boldsymbol{L}\boldsymbol{x}_n-\boldsymbol{F}(\boldsymbol{x}_n,t)]\mathrm{d}\tau. \tag{5.21}$$

Set $\boldsymbol{\lambda}$ as $diag[-1,-1,\ldots]$ and follow the manipulations we made to the one dimensional case. Suppose $\boldsymbol{x}_0=\overline{\boldsymbol{x}}_0$, from the correctional formula of VIM we have

$$\boldsymbol{x}_1=\overline{\boldsymbol{x}}_0+\boldsymbol{L}^{-1}\boldsymbol{F}(\overline{\boldsymbol{x}}_0,\tau)=\overline{\boldsymbol{x}}_0+\boldsymbol{L}^{-1}\boldsymbol{A}_0=\overline{\boldsymbol{x}}_0+\overline{\boldsymbol{x}}_1. \tag{5.22}$$

Accordingly,

$$\boldsymbol{x}_2=\overline{\boldsymbol{x}}_0+\boldsymbol{L}^{-1}\boldsymbol{F}[(\overline{\boldsymbol{x}}_0+\overline{\boldsymbol{x}}_1),\tau]. \tag{5.23}$$

Regard $\overline{\boldsymbol{x}}_1$ as a relatively small amplitude vector of functions. The nonlinear term can be rewritten as

$$\boldsymbol{F}[(\overline{\boldsymbol{x}}_0+\overline{\boldsymbol{x}}_1),\tau]=\boldsymbol{F}(\overline{\boldsymbol{x}}_0)+\boldsymbol{F}'(\overline{\boldsymbol{x}}_0)\overline{\boldsymbol{x}}_1+O(\overline{\boldsymbol{x}}_1{}^2). \tag{5.24}$$

Omitting the small terms $O(\overline{\boldsymbol{x}}_1{}^2)$, it leads to

$$\boldsymbol{x}_2 = \overline{\boldsymbol{x}}_0 + \boldsymbol{L}^{-1}[\boldsymbol{F}(\overline{\boldsymbol{x}}_0) + \boldsymbol{F}'(\overline{\boldsymbol{x}}_0)\overline{\boldsymbol{x}}_1] = \overline{\boldsymbol{x}}_0 + \boldsymbol{L}^{-1}(\boldsymbol{A}_0 + \boldsymbol{A}_1) = \overline{\boldsymbol{x}}_0 + \overline{\boldsymbol{x}}_1 + \overline{\boldsymbol{x}}_2. \tag{5.25}$$

As the iteration goes on, we have

$$\begin{aligned}\boldsymbol{x}_n &= \overline{\boldsymbol{x}}_0 + \boldsymbol{L}^{-1}\left[\boldsymbol{F}(\overline{\boldsymbol{x}}_0) + \boldsymbol{F}'(\overline{\boldsymbol{x}}_0)(\overline{\boldsymbol{x}}_1 + \overline{\boldsymbol{x}}_2 + \ldots\overline{\boldsymbol{x}}_{n-1}) + \frac{1}{2!}\boldsymbol{F}''(\overline{\boldsymbol{x}}_0)(\overline{\boldsymbol{x}}_1 + \overline{\boldsymbol{x}}_2 + \ldots\overline{\boldsymbol{x}}_{n-1})^2 + \ldots\right] \\ &= \overline{\boldsymbol{x}}_0 + \overline{\boldsymbol{x}}_1 + \overline{\boldsymbol{x}}_2 + \cdots + \overline{\boldsymbol{x}}_n + O(\overline{\boldsymbol{x}}_{m1}\overline{\boldsymbol{x}}_{m2}\ldots\overline{\boldsymbol{x}}_{mk}),\end{aligned} \tag{5.26}$$

which becomes the ADM correctional formula by omitting the small terms $O(\overline{\boldsymbol{x}}_{m1}\overline{\boldsymbol{x}}_{m2}\ldots\overline{\boldsymbol{x}}_{mk})$.

From the statements above, it is clear that the ADM and PIM can both be regarded as variants of VIM. Although they are treated as different methods in literature, and were developed into various forms such as modified ADM and MCPI method, a common principle, namely the general use of Lagrange multipliers guides all of them.

5.2 Local variational iteration method

5.2.1 Limitations of global VIM

It has been shown that the VIM is effective in approximating the periodic motions of nonlinear systems, especially the limit cycle oscillation. To verify that, an unforced Duffing equation is solved with VIM herein. The governing equation is

$$\begin{cases}\dot{x}_1 - x_2 = 0 \\ \dot{x}_2 - cx_2 + k_1x_1 + k_2x_1{}^3 = 0\end{cases} \tag{5.27}$$

where $c = 0$, $k_1 = 1$, $k_2 = 1$.

In the approach proposed by He [2], the Lagrange multiplier λ is derived with the nonlinear term being restricted from variation, and it is shown to be tenable in the unforced case with the initial conditions $x_1(t_0) = A$, $x_2(t_0) = 0$ and initial guess $x_1(t) = A\cos\omega t$. The correctional formula is

$$\begin{pmatrix} x_1 \\ x_2 \end{pmatrix}_{n+1} = \begin{pmatrix} x_1 \\ x_2 \end{pmatrix}_n + \int_{t_0}^{t}\begin{bmatrix}\lambda_{11} & \lambda_{12} \\ \lambda_{21} & \lambda_{22}\end{bmatrix}\begin{pmatrix}\dot{x}_1 - x_2 \\ \dot{x}_2 - cx_2 + k_1x_1 + k_2\tilde{x}_1{}^3\end{pmatrix}_n \mathrm{d}\tau. \tag{5.28}$$

With the nonlinear term $\tilde{x}_1{}^3$ restricted from variation, the matrix of Lagrange multipliers satisfies the conditions:

$$\delta \boldsymbol{x}(\tau)|_{\tau=t} : \boldsymbol{\lambda}(\tau)|_{\tau=t} = \begin{bmatrix} -1 & 0 \\ 0 & -1 \end{bmatrix}; \tag{5.29}$$

$$\delta \boldsymbol{x}(\tau) : \ -\dot{\boldsymbol{\lambda}}(\tau) + \boldsymbol{\lambda}(\tau)\begin{bmatrix} 0 & -1 \\ k_1 & -c \end{bmatrix} = \mathbf{0}. \tag{5.30}$$

Therefore the matrix of Lagrange multipliers $\boldsymbol{\lambda}(\tau)$ is identified as

$$\boldsymbol{\lambda}(\tau) = \begin{bmatrix} -\cos(\tau - t) & \sin(\tau - t) \\ -\sin(\tau - t) & -\cos(\tau - t) \end{bmatrix}. \tag{5.31}$$

The solution corrected only once with VIM can already give a result comparable with that of the RK4 method. The example of an unforced Duffing equation shows that the VIM could be very efficient in approximating the periodic solution of nonlinear systems by eliminating the secular term appearing in the correctional formula. The numerical results are plotted in Fig. 5.1 with the initial condition $x(0) = 1$, $\dot{x}(0) = 0$ and initial guess $x_1(t) = \cos\omega t$. To eliminate the secular term, the frequency is determined to be $\omega = 1.32774$.

However, in a more general case with $\boldsymbol{x}(t_0) = \begin{pmatrix} A & B \end{pmatrix}^{\mathrm{T}}$, the performance of VIM is far inferior to the preceding case. The initial guess is selected as

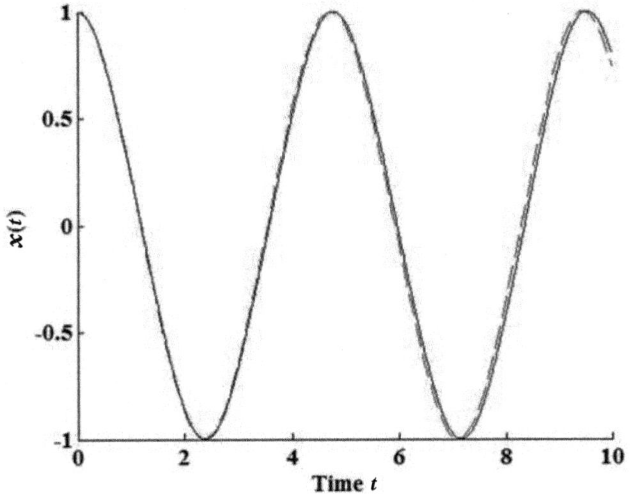

Figure 5.1 Comparison of VIM and RK4. *Solid line*: RK4. *Dashed line*: VIM.

$x(t) = A\cos\omega t + (B/\omega)\sin\omega t$. Let $A = 1$ and $B = 1$, the variational iteration formula gives the corrected solution, in which the terms of $\sin t$ and $\cos t$ are

$$\frac{-4 - 17\omega^2 + 9\omega^4}{1 - 10\omega^2 + 9\omega^4}\cos t, \quad \frac{-2 - 19\omega^2 + 9\omega^4}{1 - 10\omega^2 + 9\omega^4}\sin t. \tag{5.32}$$

It is easily identified that the coefficients of $\sin t$ and $\cos t$ cannot both be equal to zero. Here only the term of $\cos t$ is eliminated, leading to the approximation $\omega \approx 1.4493$. Substituting it into the corrected solution, we have

$$x_1(t) = 1.00598\cos(1.4493t) - 0.00597962\cos(4.34791t) - 0.111705\sin(t)$$

$$+ 0.694112\sin(1.4493t) + 0.0243163\sin(4.34791t). \tag{5.33}$$

This result is compared with that of RK4 in Fig. 5.2.

It is believed that if both terms of $\sin t$ and $\cos t$ vanish, the corrected solution will be more accurate. But with just one parameter ω in the initial guess, this will be an impossible mission. Thus we tried to bring in two parameters ω_1, ω_2 in the initial guess, which takes the form $x_1(t) = A\cos\omega_1 t + (B/\omega_2)\sin\omega_2 t$. If the resulted corrected solution can be obtained analytically, the accuracy will be much improved. But along with it, the calculation of variational iteration formula will become much more complex. We found that even with the help of Mathematica, this job is still too burdensome to be accomplished.

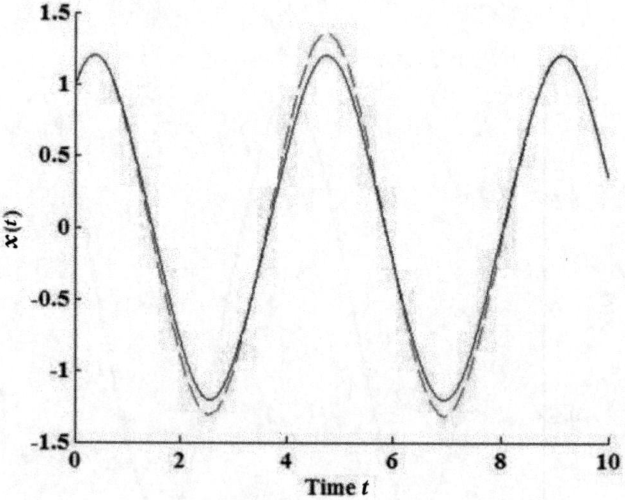

Figure 5.2 Comparison of result of Eq. (5.33) with RK4. *Solid line*: RK4. *Dashed line*: VIM.

Moreover, it is also indicated that a carefully selected initial guess is important to ensure the convergence of this method. If the approximate function takes the form as $x_1(t)=A+Bt$ rather than $x_1(t)=A\cos\omega t$, the result given by VIM will be more and more divergent from the true solution with the iteration. It is illustrated in the case $A=1$ and $B=1$. With the VIM the initial guess $x_{1,0}(t)=1+t$ is corrected for two times and the results are given as follows:

$$x_{1,1}(t)=5+3t-3t^2-t^3-4\cos t-2\sin t, \tag{5.34}$$

$$\begin{aligned}x_{2,2}(t)=\;&393349+252477t\ldots+t^9+\left(-\frac{7083587}{18}+\frac{999t}{4}\ldots-\frac{3t^7}{7}\right)\cos(t)\\&+\left(\frac{1666}{9}+6t-50t^2-6t^3\right)\cos(2t)-\frac{1}{2}\cos(3t)+\left(-\frac{505575}{2}+378t\ldots+\frac{6t^7}{7}\right)\sin t\\&+\left(32+\frac{424}{3}t-8t^3\right)\sin(2t)-\frac{11}{4}\sin(3t).\end{aligned} \tag{5.35}$$

The solution $x_{1,1}(t)$ is compared with that of RK4 in Fig. 5.3.

These above numerical examples and analysis show that VIM is fastidious to the initial condition of the problem, and also too sensitive to the initial guess function,

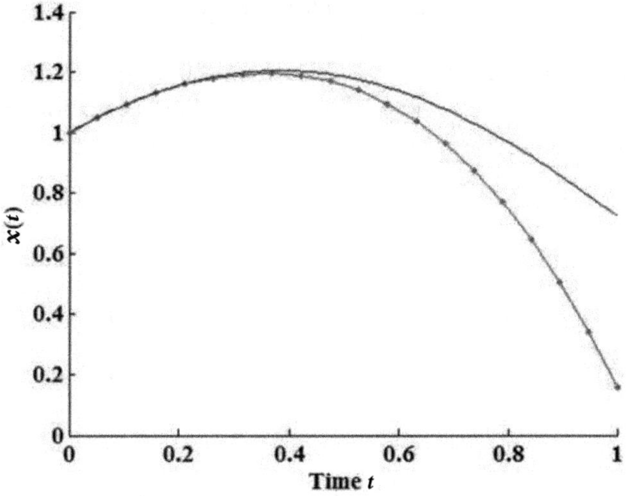

Figure 5.3 Comparison of result of Eq. (5.34) with RK4. *Solid line*: RK4. *Dotted line*: VIM.

which is analogous to the Newton's iteration method for solving nonlinear algebraic equations (NAEs). This method also fails to predict the long term responses of some complex nonlinear phenomenon, such as chaos or quasi periodic motion, because these patterns of motion are non-periodic and can hardly be approximated by any analytical functions.

5.2.2 Variational homotopy method

The proposition of VIM is inspired by the work of Inokuti, where the notion of generalized Lagrange multiplier is interpreted with the Newton method and then extended to the problem of nonlinear differential equations. The main idea of Inokuti is as follows.

If we try to obtain the solution of $F(x) = 0$, no matter it is an algebraic equation or a differential equation, we can construct an estimate expression as

$$x_{est} = x + (\lambda, F(x)), \tag{5.36}$$

where the inner product could be a simple product or an integral. The generalized Lagrange multiplier could be a number or a function accordingly. Insert an approximation $x_0 + \delta x$ nearby the true solution x_0 into the estimation.

$$x_0 = x_0 + \delta x + (\lambda, F(x_0 + \delta x)). \tag{5.37}$$

To make the correction optimal, we need to choose an appropriate λ so as to diminish the terms including δx

$$\delta x + (\lambda, F(x_0) + F'[\delta x]) + O(\delta x^2) = 0. \tag{5.38}$$

Herein an operator is introduced as $F'[]$, which stands for the first-order derivation of δx in the expansion of $F(x_0 + \delta x)$. Suppose a nonlinear equation is expressed as the sum of linear part and nonlinear part

$$F(x) = Lx + Nx, \tag{5.39}$$

the first order derivation of δx would be

$$F'[\delta x] = L\delta x + \delta Nx = L\delta x + N'[\delta x]. \tag{5.40}$$

Note that $F(x_0) = 0$, the stationary condition can be written as

$$\delta x + (\lambda, F'[\delta x]) = 0, \text{ or } \delta x = -(\lambda, F'[\delta x]). \tag{5.41}$$

So the inner product $-(\lambda, F'[*])$ can be regarded as an inverse operator to $F'[*]$. With this recognition, we will propose a variational homotopy method for solving nonlinear differential equations.

Similar to the Newton iterative method, the VIM is sensitive to initial guess. An arbitrary trial function could lead to divergent results after several loops of iteration, which has been verified earlier in this subsection. In practical, we want the initial guess to be as simple as possible, but at the same time a well convergent result can still be obtained. To achieve that, the idea of homotopy method is introduced here. Rather than finding the solution of $F(x) = 0$, a homotopic equation is considered:

$$h(x, p) = (1 - p)(x - a) + pF(x) = 0, \tag{5.42}$$

where p is a fictitious parameter and a is the initial guess. Obviously, the preceding equation is reduced to $x - a = 0$ when $p = 0$, and will evolve into the original form as p approaches to 1. The homotopic equation is mathematically equivalent to

$$\frac{\partial h}{\partial p} + \frac{\partial h}{\partial x} \cdot \frac{\partial x}{\partial p} = 0, \ x\big|_{p=0} = a. \tag{5.43}$$

If $F(x) = Lx + Nx$, we have

$$\frac{\partial h}{\partial x} \cdot \frac{\partial x}{\partial p} = (1 - p)\frac{\partial x}{\partial p} + pL\frac{\partial x}{\partial p} + pN'\left[\frac{\partial x}{\partial p}\right], \tag{5.44}$$

For simplicity, record it as $h'\left[\partial x/\partial p\right]$, thus the partial differential equation becomes

$$\frac{\partial h}{\partial p} + h'\left[\frac{\partial x}{\partial p}\right] = 0. \tag{5.45}$$

Suppose the inner product $-(\lambda, h'[*])$ is the inverse operator to $h'[*]$. The above expression can be transformed as

$$\frac{\partial x}{\partial p} = \left(\lambda, \frac{\partial h}{\partial p}\right) = \int_{t_0}^{t} \lambda \frac{\partial h}{\partial p} \mathrm{d}\tau, \tag{5.46}$$

while λ can be obtained by enforcing

$$\frac{\partial x}{\partial p} + \left(\lambda, h'\left[\frac{\partial x}{\partial p}\right]\right) = \frac{\partial x}{\partial p} + \int_{t_0}^{t} \lambda \left\{(1 - p)\frac{\partial x}{\partial p} + pL\frac{\partial x}{\partial p} + pN'\left[\frac{\partial x}{\partial p}\right]\right\} \mathrm{d}\tau = 0. \tag{5.47}$$

Take $x\big|_{p=0} = a$ as the initial condition and integrate $\partial x/\partial p$, it will arrive at the solution of $F(x) = 0$ when $p = 1$. Unfortunately, the true λ is difficult to derive because there exists the nonlinear term. In practical, an approximated λ is mostly used and combined with the restart technique. Considering that the homotopy method is globally convergent, although an approximated λ could well lead to an inexact solution, one can use the final solution as the initial guess to restart the integration. This procedure can be repeated until the convergence criteria is reached. Actually, the

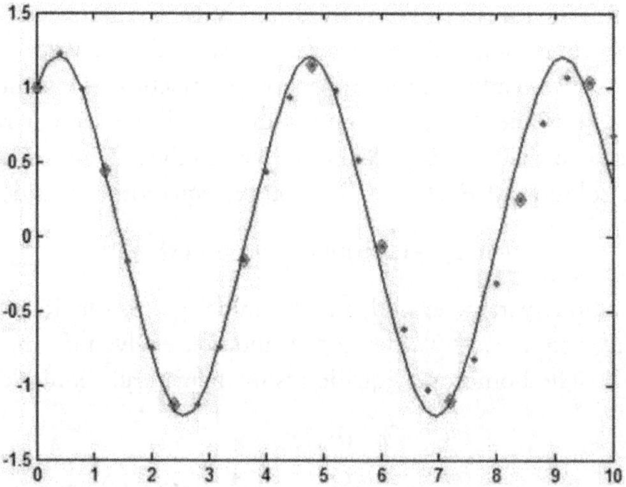

Figure. 5.4 Result generated by the variational homotopy radial basis functions method. *Red dots*: Variational homotopy collocation method. *Blue line*: RK4.

restart technique is necessary to homotopy method, because the standard integration requires a very small step size and thus needs numerous integration steps. In literature, the restart technique is combined with the integration so that one can choose an adequate step size to speed up the convergence. The efficiency and robustness of the aforementioned method will be verified and discussed later.

Fig. 5.4 shows the result generated by the variational homotopy radial basis functions method. The time step size is selected to be $\Delta t = 1.2$.

5.2.3 Methodology of LVIM

As is shown in Atluri [10] and Dong, Alotaibi, Mohiuddine, and Atluri [11], local approximations are far more effective in solving initial value and boundary value problems. We use these ideas in developing a local variational iteration method.

To remedy the drawbacks shown above, we propose to further improve the VIM. Here is the primary idea. Firstly, the nonlinear term should be kept in the derivation of Lagrange multipliers so that a better correction can be obtained for arbitrary approximation. Then the entire time domain is divided into small intervals and the LVIM is applied repeatedly in each time interval $t_{i-1} \leq \tau \leq t_i$. In each interval $t_{i-1} \leq \tau \leq t_i$ we approximate locally by an arbitrary function $\boldsymbol{x}(\tau) = \boldsymbol{A} + \boldsymbol{B}\tau$ for instance; then the corrected solution will be a function of $\boldsymbol{A}$ and $\boldsymbol{B}$, i.e., the initial condition of each interval. Thus the solution in the entire time domain can be obtained by the repetition of $\boldsymbol{A}$ and $\boldsymbol{B}$ step by step.

With this in mind, the IVP of the unforced Duffing equation is solved with the initial conditions $x_1(t_0)=1$, $x_2(t_0)=1$. The correctional formula is

$$\begin{pmatrix} x_1 \\ x_2 \end{pmatrix}_{n+1} = \begin{pmatrix} x_1 \\ x_2 \end{pmatrix}_n + \int_{t_{i-1}}^{t} \begin{bmatrix} \lambda_{11} & \lambda_{12} \\ \lambda_{21} & \lambda_{22} \end{bmatrix} \begin{pmatrix} \dot{x}_1 - x_2 \\ \dot{x}_2 - cx_2 + k_1 x_1 + k_2 x^3 \end{pmatrix}_n \mathrm{d}\tau, \tag{5.48}$$

where $t_{i-1} \le t \le t_i$. The stationary conditions are

$$\delta \boldsymbol{x}(\tau)|_{\tau=t} : \boldsymbol{\lambda}(\tau)|_{\tau=t} = \begin{bmatrix} -1 & 0 \\ 0 & -1 \end{bmatrix}; \tag{5.49}$$

$$\delta \boldsymbol{x}(\tau) : -\dot{\boldsymbol{\lambda}}(\tau) + \boldsymbol{\lambda}(\tau) \begin{bmatrix} 0 & -1 \\ k_1 + 3k_2 {x_1}^2 & -c \end{bmatrix} = \mathbf{0}, \quad t_{i-1} \le \tau \le t_i \text{ and } t_{i-1} \le t \le t_i. \tag{5.50}$$

Instead of neglecting the nonlinear term, we kept it in the derivation of $\boldsymbol{\lambda}$. This makes the differential equations more difficult to be solved analytically. But an exact analytical solution is not necessary here. The differential transform (DT) method is introduced to obtain the approximated $\boldsymbol{\lambda}$ in the form of power series. With the nonlinear term included, the approximated $\boldsymbol{\lambda}$ is actually more precise than that in the VIM of He [2].

Supposing that the initial guess of x_1 takes the form $x_{1,0}(\tau) = B\tau + A$, an approximated $\boldsymbol{\lambda}$ can be obtained using the DT method. Herein, only the first four DT terms of $\boldsymbol{\lambda}(\tau)$ are kept.

$$\begin{aligned}
\lambda_{11} &= -1 + \frac{1}{2}(-t+\tau)^2 \left(k_1 + 3(A+B\tau)^2 k_2\right) \\
&\quad + \frac{1}{3}(-t+\tau)^3 \left(6B(A+B\tau)k_2 - \frac{1}{2}c\left(k_1 + 3(A+B\tau)^2 k_2\right)\right),
\end{aligned}$$

$$\begin{aligned}
\lambda_{12} &= -t + \tau - \frac{1}{2}c(-t+\tau)^2 + \frac{1}{3}(-t+\tau)^3 \left(\frac{c^2}{2} + \frac{1}{2}\left(-k_1 - 3(A+B\tau)^2 k_2\right)\right), \\
\lambda_{21} &= (-t+\tau)\left(-k_1 - 3(A+B\tau)^2 k_2\right) + \frac{1}{2}(-t+\tau)^2\left(-6B(A+B\tau)k_2 + c\left(k_1 + 3(A+B\tau)^2 k_2\right)\right) \\
&\quad + \frac{1}{3}(-t+\tau)^3 \left(-3B^2 k_2 + 6Bc(A+B\tau)k_2 + \frac{1}{2}\left(k_1 + 3(A+B\tau)^2 k_2\right)\left(-c^2 + k_1 + 3(A+B\tau)^2 k_2\right)\right),
\end{aligned} \tag{5.51}$$

$$\lambda_{22} = -1 + c(-t+\tau) + \frac{1}{2}(-t+\tau)^2\left(-c^2 + k_1 + 3(A+B\tau)^2 k_2\right) + \frac{1}{3}(-t+\tau)^3$$
$$\left(-\frac{1}{2}c\left(-c^2 + k_1 + 3(A+B\tau)^2 k_2\right) + \frac{1}{2}\left(6B(A+B\tau)k_2 - c\left(k_1 + 3(A+B\tau)^2 k_2\right)\right)\right),$$

where $t_{i-1} \le t \le t_i$ and $t_{i-1} \le \tau \le t_i$. Using the correctional formula for one time, we have

$$x_{1,1} = A + Bt + \frac{1}{5040}t^2\left(42\left(5Bc(12+ct(4+ct)) + k_1\left(\begin{array}{c} -5A(12+ct(4+ct)) - \\ Bt(20+ct(10+ct)) + t^2(5A+Bt)k_1 \end{array}\right)\right)\right)$$
$$-6\left(35A^3(12+ct(4+ct)) + 21A^2Bt(20+ct(10+ct)) + 7AB^2t^2(30+ct(12+ct)\right)$$
$$+B^3t^3(42+ct(14+ct)) - 4t^2\left(35A^3 + 21A^2Bt + 7AB^2t^2 + B^3t^3\right)k_1)k_2$$
$$+5t^2\left(126A^5 + 126A^4Bt + 84A^3B^2t^2 + 36A^2B^3t^3 + 9AB^4t^4 + B^5t^5\right)k_2^2\Big), \quad (5.52)$$

where $t_{i-1} \le t \le t_i$.

In Fig. 5.5A the preceding solution is compared with the numerical results obtained by ODE45, for which the relative and absolute accuracy are both set as 10^{-15}. Fig. 5.5B shows the computational error of the LVIM with respect to ODE45. The time step size is set as $\Delta t = 0.01$, and the simulation is carried in $t = [0, 100]$. From Fig. 5.5B it can be seen that the error of the LVIM is less than 10^{-3} even for $t = 100$.

To further inspect the performance of the LVIM in predicting complicated responses, the forced Duffing equation is investigated.

$$\ddot{x} + c\dot{x} + k_1x + k_2x^3 = f\cos(\omega t),\ c = 0.15,\ k_1 = -1,\ k_2 = 1,\ f = 0.41,\ \omega = 0.4. \quad (5.53)$$

It is shown to be a chaotic system by RK4 and other reliable methods. Unsurprisingly, the VIM of He [2] fails to solve it because the solution is too complicated to be computable. An approximation is made to the forced term of the system. Since it is involved in the integration, a simple form of it will accelerate the calculation. For that, the forced term is expanded into power series as well. The stationary condition for λ is

$$\delta x(t){:}1 - \dot{\lambda}(\tau)\big|_{\tau=t} = 0,\ \delta\dot{x}(t){:}\lambda(\tau)\big|_{\tau=t} = 0,\ \delta x(\tau){:}\ddot{\lambda}(\tau) - c\dot{\lambda}(\tau) + k_1\lambda(\tau) + 3k_2{x_n}^2\lambda(\tau) = 0, \quad (5.54)$$

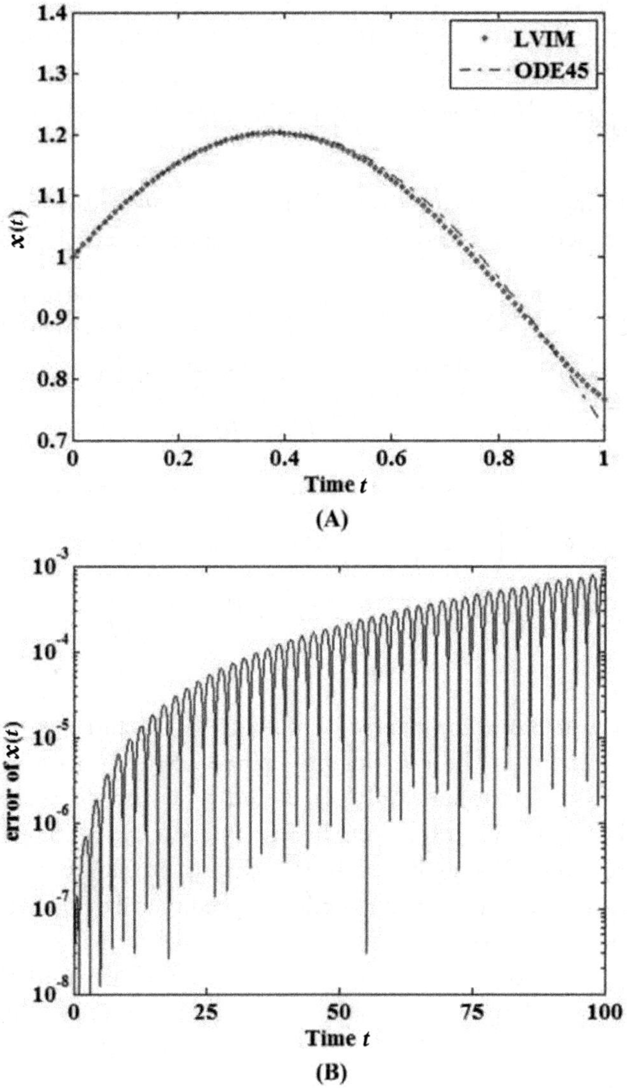

Figure 5.5 Comparison of the LVIM and the ODE45.

where $t_{i-1} \le \tau \le t_i$ and $t_{i-1} \le t \le t_i$.

Suppose the initial guess in each interval $t_{i-1} \le \tau \le t_i$ is $x_0(\tau) = A + B\tau$. With the DT method, we can get the fourth-order approximated $\lambda(\tau, t)$ in terms of A and B, which is

$$\begin{aligned}\lambda(\tau,t) = & -t+\frac{1}{2}c(t-\tau)^2+\tau+\frac{1}{6}(t-\tau)^3\left(-c^2+k_1+3(A+Bt)^2k_2\right)\\ & -\frac{1}{24}(t-\tau)^4\left(-c^3+2ck_1+6(A+Bt)(2B+Ac+Bct)k_2\right).\end{aligned} \tag{5.55}$$

Then substituting it into the correction formula, with the forced term being replaced by power series, will lead to the corrected function $x_1(t)$:

$$\begin{aligned} x_1(t) = {} & A+Bt+\frac{1}{2}\left(-Bc+f\cos(C\omega)-Ak_1-A^3k_2\right)t^2+\frac{1}{6}\begin{pmatrix} Bc^2-cf\cos(C\omega)-f\omega\sin(C\omega)\\ +(-B+Ac)k_1+A^2(-3B+Ac)k_2\end{pmatrix}t^3\\ & +\frac{1}{24}\begin{pmatrix} -Bc^3+c^2f\cos(C\omega)-f\omega^2\cos(C\omega)+cf\omega\sin(C\omega)+Ak_1^2\\ -A\left(6B^2-6ABc+A^2c^2+3Af\cos(C\omega)\right)k_2+3A^5k_2^2+\\ k_1\left(2Bc-Ac^2-f\cos(C\omega)+4A^3k_2\right)\end{pmatrix}t^4\\ & +\frac{1}{120}\begin{pmatrix} Bc^4-c^3f\cos(C\omega)+cf\omega^2\cos(C\omega)-c^2f\omega\sin(C\omega)+f\omega^3\sin(C\omega)\\ +(B-2Ac)k_1^2+\begin{pmatrix}-6B^3+24AB^2c-9A^2Bc^2\\ +A^3c^3+6A(-3B+Ac)f\cos(C\omega)\\ +3A^2f\omega\sin(C\omega)\end{pmatrix}k_2+3A^4(9B-2Ac)k_2^2\\ +k_1\begin{pmatrix}-3Bc^2+Ac^3+2cf\cos(C\omega)\\ +f\omega\sin(C\omega)-8A^2(-3B+Ac)k_2\end{pmatrix}\end{pmatrix}t^5+O[t]^6,\end{aligned} \tag{5.56}$$

where $t_{i-1}\leq t\leq t_i$. It has a compacted form with A, B, and t_{i-1} (denoted as C in the corrected function) as the coefficients. When applied to multiple time intervals, it only needs to recompute only A, B, and t_0, which correspond to the final position, velocity, and time instant of the last interval, respectively. The LVIM is very fast, precise, and efficient in solving nonlinear problems. Fig. 5.6 shows the chaotic motions of a forced Duffing oscillator predicted by both LVIM and RK4 methods. The time step size of LVIM is set as 0.2, while that of RK4 needs to be set as 0.02 to achieve the same accuracy. The numerical simulation carried out with MATLB shows that the LVIM is almost 10 times faster than RK4 in this case. The computation times of the LVIM and the RK4 are 0.013471 s and 0.135445 s, respectively.

Similar to the modified ADM, the VIM is further improved herein by bringing in some local approximation techniques, so as to be practical in predicting long-term motion and complex dynamical responses. The LVIM makes the derivation of Lagrange multipliers $\boldsymbol{\lambda}$ and the calculation of the variational iteration formula much simpler, yet still provide reliable solution in each subtime domain.

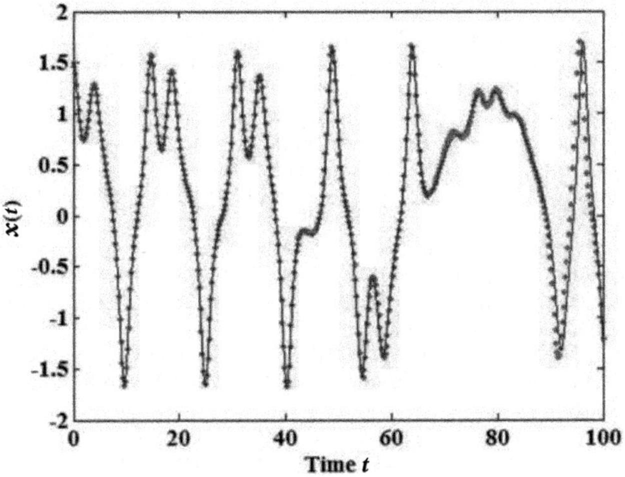

Figure 5.6 Comparison of the LVIM and the RK4 method for the forced Duffing oscillator. *Dots*: LVIM. *Solid line*: RK4.

5.3 Conclusion

The Picard iteration method, the Adomian decomposition method and the variational iteration method are unified in this chapter. From the viewpoint provided by the concept of generalized Lagrange multipliers, it is found that the three methods are rooted in the same mathematical approach of generalized Lagrange multipliers. By some simple transformations, the Picard iteration formula and Adomian decomposition formula can be derived from the variational iteration formula.

However, the global VIM is very laborious in practice and is unable to predict long term responses of nonlinear systems, although it is able to provide an analytical solution. Being aware of that, we introduced a local VIM that simplified the derivation of the matrix of generalized Lagrange multipliers $\boldsymbol{\lambda}$ with the DT method. By dividing the entire domain into small intervals and updating the initial conditions in each subinterval, the local VIM method achieves very high accuracy and very high computational speeds.

References

[1] M. Inokuti, H. Sekine, T. Mura, General use of the Lagrange multiplier in nonlinear mathematical physics, in: S. Nemat-Nasser (Ed.), Variational Method in the Mechanics of Solids, Pergamon Press, Oxford, 1978, pp. 156–162.

[2] J. He, Variational iteration method: a kind of non-linear analytical technique: some examples, International Journal of Non-linear Mechanics 34 (1999) 699–708.
[3] G. Adomian, A review of the decomposition method in applied mathematics, Journal of Mathematical Analysis and Applications 135 (1988) 501–544.
[4] T. Fukushima, Picard iteration method, Chebyshev polynomial approximation, and global numerical integration of dynamical motions, The Astronomical Journal 113 (5) (1997) 1909–1914.
[5] R.M. Woollands, A.B. Younes, J.L. Junkins, New solutions for the perturbed lambert problem using regularization and Picard iteration, Journal of Guidance, Control, and Dynamics 38 (9) (2015) 1548–1562.
[6] R. Rach, On the Adomian decomposition method and comparisons with Picard iteration, Journal of Mathematical Analysis and Applications 128 (1987) 480–483.
[7] J. He, Some asymptotic methods for strongly nonlinear equations, International Journal of Modern Physics B 20 (10) (2006) 1141–1199.
[8] S. Ghosh, A. Roy, D. Roy, An adaption of Adomian decomposition for numerical-analytic integration of strongly nonlinear and chaotic oscillators, Computer Methods in Applied Mechanics and Engineering 196 (2007) 1133–1153.
[9] M. Jang, C. Chen, Y. Liy, On solving the initial-value problems using the differential transformation method, Applied Mathematics and Computation 115 (2000) 145–160.
[10] S.N. Atluri, Methods of Computer Modeling in Engineering & The Sciences, Volume I, Tech Science Press, Forsyth, 2005.
[11] L. Dong, A. Alotaibi, S.A. Mohiuddine, S.N. Atluri, Computational methods in engineering: a variety of primal & mixed methods, with global & local interpolations, for well-posed or Ill-posed BCs, Computer Modeling in Engineering & Sciences 99 (1) (2014) 1–85.

CHAPTER 6

Collocation in Conjunction with the Local Variational Iteration Method

SUMMARY

For nonlinear dynamical systems, long term responses are often desired. We have introduced HDHB and TDC methods for approximating periodic solutions. However, the non-periodic responses, including chaotic motions and transient chaotic motions, are more common in nonlinear dynamical systems. Due to the lack of periodicity, the accurate prediction of long term non-periodic responses for the whole time domain is difficult. Naturally, to achieve long-term prediction of nonlinear systems, the local approximation methods are much more efficient and reliable [1,2] than their global counterparts. In the works of Atluri [1] and Dong, Alotaibi, Mohiuddine, and Atluri [2], both global and local weighted residual methods are illustrated clearly and systematically, including the Finite Element Method(FEM), the Finite Volume Method (FVM), the Boundary Element Method (BEM) and the Meshless Local Petrov Galerkin (MLPG) method, etc. These methods are developed using different test functions and trial functions. Among them, the collocation method is the simplest, using the Dirac Delta functions as the test functions. The selection of trial functions is flexible. Depending on the problem, one can use harmonics, polynomials, Radial Basis Functions (RBFs) and Moving Least Square (MLS) functions. But no matter which trial function is used, it will eventually lead to a system of nonlinear algebraic equations (NAEs). Therefore, the problem ultimately becomes how to solve the NAEs. For that, many NAE solvers are proposed. The most popular Newton-like methods mostly involve the inverse of the Jacobian matrix, and can be very sensitive to the initial guess of the solution. Many efforts have been made to solve the NAEs without inverting the Jacobian matrix and without being sensitive to the initial guess. Among them are the scalar-homotopy iterative algorithms that were developed by Liu, Yeih, Kuo, and Atluri [3] and Liu and Atluri [4]. For large deformation dynamic responses of beams, plates, and shells which may undergo non-periodic and chaotic motions, one needs efficient and accurate time-integrators for integrating the second-order semi-discrete differential equations in time.

In this chapter, we combine the concept of collocation method with the basic ideas of the Local Variational Iteration Method (LVIM) [5], which is one of the asymptotic methods that also include the Homotopy Perturbation Method [6], the Picard Iteration

Computational Methods for Nonlinear Dynamical Systems
DOI: https://doi.org/10.1016/B978-0-323-99113-1.00006-6

Method (PIM) [7], and the Adomian Decomposition Method (ADM), and so on [8]. These asymptotic methods start from the solution of a linearized problem and iteratively correct the initial guess so that it approaches the real solution of the nonlinear problem. It implies that the real solution can be obtained by constructing an iterative formula involving a functional of the original equations, instead of transforming the nonlinear differential equations into NAEs to be solved. However, the LVIM comes with a heavy burden of symbolic calculations. In the concepts of the weighted residual methods such as the HB method, the TDC method, and the various collocation methods, they approximate the solution in a weak sense, i.e. the original system is approximated by a set of weighted residual formulas. Thereby, the problem can be solved in a semi-analytical way which is much more convenient than a purely analytical approach. With this in mind, herein we apply the weighted residual principle to the functional iterative formulas we derive from the LVIM, and transform them into algebraic iterative formulas.

In order to obtain applicable algebraic iterative formulas, three algorithms are derived in this chapter. The first algorithm, denoted here as Algorithm-1, is obtained by transforming the original integral form of the Local Variational Iteration Method (LVIM) into a differential form. Then it is found that the matrix of generalized Lagrange multipliers can be eliminated in Algorithm-1, using the constraints on the Lagrange multipliers. The resulting formula is very concise, since it is actually the generalization of the Newton's iterative formula in a function space. The other two algorithms, denoted here as Algorithms-2 and Algorithm-3, are obtained by approximating the generalized Lagrange multipliers with power series and exponential series respectively. Further, we approximate the solution of the nonlinear problem with trial functions in a finite time interval. Thereby, Algorithms-1,2 and 3 can be transformed into three different algebraic iterative formulas for the values of the trial functions at collocation points, by using Dirac Delta functions as test functions. Once the values of trial functions at collocation points are solved for, the solutions can be obtained by interpolation. Unlike the conventional collocation method, the present approach is free from constructing the NAEs. As will be shown in the following sections, the proposed iterative formulas are also free from inverting the Jacobian matrix. The proposed iterative algorithms in this chapter are very accurate and efficient. Although they are local methods, their convergence domains are much larger than those of the simple finite difference methods. Moreover, the accumulation of computational error is relatively very small because in each finite time interval $t_{i+1} - t_i$, the approximation is made at all the collocated points, while the finite difference methods only approximate the end point. In all, the application of weighted residual principle to Algorithms-1,2 and 3 provides a class of powerful tools for solving strongly nonlinear dynamical systems and for investigating chaos and other nonperiodic responses.

The structure of this chapter is as follows: In Section 2 the methodology and the main features of the VIM are briefly introduced. In Section 3, original integral form of the LVIM [9] is transformed into a differential form first, and the generalized Lagrange multipliers are eliminated using some properties, which leads to Algorithm-1. Then the generalized Lagrange multipliers are alternatively approximated by either polynomials or exponential series respectively. The two approximation techniques lead to Algorithm-2 and Algorithm-3, respectively. In Section 4 the collocation method is introduced and some notions about weighted residual methods are explained. Using the first kind of Chebyshev polynomials as trial functions, the collocation forms of the Algorithms-1,2 and 3 are derived in Section 5. Three strongly nonlinear problems (the forced Duffing equation, the Lorenz system, and the multiple coupled Duffing equations) are used as examples to illustrate the proposed methods in Section 6.

6.1 Modifications of LVIM

To remedy the drawbacks of VIM, the LVIM was proposed as an alternative in chapter 5. With the LVIM, the solution of the nonlinear system is approximated locally in a sub-interval of time, t_i to t_{i+1}, by the correctional formula

$$\boldsymbol{x}_{n+1}(t) = \boldsymbol{x}_n(t) + \int_{t_i}^{t} \boldsymbol{\lambda}(\tau)\{\boldsymbol{L}\boldsymbol{x}_n(\tau) - \boldsymbol{F}[\boldsymbol{x}_n(\tau), \tau]\}\mathrm{d}\tau, \quad t\in[t_i, t_{i+1}] \tag{6.1}$$

where $\boldsymbol{\lambda}(\tau)$ is the matrix of the Lagrange multipliers. Based on the variational principle, the optimal $\boldsymbol{\lambda}(\tau)$ needs to satisfy the following constraints [9]:

$$\begin{cases} \delta\boldsymbol{x}_n(\tau)|_{\tau=t}\colon \boldsymbol{I} + \boldsymbol{\lambda}(\tau)|_{\tau=t} = \boldsymbol{0}\ , \\ \delta\boldsymbol{x}_n(\tau)\colon \boldsymbol{L}\boldsymbol{\lambda}(\tau) + \boldsymbol{\lambda}(\tau)\dfrac{\partial \boldsymbol{F}(\boldsymbol{x}_n, \tau)}{\partial \boldsymbol{x}_n} = \boldsymbol{0}, \end{cases} \quad \tau\in[t_i, t] \tag{6.2}$$

In LVIM, the term $\partial\boldsymbol{F}/\partial\boldsymbol{x}$ is always retained in the evolution equation for $\boldsymbol{\lambda}$ as in Eq. (6.2), so that a better correction can be obtained for an arbitrary initial guess. Furthermore, the entire time domain is divided into small intervals and the LVIM is applied repeatedly in each finite time interval $\Delta t = t_{i+1} - t_i$.

The derivation of the generalized Lagrange multipliers is accomplished by using an approximation technique, that is, the differential transform (DT) method [10]. In each interval $t_i \le \tau \le t_{i+1}$ we approximate $\boldsymbol{x}(\tau)$ locally, using a very simple function $\boldsymbol{x}(\tau) = \boldsymbol{A} + \boldsymbol{B}\tau$ for instance; then the corrected solution will be a function of $\boldsymbol{A}$ and $\boldsymbol{B}$, that is, the initial condition of each interval, thus the solution in the entire time domain can be obtained by the repetitive solution in terms of $\boldsymbol{A}$ and $\boldsymbol{B}$ step by step. In this way, the LVIM enables the initial guess function to take a very simple form and provides a convenient approach for the implementation of the correctional formula. Note that the corrected approximation in each interval takes the same expression. Therefore, it only needs to go through the symbolic calculations only one time.

In spite of all the merits of the LVIM, we found that the resulting analytical solution remains to be a very lengthy expression. If a highly accurate solution is desired, the correction would have to be conducted multiple times. In that case, the symbolic calculation would still be troublesome and the solution would be too complicated to be interpreted. Considering that, some further modifications are made to the LVIM in the following sections.

6.1.1 Algorithm-1

By differentiating the correctional formula of the LVIM in Eq. (6.1) and considering the constraints on $\boldsymbol{\lambda}(\tau)$ in Eq. (6.2), we have

$$\frac{\mathrm{d}\boldsymbol{x}_{n+1}}{\mathrm{d}t} = \frac{\mathrm{d}\boldsymbol{x}_n}{\mathrm{d}t} + \boldsymbol{\lambda}(\tau)|_{\tau=t}\left[\frac{\mathrm{d}\boldsymbol{x}_n}{\mathrm{d}t} - \boldsymbol{F}(\boldsymbol{x}_n, t)\right] + \int_{t_i}^{t}\frac{\partial \boldsymbol{\lambda}}{\partial t}[\boldsymbol{L}\boldsymbol{x}_n - \boldsymbol{F}(\boldsymbol{x}_n, \tau)]\mathrm{d}\tau, \quad t\in[t_i, t_{i+1}]$$

$$= \boldsymbol{F}(\boldsymbol{x}_n, t) + \int_{t_i}^{t}\frac{\partial \boldsymbol{\lambda}}{\partial t}[\boldsymbol{L}\boldsymbol{x}_n - \boldsymbol{F}(\boldsymbol{x}_n, \tau)]\mathrm{d}\tau \tag{6.3}$$

To further simplify the preceding formula, the following property of $\boldsymbol{\lambda}(\tau)$ is used.

Assertion:

The matrix of generalized Lagrange multipliers $\boldsymbol{\lambda}(\tau)$ in Eq. (6.2) happens to be the solution $\overline{\boldsymbol{\lambda}}(t)$ of the following ordinary differential equations:

$$\begin{cases} \boldsymbol{I} + \overline{\boldsymbol{\lambda}}(t)|_{t=\tau} = \boldsymbol{0} \\ \dfrac{\partial \overline{\boldsymbol{\lambda}}(t)}{\partial t} - \boldsymbol{J}(t)\overline{\boldsymbol{\lambda}}(t) = \boldsymbol{0} \end{cases}, \quad t\in[\tau, t_{i+1}] \tag{6.4}$$

Proof:

If we let $\boldsymbol{J}(\boldsymbol{x}_n, \tau) = \partial \boldsymbol{F}(\boldsymbol{x}_n, \tau)/\partial \boldsymbol{x}_n$, the generalized Lagrange multipliers can be identified in the Magnus approach, through which the solution of a linear ordinary differential equations with variable coefficients:

$$\boldsymbol{Y}'(t) = \boldsymbol{A}(t)\boldsymbol{Y}(t), \quad \boldsymbol{Y}(t_0) = \boldsymbol{Y}_0 \tag{6.5}$$

can be expressed as follows:

$$\boldsymbol{Y}(t) = \{\exp[\boldsymbol{\Omega}(t, t_0)]\}\boldsymbol{Y}_0 = \left\{\exp\left[\sum_{k=1}^{\infty}\boldsymbol{\Omega}_k(t)\right]\right\}\boldsymbol{Y}_0 \tag{6.6}$$

The first three terms of Magus expansion $\boldsymbol{\Omega}_k(t)$ read

$$\begin{aligned}
&\boldsymbol{\Omega}_1(t)=\int_0^t \boldsymbol{A}(t_1)\mathrm{d}t_1,\quad \boldsymbol{\Omega}_2(t)=\frac{1}{2}\int_0^t \mathrm{d}t_1\int_0^{t_1}\mathrm{d}t_2[\boldsymbol{A}(t_1),\boldsymbol{A}(t_2)]\\
&\boldsymbol{\Omega}_3(t)=\frac{1}{6}\int_0^t \mathrm{d}t_1\int_0^{t_1}\mathrm{d}t_2\int_0^{t_2}\mathrm{d}t_3([\boldsymbol{A}(t_1),[\boldsymbol{A}(t_2),\boldsymbol{A}(t_3)]]+[\boldsymbol{A}(t_3),[\boldsymbol{A}(t_2),\boldsymbol{A}(t_1)]])
\end{aligned} \tag{6.7}$$

where $[A,B]\equiv AB-BA$ is the commutator of A and B. By transposing the constrains of $\boldsymbol{\lambda}(\tau)$ in Eq. (6.2) as

$$\begin{cases}\delta\boldsymbol{x}_n(\tau)|_{\tau=t}\colon \boldsymbol{I}+\boldsymbol{\lambda}^{\mathrm{T}}(\tau)|_{\tau=t}=\boldsymbol{0}\\ \delta\boldsymbol{x}_n(\tau)\colon \boldsymbol{L}\boldsymbol{\lambda}^{\mathrm{T}}(\tau)+\boldsymbol{J}^{\mathrm{T}}\boldsymbol{\lambda}^{\mathrm{T}}(\tau)=\boldsymbol{0}\end{cases},\quad \tau\in[t_i,t] \tag{6.8}$$

we have

$$\boldsymbol{\lambda}^{\mathrm{T}}(\tau)=\exp\left\{\int_t^{\tau}(-\boldsymbol{J}^{\mathrm{T}})\mathrm{d}\tau_1+\frac{1}{2}\int_t^{\tau}\mathrm{d}\tau_1\int_t^{\tau_1}\mathrm{d}\tau_2\left[-\boldsymbol{J}^{\mathrm{T}}(\tau_1),\ -\boldsymbol{J}^{\mathrm{T}}(\tau_2)\right]+\ldots\right\}\boldsymbol{\lambda}^{\mathrm{T}}(t),\quad \tau\in[t_i,t] \tag{6.9}$$

Inverting the expression leads to

$$\begin{aligned}
\boldsymbol{\lambda}(\tau)&=\boldsymbol{\lambda}(\tau)\Big|_{\tau=t}\exp\left\{\int_t^{\tau}(-\boldsymbol{J})\mathrm{d}\tau_1+\frac{1}{2}\int_t^{\tau}\mathrm{d}\tau_1\int_t^{\tau_1}\mathrm{d}\tau_2[-\boldsymbol{J}(\tau_2),\ -\boldsymbol{J}(\tau_1)]+\ldots\right\}\\
&=-\exp\left\{-\int_t^{\tau}(\boldsymbol{J})\mathrm{d}\tau_1-\frac{1}{2}\int_t^{\tau}\mathrm{d}\tau_1\int_t^{\tau_1}\mathrm{d}\tau_2[-\boldsymbol{J}(\tau_1),\ -\boldsymbol{J}(\tau_2)]+\ldots\right\}\\
&=-\exp\left\{\int_{\tau}^{t}(\boldsymbol{J})\mathrm{d}\tau_1+\frac{1}{2}\int_{\tau}^{t}\mathrm{d}\tau_1\int_t^{\tau_1}\mathrm{d}\tau_2[\boldsymbol{J}(\tau_1),\boldsymbol{J}(\tau_2)]+\ldots\right\}
\end{aligned} \tag{6.10}$$

We notice that

$$\begin{aligned}
&\int_{\tau}^{t}\mathrm{d}\tau_1\int_t^{\tau_1}\mathrm{d}\tau_2[\boldsymbol{J}(\tau_1),\boldsymbol{J}(\tau_2)]-\int_{\tau}^{t}\mathrm{d}\tau_1\int_{\tau}^{\tau_1}\mathrm{d}\tau_2[\boldsymbol{J}(\tau_1),\boldsymbol{J}(\tau_2)]\\
&=\int_{\tau}^{t}\mathrm{d}\tau_1\int_t^{\tau}\mathrm{d}\tau_2[\boldsymbol{J}(\tau_1),\boldsymbol{J}(\tau_2)]\\
&=\int_{\tau}^{t}\mathrm{d}\tau_1\int_t^{\tau}\mathrm{d}\tau_2[\boldsymbol{J}(\tau_1)\boldsymbol{J}(\tau_2)-\boldsymbol{J}(\tau_2)\boldsymbol{J}(\tau_1)]\\
&=\boldsymbol{0}
\end{aligned} \tag{6.11}$$

and the other terms in the Magnus series also possess a similar characteristic. The expression of $\boldsymbol{\lambda}(\tau)$ can therefore be rewritten as follows:

$$\boldsymbol{\lambda}(\tau) = -\exp\left\{\int_{\tau}^{t}(\boldsymbol{J})\mathrm{d}\tau_1 + \frac{1}{2}\int_{\tau}^{t}\mathrm{d}\tau_1\int_{\tau}^{\tau_1}\mathrm{d}\tau_2[\boldsymbol{J}(\tau_1),\boldsymbol{J}(\tau_2)] + \ldots\right\},\quad \tau\in[t_i, t] \tag{6.12}$$

It is exactly $\overline{\boldsymbol{\lambda}}(t)$ in the form of Magnus series. Using the equivalence between $\boldsymbol{\lambda}(\tau)$ and $\overline{\boldsymbol{\lambda}}(t)$, we obtain the following from Eqs. (6.3) and (6.4):

$$\begin{aligned}\frac{\mathrm{d}\boldsymbol{x}_{n+1}}{\mathrm{d}t} &= \boldsymbol{F}(\boldsymbol{x}_n, t) + \int_{t_i}^{t}\frac{\partial\boldsymbol{\lambda}(\tau)}{\partial t}[\boldsymbol{L}\boldsymbol{x}_n - \boldsymbol{F}(\boldsymbol{x}_n, \tau)]\mathrm{d}\tau \\ &= \boldsymbol{F}(\boldsymbol{x}_n, t) + \int_{t_i}^{t}\frac{\partial\overline{\boldsymbol{\lambda}}(t)}{\partial t}[\boldsymbol{L}\boldsymbol{x}_n - \boldsymbol{F}(\boldsymbol{x}_n, \tau)]\mathrm{d}\tau \\ &= \boldsymbol{F}(\boldsymbol{x}_n, t) + \boldsymbol{J}(\boldsymbol{x}_n, t)\int_{t_i}^{t}\overline{\boldsymbol{\lambda}}(t)[\boldsymbol{L}\boldsymbol{x}_n - \boldsymbol{F}(\boldsymbol{x}_n, \tau)]\mathrm{d}\tau \\ &= \boldsymbol{F}(\boldsymbol{x}_n, t) + \boldsymbol{J}(\boldsymbol{x}_n, t)\int_{t_i}^{t}\boldsymbol{\lambda}(\tau)[\boldsymbol{L}\boldsymbol{x}_n - \boldsymbol{F}(\boldsymbol{x}_n, \tau)]\mathrm{d}\tau \\ &= \boldsymbol{F}(\boldsymbol{x}_n, t) + \boldsymbol{J}(\boldsymbol{x}_n, t)(\boldsymbol{x}_{n+1} - \boldsymbol{x}_n)\end{aligned} \tag{6.13}$$

This recursive formula for $\dot{\boldsymbol{x}}$ can be rewritten as follows:

Algorithm-1:

$$\frac{\mathrm{d}\boldsymbol{x}_{n+1}}{\mathrm{d}t} - \boldsymbol{J}(\boldsymbol{x}_n, t)\boldsymbol{x}_{n+1} = \boldsymbol{F}(\boldsymbol{x}_n, t) - \boldsymbol{J}(\boldsymbol{x}_n, t)\boldsymbol{x}_n,\quad t\in[t_{i-1}, t_i] \tag{6.14}$$

In this way, the Lagrange multipliers are completely eliminated in the recursive formula for the analytical expression for $\dot{\boldsymbol{x}}$.

6.1.2 Algorithm-2

It is difficult to explicitly derive the generalized Lagrange multipliers, since the constraints include $\boldsymbol{x}_n$. However, with the help of the differential transform method, it is convenient to approximate the generalized Lagrange multipliers in a series of polynomials [9] as follows:

$$\boldsymbol{\lambda}(\tau) \approx \boldsymbol{T}_0[\boldsymbol{\lambda}] + \boldsymbol{T}_1[\boldsymbol{\lambda}](\tau - t) + \cdots + \boldsymbol{T}_K[\boldsymbol{\lambda}](\tau - t)^K \tag{6.15}$$

where $\boldsymbol{T}_k[\boldsymbol{\lambda}]$ is the kth-order differential transformation of $\boldsymbol{\lambda}(\tau)$, that is,

$$\boldsymbol{T}_k[\boldsymbol{\lambda}] = \frac{1}{k!}\frac{\mathrm{d}^k\boldsymbol{\lambda}(\tau)|_{\tau=t}}{\mathrm{d}\tau^k} \tag{6.16}$$

From the constraints on $\boldsymbol{\lambda}(\tau)$, $\boldsymbol{T}_k[\boldsymbol{\lambda}]$ can be determined in an iterative way as following:

$$\boldsymbol{T}_0[\boldsymbol{\lambda}] = diag\,[-1, -1, \ldots], \ \boldsymbol{T}_{k+1}[\boldsymbol{\lambda}] = -\frac{\boldsymbol{T}_k[\boldsymbol{\lambda J}]}{k+1}, \ 0 \le k \le K+1 \tag{6.17}$$

In the VIM of He [5] the nonlinear term in the dynamical system is normally restricted from variation to simplify the derivation of $\boldsymbol{\lambda}(\tau)$. If we also restrict the nonlinear term $\boldsymbol{F}(\boldsymbol{x}, \tau)$ from variation herein, the matrix of generalized Lagrange multipliers can be easily determined to be $\boldsymbol{\lambda}(\tau) = -\boldsymbol{I}$. Substituting it into the correctional formula of the LVIM leads to the Picard iteration method

$$\boldsymbol{x}_{n+1}(t) = \boldsymbol{x}(t_i) + \int_{t_i}^{t} \boldsymbol{F}[\boldsymbol{x}_n(\tau), \tau]\mathrm{d}\tau \tag{6.18}$$

If we differentiate this formula, we have

$$\frac{\mathrm{d}\boldsymbol{x}_{n+1}}{\mathrm{d}t} = \boldsymbol{F}(\boldsymbol{x}_n, t), \ \ t \in [t_i, t_{i+1}] \tag{6.19}$$

However, if the nonlinear term is included in the variation as in present Chapter, the resulting $\boldsymbol{\lambda}(\tau)$ should be a function of τ and t. Making differentiations to the correctional formula of the LVIM for $K+1$ times will lead to

$$\boldsymbol{x}_{n+1}^{(K+1)} = \boldsymbol{x}_n^{(K+1)} + (\boldsymbol{\lambda}\mathbf{G}|_{\tau=t})^{(K)} + \left(\frac{\partial \boldsymbol{\lambda}}{\partial t}\mathbf{G}|_{\tau=t}\right)^{(K-1)} + \cdots + \left(\frac{\partial^K \boldsymbol{\lambda}}{\partial t^K}\mathbf{G}|_{\tau=t}\right) + \int_{t_i}^{t}\frac{\partial^{K+1}\boldsymbol{\lambda}}{\partial t^{K+1}}\mathbf{G}\mathrm{d}\tau \tag{6.20}$$

where $\mathbf{G} = \boldsymbol{L}\boldsymbol{x}_n(\tau) - \boldsymbol{F}[\boldsymbol{x}_n(\tau), \tau]$.

If we simply enforce $\boldsymbol{x}(\tau)$ to be a constant $\boldsymbol{x}(t_i)$, then $\boldsymbol{\lambda}(\tau)$ can be approximated by a truncated polynomial series in which the highest order of t is K. The last term of the preceding expression can be omitted, and we will obtain a correctional formula without the integral operator:

$$\begin{aligned}\boldsymbol{x}_{n+1}^{(K+1)} &= \boldsymbol{x}_n^{(K+1)} + (\boldsymbol{\lambda}\mathbf{G}|_{\tau=t})^{(K)} + \left(\frac{\partial \boldsymbol{\lambda}}{\partial t}\mathbf{G}|_{\tau=t}\right)^{(K-1)} + \cdots + \left(\frac{\partial^K \boldsymbol{\lambda}}{\partial t^K}\mathbf{G}|_{\tau=t}\right)\\ &= \boldsymbol{x}_n^{(K+1)} + \boldsymbol{T}_0[\boldsymbol{\lambda}]\mathbf{G}\big|_{\tau=t}^{(K)} - \boldsymbol{T}_1[\boldsymbol{\lambda}]\mathbf{G}\big|_{\tau=t}^{(K-1)} + \cdots + (-1)^K(K!)\boldsymbol{T}_K[\boldsymbol{\lambda}]\mathbf{G}|_{\tau=t}\end{aligned} \tag{6.21}$$

However, this formula is not complete. It should be noted that each time that we differentiate the correctional formula, the information of constant terms in it is lost. For that reason, before each differentiation is made, the initial value $\boldsymbol{x}_{n+1}^{(k)}(t_0)$, $k=0,1,2,\ldots,K$, is kept as additional constrains to the correctional formula. In the expression above, $\boldsymbol{T}_k[\boldsymbol{\lambda}]$ $(k=0,1,2,\ldots,K)$ are constants. So $\boldsymbol{x}_{n+1}^{(K+1)}$ is a simple linear combination of $\boldsymbol{x}_n^{(K+1)}$ and $\boldsymbol{G}$. Clearly, if the iteration become static, that is, $\boldsymbol{x}_{n+1}=\boldsymbol{x}_n$, the correctional formula will guarantee

$$\mathbf{G}=\boldsymbol{L}\boldsymbol{x}_n(\tau)-\boldsymbol{F}[\boldsymbol{x}_n(\tau),\tau]=\mathbf{0} \tag{6.22}$$

By approximating the generalized Lagrange multipliers in a series of polynomials, the LVIM can also be altered in the following way for $\boldsymbol{x}_{n+1}$:

$$\begin{aligned}\boldsymbol{x}_{n+1}(t)&=\boldsymbol{x}_n(t)+\int_{t_i}^{t}\boldsymbol{\lambda}(\tau)\{\boldsymbol{L}\boldsymbol{x}_n(\tau)-\boldsymbol{F}[\boldsymbol{x}_n(\tau),\tau]\}\mathrm{d}\tau\\ &=\boldsymbol{x}_n(t)+\int_{t_i}^{t}\{\boldsymbol{T}_0[\boldsymbol{\lambda}]+\boldsymbol{T}_1[\boldsymbol{\lambda}](\tau-t)+\cdots+\boldsymbol{T}_K[\boldsymbol{\lambda}](\tau-t)^K\}\mathbf{G}\mathrm{d}\tau\end{aligned} \tag{6.23}$$

Considering that $\boldsymbol{T}_k[\boldsymbol{\lambda}]$, $k=0,1,\ldots K$ are functions of $\boldsymbol{x}_n(t)$ and t, the preceding expression can be rewritten as

---

Algorithm-2:

$$\boldsymbol{x}_{n+1}(t)=\boldsymbol{x}_n(t)+\boldsymbol{A}_0(t)\int_{t_i}^{t}\mathbf{G}\mathrm{d}\tau+\boldsymbol{A}_1(t)\int_{t_i}^{t}\tau\mathbf{G}\mathrm{d}\tau+\cdots+\boldsymbol{A}_K(t)\int_{t_i}^{t}\tau^K\mathbf{G}\mathrm{d}\tau \tag{6.24}$$

---

where $t\in[t_i,t_{i+1}]$. The coefficient matrices $\boldsymbol{A}_k(t)$, $k=0,1,\ldots K$, are combinations of $\boldsymbol{T}_k[\boldsymbol{\lambda}]$ and t. It is worth to note that the Algorithm-2 can be easily generalized into higher order systems, although it is derived in the case of first-order differential equations.

6.1.3 Algorithm-3

As is stated in subsection 2.1, the matrix of generalized Lagrange multipliers can be expressed in the form of exponential functions. Since the exponential functions can be equivalently defined by power series, we have

$$\boldsymbol{\lambda}(\tau)=-\exp\left[\int_{\tau}^{t}\tilde{\mathbf{G}}(\boldsymbol{x}_n,\varsigma)\mathrm{d}\varsigma\right]=-\left\{diag\,[1,1,\ldots]+\int_{\tau}^{t}\tilde{\mathbf{G}}(\boldsymbol{x}_n,\varsigma)\mathrm{d}\varsigma+\frac{1}{2!}\left[\int_{\tau}^{t}\tilde{\mathbf{G}}(\boldsymbol{x}_n,\varsigma)\mathrm{d}\varsigma\right]^2+\cdots\right\} \tag{6.25}$$

We then substitute it into $\partial\boldsymbol{\lambda}(\tau)/\partial t = \boldsymbol{J}(\boldsymbol{x}_n, t)\boldsymbol{\lambda}(\tau)$. If the higher-order terms of $\boldsymbol{\lambda}(\tau)$ are ignored, $\partial\boldsymbol{\lambda}(\tau)/\partial t$ can be approximated by $\partial\boldsymbol{\lambda}(\tau)/\partial t = -\boldsymbol{J}(\boldsymbol{x}_n, t)$. Accordingly, it gives rise to the following iterative formula:

$$
\begin{aligned}
\frac{\mathrm{d}\boldsymbol{x}_{n+1}}{\mathrm{d}t} &= \frac{\mathrm{d}\boldsymbol{x}_n}{\mathrm{d}t} + \boldsymbol{\lambda}(t)\left[\frac{\mathrm{d}\boldsymbol{x}_n}{\mathrm{d}t} - \boldsymbol{F}(\boldsymbol{x}_n, t)\right] + \int_{t_i}^{t} \frac{\partial\boldsymbol{\lambda}}{\partial t}[\boldsymbol{L}\boldsymbol{x}_n - \boldsymbol{F}(\boldsymbol{x}_n, \tau)]\mathrm{d}\tau \\
&= \boldsymbol{F}(\boldsymbol{x}_n, t) - \boldsymbol{J}(\boldsymbol{x}_n, t)\int_{t_i}^{t}[\boldsymbol{L}\boldsymbol{x}_n - \boldsymbol{F}(\boldsymbol{x}_n, \tau)]\mathrm{d}\tau \\
&= \boldsymbol{F}(\boldsymbol{x}_n, t) - \boldsymbol{J}(\boldsymbol{x}_n, t)\left[\boldsymbol{x}_n - \int_{t_i}^{t}\boldsymbol{F}(\boldsymbol{x}_n, \tau)\mathrm{d}\tau\right]
\end{aligned}
\tag{6.26}
$$

or simply

Algorithm-3:

$$
\frac{\mathrm{d}\boldsymbol{x}_{n+1}}{\mathrm{d}t} = \boldsymbol{F}(\boldsymbol{x}_n, t) - \boldsymbol{J}(\boldsymbol{x}_n, t)\left[\boldsymbol{x}_n - \int_{t_i}^{t}\boldsymbol{F}(\boldsymbol{x}_n, \tau)\mathrm{d}\tau\right]
\tag{6.27}
$$

---

where $t \in [t_i, t_{i+1}]$. In summary, the three proposed modifications of the Local Variational Iteration Method are listed as below.

It is not hard to see that direct applications of the Algorithms-1,2, and 3 with an initial analytical guess still needs very lengthy symbolic computations. Instead of using analytical initial guess functions, we satisfy the iterative equations of Algorithms-1,2, and 3 in a weak form by using trial functions and test functions in the time interval $t_i \le t \le t_{i+1}$. Depending on the selection of test functions and trial functions, there are various weak forms, of which the collocation form is the simplest. To explain the idea of weak forms of the three modifications clearly, the collocation method is introduced first (Table 6.1).

Table 6.1 Three modifications of the LVIM with $t \in [t_i, t_{i+1}]$.

Algorithm-1	$\frac{\mathrm{d}\boldsymbol{x}_{n+1}}{\mathrm{d}t} - \boldsymbol{J}(\boldsymbol{x}_n, t)\boldsymbol{x}_{n+1} = \boldsymbol{F}(\boldsymbol{x}_n, t) - \boldsymbol{J}(\boldsymbol{x}_n, t)\boldsymbol{x}_n$
Algorithm-2	$\boldsymbol{x}_{n+1}(t) = \boldsymbol{x}_n(t) + \boldsymbol{A}_0(t)\int_{t_i}^{t}\mathbf{G}\mathrm{d}\tau + \boldsymbol{A}_1(t)\int_{t_i}^{t}\tau\mathbf{G}\mathrm{d}\tau + \cdots + \boldsymbol{A}_K(t)\int_{t_i}^{t}\tau^K\mathbf{G}\mathrm{d}\tau$
Algorithm-3	$\frac{\mathrm{d}\boldsymbol{x}_{n+1}}{\mathrm{d}t} = \boldsymbol{F}(\boldsymbol{x}_n, t) - \boldsymbol{J}(\boldsymbol{x}_n, t)\left[\boldsymbol{x}_n - \int_{t_i}^{t}\boldsymbol{F}(\boldsymbol{x}_n, \tau)\mathrm{d}\tau\right]$

6.2 Implementation of LVIM

6.2.1 Discretization using collocation

Considering a vector of trial functions $\boldsymbol{u}$, the residual error of the first-order differential equations considered in this chapter is

$$\boldsymbol{R} = \boldsymbol{L}\boldsymbol{u} - \boldsymbol{F}(\boldsymbol{u}, t) \neq \boldsymbol{0}, \quad t \in [t_i, t_{i+1}] \tag{6.28}$$

With a diagonal matrix of test functions $\boldsymbol{v} = diag\,[v, v, \ldots]$, the local weighted residual weak form of the formulation is written as

$$\int_{t_i}^{t_{i+1}} \boldsymbol{v}\boldsymbol{R}\mathrm{d}t = \int_{t_i}^{t_{i+1}} \boldsymbol{v}[\boldsymbol{L}\boldsymbol{u} - \boldsymbol{F}(\boldsymbol{u}, t)]\mathrm{d}t = \boldsymbol{0} \tag{6.29}$$

Let the trial function u_e be the linear combinations of basis functions $\phi_{e,nb}(t)$:

$$u_e = \sum_{nb=1}^{N} \alpha_{e,nb}\phi_{e,nb}(t) = \boldsymbol{\Phi}_e(t)\boldsymbol{A}_e \tag{6.30}$$

where u_e represents an element of $\boldsymbol{u}$, the columns of $\boldsymbol{\Phi}_e$ represent each of the independent basis functions, and the vector $\boldsymbol{A}_e$ contains undetermined coefficients.

We use the Dirac delta functions as the test function, i.e., $v = \delta(t - t_m)$ for a group of preselected points t_m $(m = 1, 2, \ldots, M)$ in the local domain $t \in [t_i, t_{i+1}]$. It leads to the local collocation method:

$$\boldsymbol{L}\boldsymbol{u}(t_m) - \boldsymbol{F}[\boldsymbol{u}(t_m), t_m] = \boldsymbol{0}, \quad t_m \in [t_i, t_{i+1}] \tag{6.31}$$

The value of u_e and the differentiation of u_e at time point t_m can be expressed as

$$u_e(t_m) = \sum_{nb=1}^{N} \alpha_{e,nb}\phi_{e,nb}(t_m) = \boldsymbol{\Phi}_e(t_m)\boldsymbol{A}_e \tag{6.32}$$

and

$$Lu_e(t_m) = \sum_{nb=1}^{N} \alpha_{e,nb}L\phi_{e,nb}(t_m) = \boldsymbol{L}\boldsymbol{\Phi}_e(t_m)\boldsymbol{A}_e \tag{6.33}$$

In matrix form, they can be written as $\boldsymbol{U}_e = \boldsymbol{B}_e\boldsymbol{A}_e$ and $\boldsymbol{L}\boldsymbol{U}_e = \boldsymbol{L}\boldsymbol{B}_e\boldsymbol{A}_e$ separately, where $\boldsymbol{U}_e = [u_e(t_1), u_e(t_2), \ldots, u_e(t_M)]^{\mathrm{T}}$ and $\boldsymbol{B}_e = [\boldsymbol{\Phi}_e(t_1)^{\mathrm{T}}, \boldsymbol{\Phi}_e(t_2)^{\mathrm{T}}, \ldots, \boldsymbol{\Phi}_e(t_M)^{\mathrm{T}}]^{\mathrm{T}}$. By simple transformations, we have $\boldsymbol{L}\boldsymbol{U}_e = \boldsymbol{L}\boldsymbol{B}_e\boldsymbol{A}_e = (\boldsymbol{L}\boldsymbol{B}_e)\boldsymbol{B}_e^{-1}\boldsymbol{U}_e$, thus $\boldsymbol{L}\boldsymbol{U}_e$ is expressed in the form of $\boldsymbol{U}_e$. Further we can derive the collocation of higher order differentiations as

$$\boldsymbol{L}^k\boldsymbol{U}_e = \boldsymbol{L}^k\boldsymbol{B}_e\boldsymbol{A}_e = (\boldsymbol{L}^k\boldsymbol{B}_e)\boldsymbol{B}_e^{-1}\boldsymbol{U}_e \tag{6.34}$$

Now we can rewrite the formulation of local collocation method as a system of NAEs of $\boldsymbol{U}$:

$$\boldsymbol{EU} - \boldsymbol{F}(\boldsymbol{U}, \boldsymbol{t}) = \boldsymbol{0} \tag{6.35}$$

where $\boldsymbol{t} = [t_1, t_2, \ldots, t_M]$, $\boldsymbol{E} = diag\,[(\boldsymbol{LB}_1)\boldsymbol{B}_1^{-1}, \ldots, (\boldsymbol{LB}_e)\boldsymbol{B}_e^{-1}, \ldots]$, and $\boldsymbol{U} = [\boldsymbol{U}_1^{\mathrm{T}}, \ldots, \boldsymbol{U}_e^{\mathrm{T}}, \ldots]^{\mathrm{T}}$.

Following the concept of the collocation method, we approximate the guess function $\mathbf{x}_n(t)$ and the corrected function $\boldsymbol{x}_{n+1}(t)$ in the three modifications of the LVIM respectively by $\boldsymbol{u}_n(t)$ and $\boldsymbol{u}_{n+1}(t)$, respectively, which are composed of the same set of basis functions $\boldsymbol{\Phi}$. By collocating at a set of points in the local time domain of the system, the iterative formulas of Algorithm-1,2, and 3 are then reduced to algebraic iterative formulas for the values of the solution at collocation points.

6.2.2 Collocation of algorithm-1

Suppose that the solution of the nonlinear problem $\boldsymbol{Lx} = \boldsymbol{F}(\boldsymbol{x}, t)$ in a subinterval $t \in [t_i, t_{i+1}]$ is approximated by $\boldsymbol{u}$. From the MLVIM-1, we have

$$\boldsymbol{Lu}_{n+1}(t_m) - \boldsymbol{J}[\boldsymbol{u}_n(t_m), t_m]\boldsymbol{u}_{n+1}(t_m) = \boldsymbol{F}[\boldsymbol{u}_n(t_m), t_m] - \boldsymbol{J}[\boldsymbol{u}_n(t_m), t_m]\boldsymbol{u}_n(t_m) \tag{6.36}$$

where t_m $(m = 1, 2, \ldots, M)$ are the collocation points in the time interval t_i to t_{i+1}. Following the transformations made in section 3, this formulation can be rewritten in matrix form as

$$\boldsymbol{EU}_{n+1} - \boldsymbol{J}(\boldsymbol{U}_n, \boldsymbol{t})\boldsymbol{U}_{n+1} = \boldsymbol{F}(\boldsymbol{U}_n, \boldsymbol{t}) - \boldsymbol{J}(\boldsymbol{U}_n, \boldsymbol{t})\boldsymbol{U}_n \tag{6.37}$$

By rearranging the matrices, it can be further expressed as

$$\boldsymbol{U}_{n+1} = \boldsymbol{U}_n - [\boldsymbol{E} - \boldsymbol{J}(\boldsymbol{U}_n, \boldsymbol{t})]^{-1}[\boldsymbol{EU}_n - \boldsymbol{F}(\boldsymbol{U}_n, \boldsymbol{t})] \tag{6.38}$$

Interestingly, this recursive formula happens to be the Newton-Raphson iterative algorithm for solving the NAEs that are generated by the conventional collocation method. It gives a straightforward illustration of the relationship between Newton's method and the variational iteration method, which is implied in the work of Inokuti, Sekine, and Mura [11].

6.2.3 Collocation of algorithm-2

Firstly, the Picard iteration method is derived here as a special case of the LVIM to give a brief explanation. Considering the first order differential equations, the sequence of approximations given by Picard iteration is

$$\frac{\mathrm{d}\boldsymbol{x}_{n+1}}{\mathrm{d}t} = \boldsymbol{F}(\boldsymbol{x}_n, t), \quad \boldsymbol{x}_{n+1}(t_i) = \boldsymbol{x}(t_i) \tag{6.39}$$

or

$$\boldsymbol{x}_{n+1}(t) = \boldsymbol{x}(t_i) + \int_{t_i}^{t} \boldsymbol{F}[\boldsymbol{x}_n(\tau), \tau]\mathrm{d}\tau, \quad t \in [t_i, t_{i+1}] \tag{6.40}$$

The adoption of collocation method will then generate an iteration formula for values of the solution at collocation points t_m ($m = 1, 2, \ldots, M$), which is

$$\boldsymbol{E}\boldsymbol{U}_{n+1} = \boldsymbol{F}(\boldsymbol{U}_n, \boldsymbol{t}), \quad \boldsymbol{u}_{n+1}(t_i) = \boldsymbol{x}(t_i)$$

or

$$\boldsymbol{U}_{n+1} = \boldsymbol{U}_n - \tilde{\boldsymbol{E}}[\boldsymbol{E}\boldsymbol{U}_n - \boldsymbol{F}(\boldsymbol{U}_n, \boldsymbol{t})] \tag{6.41}$$

The coefficient matrix $\tilde{\boldsymbol{E}}$ is composed by $\tilde{\boldsymbol{E}} = diag[(\boldsymbol{L}^{-1}\boldsymbol{B}_1)\boldsymbol{B}_1{}^{-1}, \ldots, (\boldsymbol{L}^{-1}\boldsymbol{B}_e)\boldsymbol{B}_e{}^{-1}, \ldots]$. It should be noted that these two iterative formulas are not the same in practice. In the first formula, the boundary conditions are enforced by adding the constraints $\boldsymbol{u}_{n+1}(t_i) = \boldsymbol{x}(t_i)$, which will introduce an extra computational error. Converstly, the second formula is not plagued by this problem because the boundary conditions are naturally satisfied.

Obviously, if the basis functions are selected as the Chebyshev polynomials of the first kind, the second iterative formula stated above will give rise to the modified Chebyshev Picard iteration method (MCPI) method [12]. The iteration formula works if only the collocations $\boldsymbol{U}_n$ is updated in each step, so it is very simple and direct.

Remark 1

The MCPI method is exactly the same as the zeroth order approximation of Algorithm-2, using the collocation method with Chebyshev polynomials as basis functions.

In a general case, if $\boldsymbol{x}(\tau)$ is approximated by $\boldsymbol{x}(t_i)$, we have

$$\boldsymbol{x}_{n+1}^{(K+1)} = \boldsymbol{x}_n^{(K+1)} + \boldsymbol{T}_0[\boldsymbol{\lambda}]\mathbf{G}|_{\tau=t}^{(K)} - \boldsymbol{T}_1[\boldsymbol{\lambda}]\mathbf{G}|_{\tau=t}^{(K-1)} + \cdots + (-1)^K(K!)\boldsymbol{T}_K[\boldsymbol{\lambda}]\mathbf{G}|_{\tau=t} \tag{6.42}$$

By collocating points in the local domain, it leads to

$$\boldsymbol{u}_{n+1}^{(K+1)}(t_m) = \boldsymbol{u}_n^{(K+1)}(t_m) + \boldsymbol{T}_0\mathbf{G}^{(K)}(t_m) - \boldsymbol{T}_1\mathbf{G}^{(K-1)}(t_m) + \cdots + (-1)^K(K!)\boldsymbol{T}_K\mathbf{G}(t_m) \tag{6.43}$$

where $\mathbf{G}|_{\tau=t}$ and $\boldsymbol{T}_k[\boldsymbol{\lambda}]$ are denoted as $\mathbf{G}$ and $\boldsymbol{T}_k$ for simplicity. It can be rewritten in matrix form as

$$\boldsymbol{E}_{K+1}\boldsymbol{U}_{n+1} = \boldsymbol{E}_{K+1}\boldsymbol{U}_n + \boldsymbol{T}_0\mathbf{G}^{(K)}(\boldsymbol{t}) - \boldsymbol{T}_1\mathbf{G}^{(K-1)}(\boldsymbol{t}) + \cdots + (-1)^K(K!)\boldsymbol{T}_K\mathbf{G}(\boldsymbol{t}) \tag{6.44}$$

of which

$$\boldsymbol{E}_k = diag[(\boldsymbol{LB}_1)\boldsymbol{B}_1^{-k}, \ldots, (\boldsymbol{LB}_e)\boldsymbol{B}_e^{-k}, \ldots].$$

This iterative formula is very simple since there is no need to calculate the Jacobian matrix of nonlinear terms and its inverse, and all the matrices of coefficients are constant. However, the convergence domain of it could be very limited because $\boldsymbol{x}(\tau)$ is approximated too roughly.

Normally, it is more reasonable to use the integral form of Algorithm-2 (i.e., Eq. 6.24):

$$\boldsymbol{x}_{n+1}(t) = \boldsymbol{x}_n(t) + \boldsymbol{A}_0(t)\int_{t_i}^{t} \mathbf{G}\mathrm{d}\tau + \boldsymbol{A}_1(t)\int_{t_i}^{t} \tau\mathbf{G}\mathrm{d}\tau + \cdots + \boldsymbol{A}_K(t)\int_{t_i}^{t} \tau^K\mathbf{G}\mathrm{d}\tau \tag{6.45}$$

The adoption of collocation method leads to

$$\boldsymbol{U}_{n+1} = \boldsymbol{U}_n + \boldsymbol{A}_0(\boldsymbol{t})\tilde{\boldsymbol{E}}\boldsymbol{G}(\boldsymbol{t}) + \boldsymbol{A}_1(\boldsymbol{t})\tilde{\boldsymbol{E}}[\boldsymbol{t}\cdot\mathbf{G}(\boldsymbol{t})] + \cdots + \boldsymbol{A}_K(\boldsymbol{t})\tilde{\boldsymbol{E}}[\boldsymbol{t}^K\cdot\mathbf{G}(\boldsymbol{t})] \tag{6.46}$$

The integral form is superior to the differential form in several aspects:

1. The integral form is mathematically correct. It considers the change of $\boldsymbol{\lambda}(\tau)$ along with $\boldsymbol{x}_n(\tau)$ in each iteration step, while the differential form fails to achieve that.
2. The integral form satisfies the initial conditions inherently. In contrast, the differential form needs additional constraints to guarantee that, which will bring in extra computational errors.
3. In the integral form, there is no need to calculate the inverse of $\boldsymbol{E}_k$, which could get ill-conditioned easily for high order interpolations [2].

Overall, the integral form is more robust than the differential form, although the computational burden is slightly heavier. Unlike the collocation form of Algorithm-1 (i.e., Eq. 6.38), the collocation form of Algorithm-2 (i.e., Eq. 6.46) does not need to calculate the Jacobian matrix of nonlinear terms or its inverse in the computation. For high dimensional systems with complex structures, the simple operations of this formula can speed up the computation.

With $\boldsymbol{T}_0[\boldsymbol{\lambda}]$, $\boldsymbol{T}_1[\boldsymbol{\lambda}]$, and $\boldsymbol{T}_2[\boldsymbol{\lambda}]$, the first three low order correctional formulas in differential form can be obtained by using the collocation formula of Algorithm-2 (i.e., Eq. 6.44), as shown in Table 6.2.

Similarly, the integral forms are obtained as shown in Table 6.3.

6.2.4 Collocation of algorithm-3

Since there exists the integration of nonlinear terms, we need to approximate the nonlinear terms with a set of basis functions first, so that it can be written as

Table 6.2 Differential forms of collocated Algorithm-2.

Approximation of $\boldsymbol{\lambda}$	Correctional formula in differential form
0th order: $\boldsymbol{\lambda} = \boldsymbol{T}_0$	$\boldsymbol{E}_1\boldsymbol{U}_{n+1} = \boldsymbol{E}_1\boldsymbol{U}_n + \boldsymbol{T}_0\mathbf{G}(t)$
1st order: $\boldsymbol{\lambda} = \boldsymbol{T}_0 + \boldsymbol{T}_1(\tau - t)$	$\boldsymbol{E}_2\boldsymbol{U}_{n+1} = \boldsymbol{E}_2\boldsymbol{U}_n + \boldsymbol{T}_0\mathbf{G}^{(1)}(t) - \boldsymbol{T}_1\mathbf{G}(t)$
2nd order: $\boldsymbol{\lambda} = \boldsymbol{T}_0 + \boldsymbol{T}_1(\tau - t) + \boldsymbol{T}_2(\tau - t)^2$	$\boldsymbol{E}_3\boldsymbol{U}_{n+1} = \boldsymbol{E}_3\boldsymbol{U}_n + \boldsymbol{T}_0\mathbf{G}^{(2)}(t) - \boldsymbol{T}_1\mathbf{G}^{(1)}(t) + 2\boldsymbol{T}_2\mathbf{G}(t)$

Table 6.3 Integral forms of collocated Algorithm-2.

Approximation of $\boldsymbol{\lambda}$	Correctional formula in integral form
0th order: $\boldsymbol{\lambda} = \boldsymbol{T}_0$	$\boldsymbol{U}_{n+1} = \boldsymbol{U}_n + \boldsymbol{T}_0\tilde{\mathbf{E}}\mathbf{G}$
1st order: $\boldsymbol{\lambda} = \boldsymbol{T}_0 + \boldsymbol{T}_1(\tau - t)$	$\boldsymbol{U}_{n+1} = \boldsymbol{U}_n + (\boldsymbol{T}_0 - [\boldsymbol{T}_1 \cdot t])\tilde{\mathbf{E}}\mathbf{G} + \boldsymbol{T}_1\tilde{\mathbf{E}}[t \cdot \mathbf{G}]$
2nd order: $\boldsymbol{\lambda} = \boldsymbol{T}_0 + \boldsymbol{T}_1(\tau - t) + \boldsymbol{T}_2(\tau - t)^2$	$\boldsymbol{U}_{n+1} = \boldsymbol{U}_n + (\boldsymbol{T}_0 - [\boldsymbol{T}_1 \cdot t] + [\boldsymbol{T}_2 \cdot t^2])\tilde{\mathbf{E}}\mathbf{G} + (\boldsymbol{T}_1 - 2[\boldsymbol{T}_2 \cdot t])\tilde{\mathbf{E}}[t \cdot \mathbf{G}] + \boldsymbol{T}_2\tilde{\mathbf{E}}[t^2 \cdot \mathbf{G}]$

$$F_e(\boldsymbol{x}_n, t) \approx \sum_{n=1}^{N} \eta_{en}\phi_{en}(t) = \boldsymbol{\Phi}_e(t)\boldsymbol{Y}_e \tag{6.47}$$

where $\boldsymbol{Y}_e$ can be determined from the relationship $F_e[\boldsymbol{u}(t_m), t_m] = \boldsymbol{\Phi}_e(t_m)\boldsymbol{Y}_e$, $m = 1, 2, \ldots, M$. Therefore, the integral terms at collocation point t_m can be expressed as

$$\int_{t_i}^{t_m} F_e(\boldsymbol{u}_n, \tau)\mathrm{d}\tau = \boldsymbol{L}^{-1}\boldsymbol{\Phi}_e(t_m)\boldsymbol{Y}_e = [\boldsymbol{L}^{-1}\boldsymbol{\Phi}_e(t_m)]\boldsymbol{B}_e^{-1}F_e(\boldsymbol{U}_{ne}, \boldsymbol{t}) \tag{6.48}$$

where $\boldsymbol{B}_e = [\boldsymbol{\Phi}_e(t_1)^{\mathrm{T}}, \boldsymbol{\Phi}_e(t_2)^{\mathrm{T}}, \ldots, \boldsymbol{\Phi}_e(t_M)^{\mathrm{T}}]^{\mathrm{T}}$ and $\boldsymbol{\Phi}^{-1}$ is the integration of the basis functions. Substituting it into the collocation form of Algorithm-3 leads to

$$\boldsymbol{EU}_{n+1} = \boldsymbol{F}(\boldsymbol{U}_n, \boldsymbol{t}) - \boldsymbol{J}(\boldsymbol{U}_n, \boldsymbol{t})\tilde{\boldsymbol{E}}[\boldsymbol{EU}_n - \boldsymbol{F}(\boldsymbol{U}_n, \boldsymbol{t})] \tag{6.49}$$

Compared to the collocation form of Algorithm-2 (i.e., Eq. 6.46), this iteration formula (i.e., Eq. 6.49) includes the Jacobian matrix of nonlinear terms, but there is still no need to calculate the inverse of the Jacobian matrix.

6.3 Numerical examples

Similar to the conventional collocation methods, the basis functions will determine the specific forms of the iterative schemes proposed in Section 5, while the selection of basis functions could be fairly flexible. The common types of basis functions used in collocation method include harmonics, polynomials, Radial Basis Functions (RBFs), Moving Least Square (MLS) functions. For the sake of brevity and clarity, only one kind of basis functions is used in this chapter to illustrate the methods of section 6, which is the first kind of Chebyshev polynomials. All the discussions made in the following are based on this type of basis functions. For clarity, we use the name Chebyshev Local Iterative Collocation (CLIC) method to denote the proposed methods with Chebyshev polynomials as basis functions. Thus as counterparts of Eqs. (6.38), (6.46), and (6.49), we have CLIC-1, CLIC-2, and CLIC-3 respectively.

With the first kind of Chebyshev polynomials as the basis functions, the trial functions are expressed as follows:

$$u_i(\xi) = \sum_{n=0}^{N} \alpha_{in} T_n(\xi), \quad \xi = \frac{2t - (t_i + t_{i+1})}{t_{i+1} - t_i} \tag{6.50}$$

where $T_n(t)$ denotes the nth Chebyshev polynomial. A rescaled time ξ is introduced herein so that the Chebyshev polynomials are defined in a valid range $-1 \leq \xi \leq 1$.

The Chebyshev polynomials of the first kind are defined by the recurrence relation

$$T_0(\xi) = 1, \quad T_1(\xi) = \xi, \quad T_{n+1}(\xi) = 2\xi T_n(\xi) - T_{n-1}(\xi) \tag{6.51}$$

or by trigonometric functions

$$T_n(\xi) = \cos[n \arccos(\xi)], \quad -1 \leq \xi \leq 1 \tag{6.52}$$

The differentiations of $T_n(\xi)$ can be obtained by the properties of Chebyshev polynomials or simply by

$$\begin{cases} LT_n(\xi) = \dfrac{n \sin[n \arccos(\xi)]}{\sqrt{1-\xi^2}} \\ LT_n(\xi)|_{\xi=\pm 1} = (\pm 1)^{n+1} n^2 \end{cases} \tag{6.53}$$

For higher order differentiations of Chebyshev polynomials, one may seek the help of Mathematica. With the properties of Chebyshev polynomials, the integration can be calculated conveniently by

$$\int T_n(\xi) \mathrm{d}\xi = \frac{1}{2}\left(\frac{T_{n+1}}{n+1} - \frac{T_{n-1}}{n-1}\right) \tag{6.54}$$

For better approximation, the collocation points are selected as the Chebyshev-Gauss-Lobatto (CGL) nodes, which are calculated from

$$\xi_m = \cos[m-1]\pi/(M-1), \quad m = 1, \ldots, M$$

6.3.1 The forced Duffing equation

Consider a Duffing-Holmes's oscillator, of which the governing equation is

$$\ddot{x} + c\dot{x} + k_1 x + k_2 x^3 = f\cos(\omega t), \quad x(0) = a, \quad \dot{x}(0) = b \tag{6.55}$$

It can be rewritten as a system of first-order differential equations

$$\begin{cases} \dot{x}_1 = x_2 \\ \dot{x}_2 = -cx_2 - k_1 x_1 - k_2 {x_1}^3 + f\cos(\omega t) \end{cases} \tag{6.56}$$

in which the variables and nonlinear terms are

$$\mathbf{X} = [x_1 \ x_2]T, \ \boldsymbol{F}(\mathbf{X}, t) = \left[x_2 - cx_2 - k_1 x_1 - k_2 {x_1}^3 + f\cos(\omega t)\right]^{\mathrm{T}} \tag{6.57}$$

Accordingly, the Jacobian matrix is

$$\boldsymbol{J}(\mathbf{X}, t) = \partial \boldsymbol{F}(\mathbf{X}, t)/\partial \mathrm{X} = \begin{bmatrix} 0 & 1 \\ -k_1 - 3k_2 {x_1}^2 & -c \end{bmatrix} \tag{6.58}$$

The Duffing-Holmes's equation is solved with the proposed CLIC methods. Both the 0th order and the 1st order correctional formulas of CLIC-2 are used. In each local domain, the iteration of the proposed methods stops when $\|\boldsymbol{U}_n - \boldsymbol{U}_{n+1}\| \le 10^{-12}$.

The results of ODE45 are used as benchmarks to measure the computational errors. The relative accuracy and the absolute accuracy of ODE45 are set to be 10^{-15}.

The chaotic phase portrait and time response curves obtained by the CLIC methods and the ODE45 are plotted in Fig. 6.1, in which the Chebyshev polynomials of the first kind are used as the basis functions. The repetitive time interval for the proposed methods is set as $\Delta t = 2$. The number of basis functions and collocation points are $N = M = 32$. Obviously, the results are consistent even for $t > 200$. For comparison the RK4 method is also used, and the step length of RK4 is selected as 0.001. The computational errors of the methods, including RK4, are recorded and plotted in Fig. 6.2.

As is shown in Fig. 6.2, the accuracy of the proposed methods is very high. Although the time intervals of the proposed methods are 2000 times larger than the time step size of RK4, the computational errors of them are 10^5 times smaller than that of RK4 on the whole. The accuracy can be improved further by including more Chebyshev polynomials in the interpolation or by reducing the size of the time intervals of the present algorithms.

For a comprehensive evaluation of the proposed methods, the indices of computational efficiency, including the iterative steps, the computational time, and the maximal errors are compared in Fig. 6.3 and Table 6.4. It can be seen that CLIC-1, is the fastest method in this case. The CLIC-2 of the first order and CLIC-3 behave very

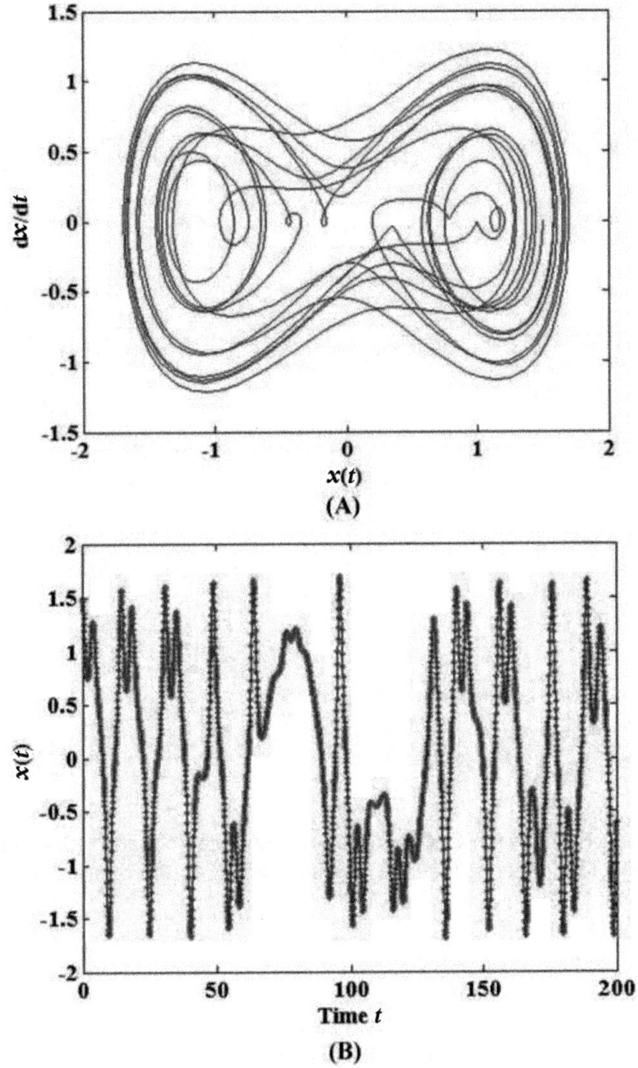

Figure 6.1 (A) Phase portrait obtained by ODE45. (B) Response curves obtained by the CLIC methods (*dots*) and ODE45 (*solid line*).

similarly in both the convergence speed and the computational speed. From Fig. 6.3 it is found that the iterative steps of CLIC-2 of the 0th order, which can also be referred to as MCPI method, is almost twice those of the CLIC-3 or the CLIC-2 of the first order and almost four times those of the CLIC-1.

Compared with RK4 method, CLIC-1 and CLIC-2 (0th order) are superior in all aspects, especially CLIC-1. The other two proposed methods, CLIC-2 (1st order) and CLIC-3, are also very efficient considering their high accuracy, although they cost a

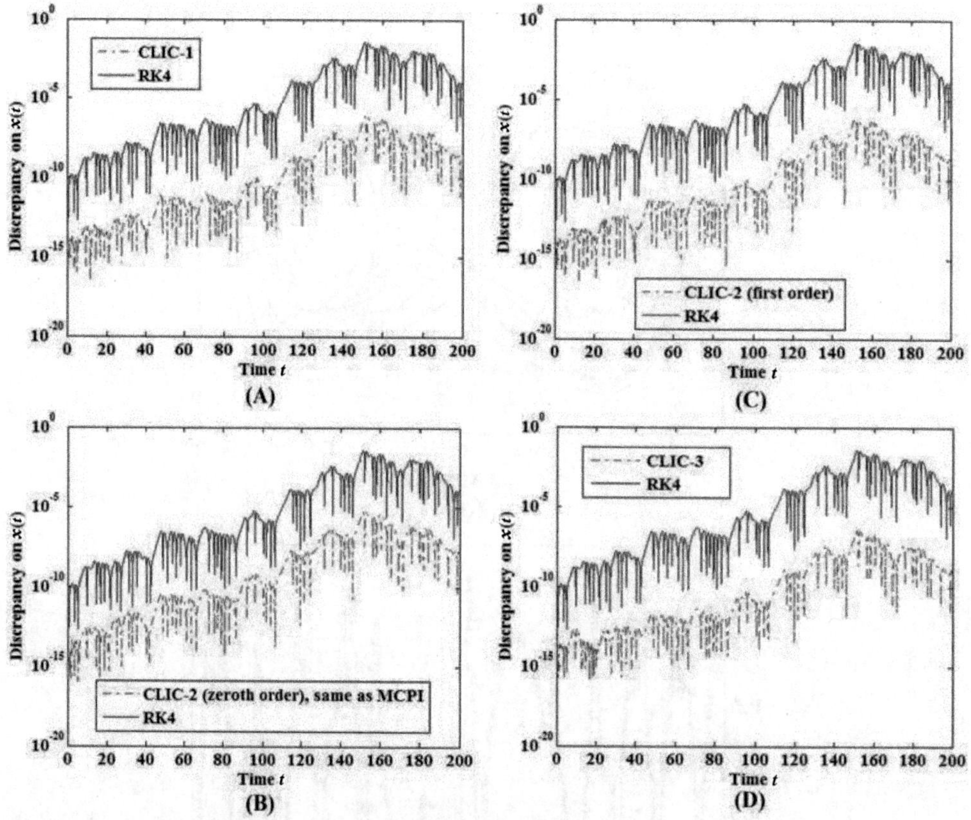

Figure 6.2 Discrepancies between CLIC methods, the RK4 method, and ODE45 on $x(t)$. (A) Results for CLIC-1. (B) Results for CLIC-2(1st order). (C) Results for CLIC-2 (0th order). (D) Results for CLIC-3.

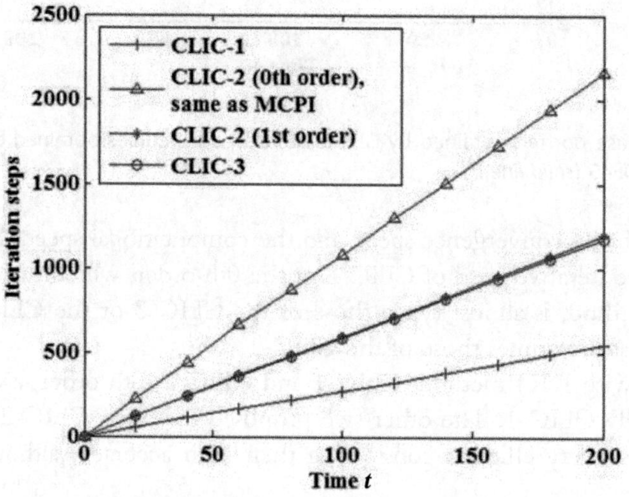

Figure 6.3 Iterative steps of CLIC methods.

Table 6.4 Performance of the methods on solving the forced Duffing equation.

Methods	CLIC-1	CLIC-2 (zeroth order) or MCPI method	CLIC-2 (first order)	CLIC-3	RK4	ODE45
Iterative steps	546	2159	1195	1174	200000	188625
Total computational time (s)	0.389253	0.713563	1.315920	1.160566	0.814100	3.705868
Maximal error	7.602e − 07	5.178e − 06	7.006e − 07	3.414e − 07	0.03761	–

bit more time than RK4 method. It is also noticed that the accuracy of the CLIC-3 is the highest while that of CLIC-2 (zeroth order) is the lowest among all the proposed methods.

6.3.2 The Lorenz system

This three-dimensional nonlinear system is expressed as

$$\left\{\begin{aligned} x &= \sigma(y-x) \\ y &= rx - y - xz \\ z &= -bz + xy \end{aligned}\right\} \tag{6.59}$$

The nonlinear term is $\boldsymbol{F}(x, y, z, t) = [\sigma(y-x) rx-y-xz-bz+xy]^{\mathrm{T}}$, and the corresponding Jacobian matrix is

$$\boldsymbol{J} = \begin{bmatrix} -\sigma & \sigma & 0 \\ r-z & -1 & -x \\ y & x & -b \end{bmatrix} \tag{6.60}$$

The Lorenz system is also solved with all the CLIC methods, the RK4 method and the ODE45. The chaotic phase portrait and time-response curves obtained by the CLIC methods and ODE45 are plotted in Fig. 6.4. The size of the time intervals of the CLIC methods is set as $\Delta t = 0.2$. The number of basis functions and collocation points are $N = M = 32$. The simulation covers a time period $[0, 40]$. For comparison, the RK4 is also used and the step length of RK4 is selected as 0.0001. The computational errors of the methods, including RK4, are recorded and plotted in Fig. 6.5.

For the reason that the CLIC methods give similar results of computational errors in the macro scope, only one figure (Fig. 6.5) is plotted here. From Fig. 6.5 it can be found that the solution by RK4 diverges from the more accurate solution by ODE45 at an early time $t = 20$, while the solution of the proposed methods are consistent with that of ODE45 until $t = 40$. It can also be seen that the computational errors of the proposed methods accumulate much more slower than those of the RK4 method. Considering that the chaotic system is sensitive to the current state, any small

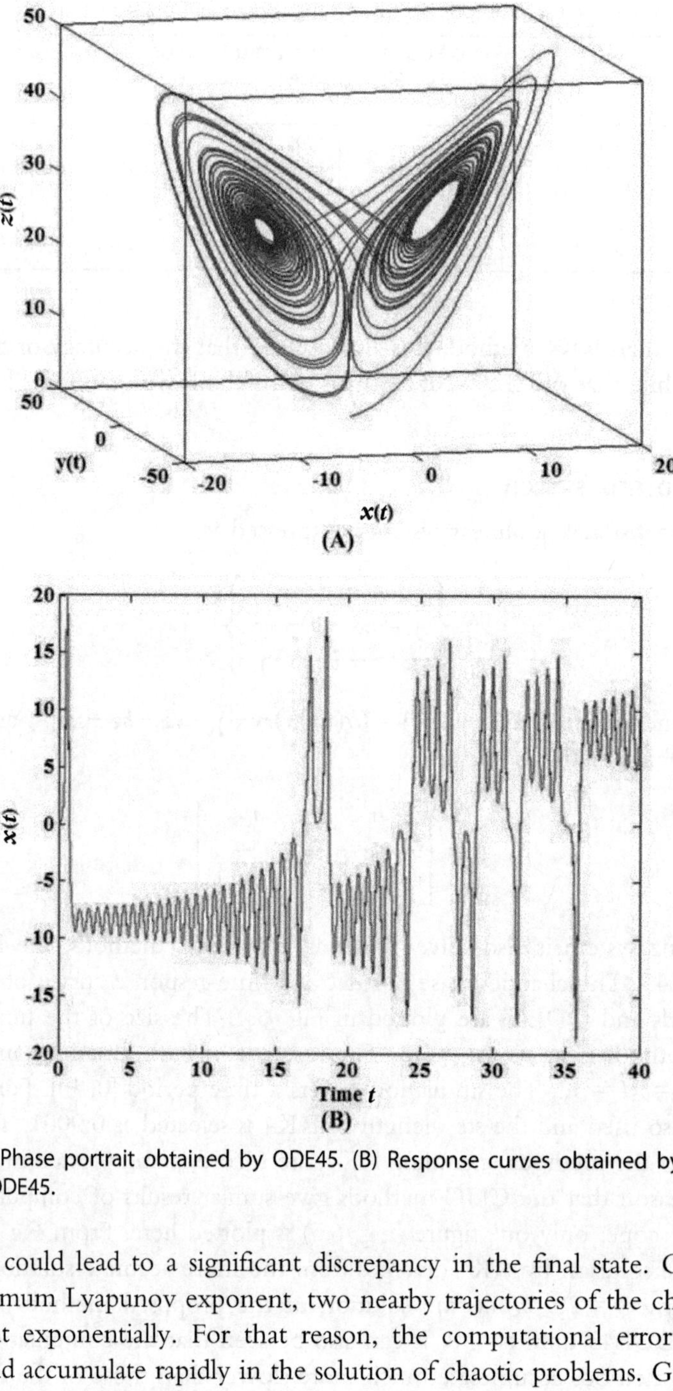

Figure 6.4 (A) Phase portrait obtained by ODE45. (B) Response curves obtained by the proposed methods and ODE45.

interruptions could lead to a significant discrepancy in the final state. Owing to the positive maximum Lyapunov exponent, two nearby trajectories of the chaotic motion will divergent exponentially. For that reason, the computational error of common methods could accumulate rapidly in the solution of chaotic problems. Given that, the result in Fig. 6.5 is fairly satisfying.

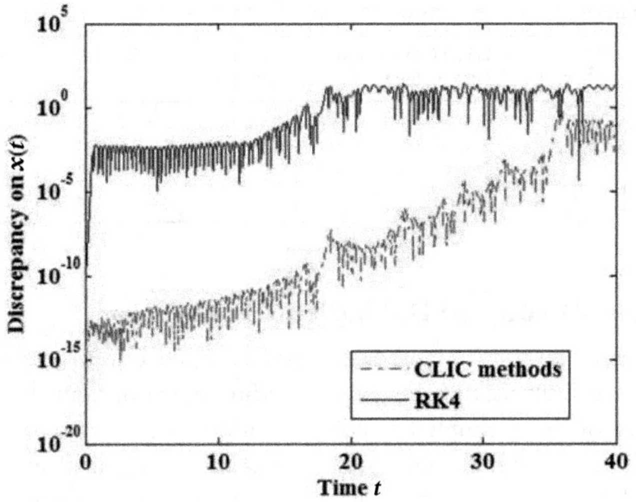

Figure 6.5 The discrepancies between the CLIC methods, the RK4 method, and ODE45 in the solution of $x(t)$.

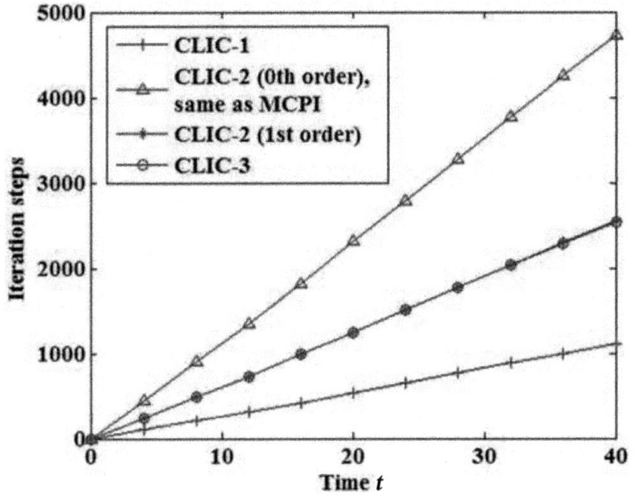

Figure 6.6 Iterative steps.

The results in Fig. 6.6 and Table 6.5 verify the analysis made in the case of the forced Duffing equation. Overall, CLIC-1 is the fastest, and CLIC-3 is the most accurate. Besides, the differences between CLIC-2 (1st order) and CLIC-3 are still trivial, which is the same as the previous example. Table 6.5 shows that the CLIC methods saves up to 77% of the computational time of ODE45, while the computational step size of CLIC is more than 1000 times larger than that of ODE45 and 2000 times larger than that of the RK4.

Table 6.5 Performance of the methods on solving the Lorenz system.

Methods	CLIC-1	CLIC-2 (0th order) or MCPI method	CLIC-2 (1st order)	CLIC-3	RK4	ODE45
Mean step size	0.2	0.2	0.2	0.2	1e-4	1.6e-04
Total computational time (sec)	1.118140	2.037919	3.854775	3.639701	1.529091	4.847664

6.3.3 The multiple coupled Duffing equations

The performance of the CLIC methods in predicting chaotic motions has been illustrated by using the examples of the forced Duffing equation and the Lorenz system. Herein, a system of three coupled Duffing equations is further studied to show that the CLIC methods are also capable of capturing the transient chaotic motion. Such coupled multiple system of Duffing equations arise as semi-discrete equations for beams, plates, and shells that are undergoing large deformations.

The governing equations of the three coupled Duffing equations considered here are expressed as

$$\begin{cases} m\ddot{x}_1 + Vx_1^3 + c\dot{x}_1 - V(x_2 - x_1)^3 - c(\dot{x}_2 - \dot{x}_1) = P\cos\omega t \\ m\ddot{x}_2 + V(x_2 - x_1)^3 + c(\dot{x}_2 - \dot{x}_1) - K_1(x_3 - x_2) = 0 \\ m\ddot{x}_3 + Vx_3^3 + c\dot{x}_3 + K_1(x_3 - x_2) = 0 \end{cases} \tag{6.61}$$

The above system can be transformed into the following autonomous system of equations:

$$\begin{cases} \dot{y}_1 = y_2 \\ \dot{y}_2 = B\cos y_3 - ky_2 - y_1^3 + (y_4 - y_1)^3 + k(y_5 - y_2) \\ \dot{y}_3 = \omega \\ \dot{y}_4 = y_5 \\ \dot{y}_5 = K_c(y_6 - y_4) - k(y_5 - y_2) - (y_4 - y_1)^3 \\ \dot{y}_6 = y_7 \\ \dot{y}_7 = -k_c(y_6 - y_4) - y_6^3 - ky_7 \end{cases} \tag{6.62}$$

where $k = c/m$, $k_c = K_l/m$, and $B = P/m$.

It is observed by the author that this oscillator possesses a transient chaotic motion when the parameters are selected as $k = 0.05$, $k_c = 10$, $\omega = 1$, and $B = 18$. Compared to chaos, the transient chaotic motion behaves irregularly at first for a limited time interval and then enters into the region of steady periodic or quasiperiodic motion. Since the period of irregular motion is often much longer than that of normal transient motion, it is not easy to determine if the investigated motion is chaotic. Although the largest Lyapunov exponent can be used as an indicator, the calculation of chaotic-transients is troublesome, and it is corrupted by the computational errors. The most straightforward approach is to simulate the responses of the coupled oscillators for long enough times. However, common numerical integration methods, such as RK4, can easily get divergent in the stage of transient chaos because of the accumulation of computational error. In that case, the simulated oscillation will behave like a real chaotic motion, which could be misleading for analyzing the dynamical behaviors of the system.

By using the CLIC methods and the RK4 method respectively, the simulated responses of the coupled Duffing oscillator are presented in Figs. 6.7 and 6.8, and the ODE45 function of MATLAB is used to provide the benchmark solution. The step size of the CLIC methods is selected to be $\Delta t = 0.5$, while a much smaller step size $\Delta t = 0.001$ is used in the RK4 method. The number of Chebyshev polynomials that are used in CLIC-1,2,3 are 21, 25, and 21, respectively. As is shown in Fig. 6.7, the results provided by the CLIC methods and ODE45 coincide very well. The enlarged partial view of the ending part of the transient chaotic motion shows that the results that were provided by the CLIC methods are of very accurate and are capable of revealing the true dynamical responses during the transient chaotic stage. As a comparison, in Fig. 6.8, the simulated responses by using the RK4 seem to be totally chaotic. From Fig. 6.8B it can be seen that although the results of the RK4 method and ODE45 stay the same at first, they are divergent obviously at the ending part of the transient chaotic motion. Thereafter, the RK4 method was tested by using step sizes of both higher and lower magnitudes. All the simulations carried on MATLAB show that the classical RK4 method fails to predict the transient chaotic motion of the system of equations.

In Fig. 6.9 discrepancies between the results of CLIC, the RK4 method, and ODE45 are plotted. It is shown that the approximated $y1(t)$ given by CLIC matches much better with that of ODE45 than that of the RK4 method from the very beginning. Although the discrepancy between CLVI and ODE45 keeps increasing during the transient chaotic motion, there is no obvious divergence throughout the simulation using these two kinds of methods. Just like the true chaos, the transient chaotic motion is sensitive to small disturbance. Despite the fact that the relatively accurate solution of ODE45 is used as a benchmark at here, it does not mean that the ODE45 is absolutely accurate. Actually, a simple conserved two body problem modeled in Hamiltonian was used by the author to test the accuracy of ODE45 and the CLIC methods in terms of the computational error of total energy, and it was found that the total energy error of the CLIC method is smaller than that of ODE45. Thus the

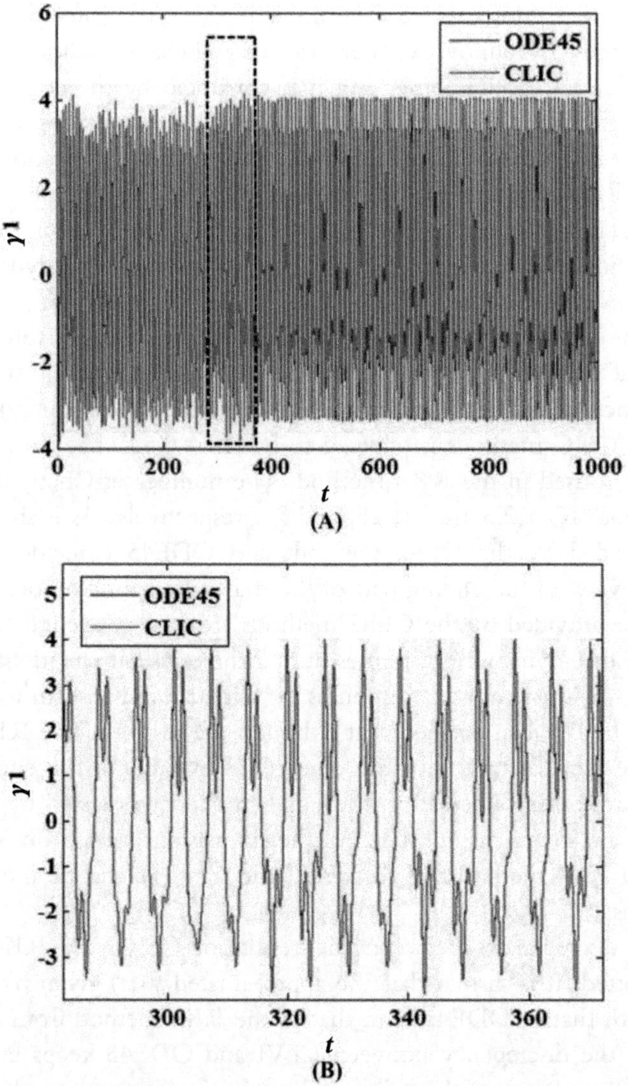

Figure 6.7 (A) The transient chaotic motion obtained by ODE45 and CLIC methods. (B) Enlarged partial view.

increasing discrepancy between the CLIC and the ODE45 does not necessarily indicate the deterioration of CLIC. The good agreement between CLIC and ODE45 is enough to verify the high accuracy of the CLIC methods and its capability of predicting transient chaotic motion.

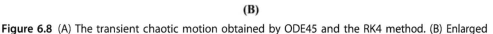

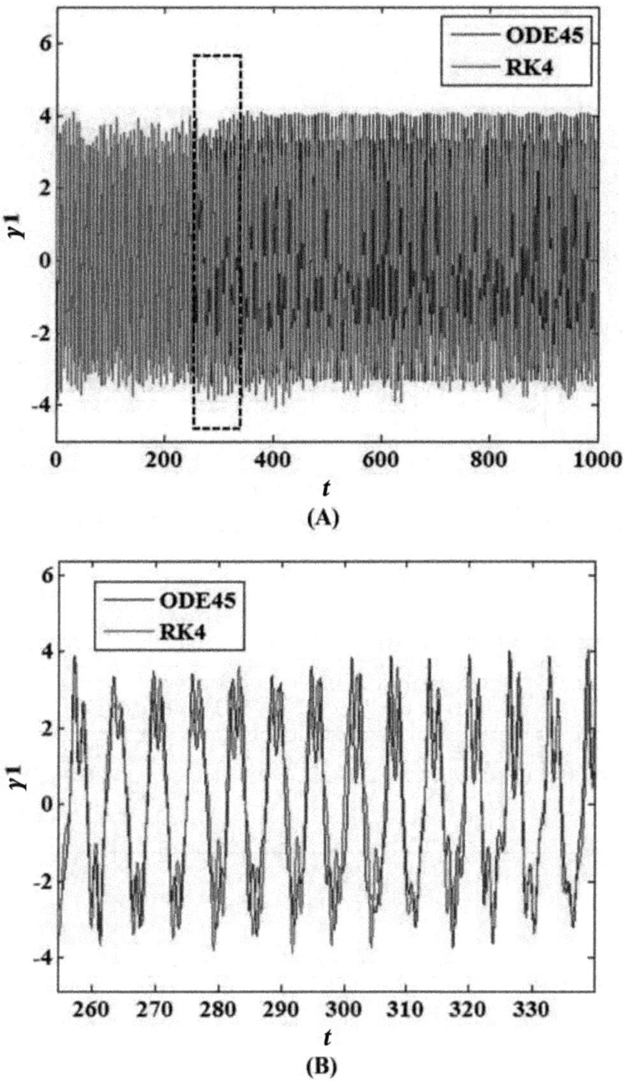

Figure 6.8 (A) The transient chaotic motion obtained by ODE45 and the RK4 method. (B) Enlarged partial view.

A brief illustration of the computational efficiency is made in Table 6.6. As is shown, the computational steps of ODE45 is almost 100–400 times higher than those of the total iterative steps of the CLIC methods. If one merely considers the mean length of time interval in one step, the time interval of CLIC is 2000 times larger than that of ODE45. It is also shown that the CLIC methods save up to 89% of the computational time of

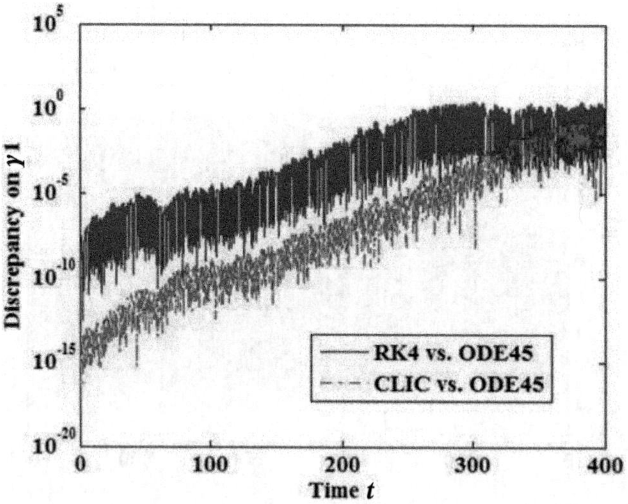

Figure 6.9 The discrepancies between the CLIC methods, the RK4 method, and ODE45 in the solution of $y1(t)$.

Table 6.6 Performance of the methods used in solving the coupled Duffing oscillator.

Methods	CLIC-1	CLIC-2 (zeroth order), or MCPI method	CLIC-2 (First Order)	CLIC-3	RK4	ODE45
Iterative steps	4034	16955	9173	9076	—	1633501
Total computational time (s)	6.174757	26.702937	15.410866	10.594174	—	59.527724

ODE45. Since the RK4 method cannot predict the transient chaotic motion correctly, the computational steps and total computational time of it are not displayed at here.

6.4 Conclusion

We presented a class of iterative semi-analytical algorithms for predicting long-term non-periodic responses of strongly nonlinear dynamical systems, based on a local application in a time interval $t_{i+1} - t_i$, of the Variation Iteration Method, wherein variations of the nonlinear term $\boldsymbol{F}(\boldsymbol{x}, t)$ are retained. We labeled this the improved sequential Local Variation Iteration Method as LVIM and then further modified it into Algorithm-1, Algorithm-2, and Algorithm-3, respectively, wherein the variational iterative formulas: (1) do not contain the Lagrange multipliers; (2) the Lagrange multipliers are approximated by polynomial series; and (3) the Lagrange multipliers are approximated by exponential functions. The weak-form solutions of the iterative formulas of Algorithms-1,2, and 3, are derived respectively, by using Chebyshev polynomials as trial functions in each time interval $(t_{i+1} - t_i)$. We denoted these Chebyshev polynomial–based collocation methods as CLIC-1, CLIC-2, and CLIC-3, respectively. It is shown in this chapter that CLIC-2, with zeroth-order polynomial approximation to the Lagrange multipliers, is identical to the Modified Chebyshev Picard Iteration Method.

In solving strongly nonlinear problems with non-periodic responses, such as the forced Duffing oscillator, the Lorenz system and the system of multiple coupled Duffing equations, we have shown that the CLIC methods, especially CLIC-1, are far superior in computational accuracy as well as time, as compared to classical RK4 method. It is also shown that the CLIC methods are much more efficient than ODE45 and can save a lot of computational time in accurately predicting long term non-periodic motions of strongly nonlinear systems. In the example of three coupled Duffing equations, it is found by the authors that there exists transient chaotic motion with a certain configuration of parameters. This transient chaotic motion can be successfully predicted by both ODE45 and the CLIC methods, while the RK4 method fails to predict it and gives a chaos-like response throughout the simulation time.

In sum, the proposed methods in this chapter are of high accuracy and efficiency. They are readily applicable to investigate long term dynamical behaviors of a broad class of strongly nonlinear systems, regardless of whether the system is high dimensional or low dimensional, autonomous or non-autonomous. The underlying applications of the proposed methods cover many different areas, such as the vibration of aeroelastic structures, the perturbed orbital propagations, and even the stochastic problems. By slightly modifying the code of the CLIC methods in this chapter, one can further use them to solve boundary value problems.

References

[1] S.N. Atluri, Methods of Computer Modeling in Engineering & the Sciences, I, Tech Science Press, Forsyth, 2005.

[2] L. Dong, A. Alotaibi, S.A. Mohiuddine, S.N. Atluri, Computational methods in engineering: a variety of primal & mixed methods, with global & local interpolations, for well-posed or Ill-posed BCs, Computer Modeling in Engineering & Sciences 99 (1) (2014) 1–85.

[3] C.S. Liu, W. Yeih, C.L. Kuo, S.N. Atluri, A scalar homotopy method for solving an over/under determined system of non-linear algebraic equations, Computer Modeling in Engineering and Sciences 53 (1) (2009) 47–71.

[4] C.S. Liu, S.N. Atluri, An iterative algorithm for solving a system of nonlinear algebraic equations, $\mathbf{F}(\mathbf{x}) = 0$, using the system of ODEs with an optimum α in $\mathrm{x} = \lambda[\alpha\mathbf{F} + (1\text{-}\alpha)\mathbf{B}^T\mathbf{F}]$, $B_{ij} = \partial F_i/\partial x_j$, Computer Modeling in Engineering and Sciences 73 (4) (2011) 395–431.

[5] J. He, Variational iteration method: a kind of non-linear analytical technique: some examples, International Journal of Non-linear Mechanics 34 (1999) 699–708.

[6] J. He, Homotopy perturbation method for solving boundary value problems, Physics Letters. A 350 (1–2) (2006) 87–88.

[7] T. Fukushima, Picard iteration method, Chebyshev polynomial approximation, and global numerical integration of dynamical motions, The Astronomical Journal 113 (5) (1997) 1909–1914.
[8] G. Adomian, A review of the decomposition method in applied mathematics, Journal of Mathematical Analysis and Applications 135 (1988) 501–544.
[9] X. Wang, S.N. Atluri, A unification of the concepts of the variational iteration, Adomian decomposition and Picard iteration methods; and a local variational iteration method, Computer Modelling in Engineering and Sciences 111 (6) (2016) 567–585.
[10] M. Jang, C. Chen, Y. Liy, On solving the initial-value problems using the differential transformation method, Applied Mathematics and Computation 115 (2000) 145–160.
[11] M. Inokuti, H. Sekine, T. Mura, General use of the Lagrange multiplier in nonlinear mathematical physics, in: S. Nemat-Nasser (Ed.), Variational Method in the Mechanics of Solids, Pergamon Press, Oxford, 1978, pp. 156–162.
[12] X. Bai, J.L. Junkins, Modified Chebyshev-Picard iteration methods for solution of boundary value problems, The Journal of Astronautical Sciences 58 (4) (2011) 615–642.

CHAPTER 7

Application of the Local Variational Iteration Method in Orbital Mechanics

SUMMARY

In orbital mechanics, the most concerned aspects of numerical methods are accuracy and efficiency. To obtain highly accurate results, the step size has to be very small in traditional methods that are based on finite-difference approximations. The collocation method is superior to traditional methods in that the solution can be semianalytically approximated in a much larger interval, and only a few coefficients of basis functions need to be saved, rather than a large amount of discrete data [1]. By choosing appropriate collocating points in the domain of either the initial value problem (IVP) or the two-point boundary value problem (TPBVP), the governing equations and the boundary conditions of the problem can be converted into algebraic equations governing the values at the collocation points. The applications of collocation method on orbital mechanics have been quite successful. The drawback is that researchers have to construct nonlinear algebraic equations and calculate the inverse of Jacobian matrix, which could be troublesome. A simpler idea is introduced by using the Picard iteration method to produce a sequence of correctional formulas, and then using a set of orthogonal basis functions to approximate the solution. This leads to a very simple algorithm without inverting a matrix, and is named as the modified Chebyshev-Picard iteration (MCPI) method [2–5].

In this chapter, we show that the local variational iteration method (LVIM) developed by the current authors can be used as a highly efficient and accurate alternative to traditional numerical methods. This method is applied on orbit propagation as well as Lambert's problem. The convergence as well as the accuracy and efficiency are verified through these examples.

In addition, for solving TPBVPs, a novel quasi-linearization method [6] is introduced to transform TPBVPs into initial value problems (IVPs). Particularly, in conservative problems, a so-called fish-scales-growing method (FSGM) is further proposed. Compared with the multiple shooting method and the modified simple shooting method, the proposed FSGM is based on the principle of least action for a Hamiltonian system, instead of using the Newton's iterative algorithm or a homotopy asymptotic strategy. In a conservative system, the solution of a TPBVP is determined

Computational Methods for Nonlinear Dynamical Systems
DOI: https://doi.org/10.1016/B978-0-323-99113-1.00007-8

by the principle of least action. Using the FSGM, a piecewise solution obtained by solving multiple TPBVPs is used as an approximation of the true solution. Then by changing the boundary conditions of the multiple TPBVPs, the piecewise solution will iteratively evolve in a pattern that looks like the growing of fish scales. In this pattern, the action along the piecewise trajectory is guaranteed to decrease. Theoretically, the piecewise solution will eventually approach the true solution in a conservative system. In the FSGM, the TPBVP to be solved is converted into multiple TPBVPs that can be solved much more easily. Since the convergence of the piecewise solutions obtained by FSGM is theoretically guaranteed by the principle of least action, the initial guess can be selected rather arbitrarily, although different initial guesses may affect the convergence speed. Moreover, in each iteration step of the FSGM, the multiple TPBVPs are independent of each other. Thus, it can be conveniently coded for parallel computation to enhance the computational efficiency. At last for each solution of the multiple TPBVPs, only one time instant and its corresponding positions and velocities need to be recorded for the next iteration, so the FSGM is very efficient and memory-saving.

7.1 Local variational iteration method and quasi-linearization method

7.1.1 Local variational iteration method

As is introduced in previous chapters, for solving a system of first-order differential equations

$$\dot{\boldsymbol{x}} = \boldsymbol{f}(\boldsymbol{x}, \tau), \quad \tau \in [t_0, t] \tag{7.1}$$

the original variational iteration method (VIM) approximates the solution with an initial approximation $\boldsymbol{x}_0(\tau)$, and the correctional iterative formula is

$$\boldsymbol{x}_{n+1}(t) = \boldsymbol{x}_n(\tau)|_{\tau=t} + \int_{t_0}^{t} \lambda(\tau)\{\dot{x}_n(\tau) - \boldsymbol{f}[\boldsymbol{x}_n(\tau), \tau]\}\mathrm{d}\tau, \quad t_0 \leq \tau \leq t \tag{7.2}$$

where $\boldsymbol{\lambda}(\tau)$ is a matrix of Lagrange multipliers to be determined. Eq. (7.2) indicates that the $(n+1)$th correction to the analytical solution $\boldsymbol{x}_{n+1}$ involves the addition of $\boldsymbol{x}_n$ and a feedback-weighted optimal error in the solution $\boldsymbol{x}_n$ up to the current time t. $\boldsymbol{\lambda}(\tau)$ can be optimally determined by making the right-hand side of Eq. (7.2) stationary about $\delta\boldsymbol{x}_n(\tau)$, which is the variation of $\boldsymbol{x}_n(\tau)$. The stationary condition is obtained as

$$\delta\boldsymbol{x}_n(\tau)|_{\tau=t} + \boldsymbol{\lambda}(\tau)\delta\boldsymbol{x}_n(\tau)|_{\tau=t_0}^{\tau=t} - \int_{t_0}^{t}\left[\dot{\lambda}(\tau) + \lambda(\tau)\frac{\partial \boldsymbol{f}(\boldsymbol{x}_n, \tau)}{\partial \boldsymbol{x}_n}\right]\delta\boldsymbol{x}_n(\tau)\mathrm{d}\tau = 0 \tag{7.3}$$

Then we collect the terms including $\delta \boldsymbol{x}_n(\tau)|_{\tau=t}$ and $\delta \boldsymbol{x}_n(\tau)$. Noting that the boundary value of $\boldsymbol{x}(\tau)$ at $\tau = t_0$ is prescribed, we have $\delta \boldsymbol{x}_n(\tau)\big|_{\tau=t_0} = 0$. Thus Eq. (7.3) leads to the constraints of $\boldsymbol{\lambda}(\tau)$ as following:

$$\begin{cases} \delta \boldsymbol{x}_n(\tau)|_{\tau=t}:\boldsymbol{I} + \boldsymbol{\lambda}(\tau)|_{\tau=t} = \boldsymbol{0} \\ \delta \boldsymbol{x}_n(\tau):\dot{\boldsymbol{\lambda}}(\tau) + \boldsymbol{\lambda}(\tau)\dfrac{\partial \boldsymbol{f}(\boldsymbol{x}_n, \tau)}{\partial \boldsymbol{x}_n} = \boldsymbol{0} \end{cases} \tag{7.4}$$

The VIM leads to two modifications. First, by differentiating Eq. (7.2) and using the constraints on $\boldsymbol{\lambda}(\tau)$ in Eq. (7.4), we have

$$\begin{aligned} \frac{\mathrm{d}\boldsymbol{x}_{n+1}}{\mathrm{d}t} &= \frac{\mathrm{d}\boldsymbol{x}_n}{\mathrm{d}t} + \boldsymbol{\lambda}(\tau)|_{\tau=t}\left[\frac{\mathrm{d}\boldsymbol{x}_n}{\mathrm{d}t} - \boldsymbol{f}(\boldsymbol{x}_n, t)\right] + \int_{t_i}^{t} \frac{\partial \lambda}{\partial t}[\dot{\boldsymbol{x}}_n - \boldsymbol{f}(\boldsymbol{x}_n, \tau)]\mathrm{d}\tau \\ &= \boldsymbol{f}(\boldsymbol{x}_n, t) + \int_{t_i}^{t} \frac{\partial \lambda}{\partial t}[\dot{\boldsymbol{x}}_n - \boldsymbol{f}(\boldsymbol{x}_n, \tau)]\mathrm{d}\tau \end{aligned} \tag{7.5}$$

It was proved that the generalized Lagrange multipliers $\boldsymbol{\lambda}(\tau)$ happen to be the solution $\overline{\boldsymbol{\lambda}}(\tau)$ of the following ordinary differential equations [7]:

$$\begin{cases} \boldsymbol{I} + \overline{\boldsymbol{\lambda}}(t)|_{t=\tau} = \boldsymbol{0} \\ \dfrac{\partial \overline{\boldsymbol{\lambda}}(t)}{\partial t} - \boldsymbol{J}(t)\overline{\boldsymbol{\lambda}}(t) = \boldsymbol{0} \end{cases} \tag{7.6}$$

where $\boldsymbol{J}(t) = \partial \boldsymbol{f}(\boldsymbol{x}_n, \tau)/\partial \boldsymbol{x}_n$. Therefore Eq. (7.5) can be further simplified by using the equivalence between $\boldsymbol{\lambda}(\tau)$ and $\overline{\boldsymbol{\lambda}}(\tau)$, which leads to

$$\frac{\mathrm{d}\boldsymbol{x}_{n+1}}{\mathrm{d}t} - \boldsymbol{J}(\boldsymbol{x}_n, t)\boldsymbol{x}_{n+1} = \boldsymbol{f}(\boldsymbol{x}_n, t) - \boldsymbol{J}(\boldsymbol{x}_n, t)\boldsymbol{x}_n \tag{7.7}$$

In this way, the Lagrange multipliers $\boldsymbol{\lambda}(\tau)$ are completely eliminated in the correctional formula of the VIM. The preceding formula can be regarded as a generalization of Newton's method in function space.

Another modification was made to the VIM by approximating $\boldsymbol{\lambda}(\tau)$ in Taylor series instead of eliminating $\boldsymbol{\lambda}(\tau)$. The Taylor series approximation of $\boldsymbol{\lambda}(\tau)$ can be readily obtained from the Eq. (7.4) in the following form:

$$\boldsymbol{\lambda}(\tau) \approx \boldsymbol{T}_0[\boldsymbol{\lambda}] + \boldsymbol{T}_1[\boldsymbol{\lambda}](\tau - t) + \ldots + \boldsymbol{T}_K[\boldsymbol{\lambda}](\tau - t)^K \tag{7.8}$$

where $\boldsymbol{T}_K[\boldsymbol{\lambda}]$ is the kth-order differential transformation [8] of $\boldsymbol{\lambda}(\tau)$, that is,

$$\boldsymbol{T}_k[\boldsymbol{\lambda}] = \frac{1}{k!}\frac{\mathrm{d}^k \boldsymbol{\lambda}(\tau)|_{\tau=t}}{\mathrm{d}\tau^k} \tag{7.9}$$

By using Eq. (7.4), $\boldsymbol{T}_k[\boldsymbol{\lambda}]$ can be determined in an iterative way:

$$\boldsymbol{T}_0[\boldsymbol{\lambda}] = diag[-1, -1, \ldots], \quad \boldsymbol{T}_{k+1}[\boldsymbol{\lambda}] = -\frac{\boldsymbol{T}_k[\boldsymbol{\lambda J}]}{k+1}, \quad 0 \leq k \leq K+1 \tag{7.10}$$

Let $\mathbf{G} = \dot{\boldsymbol{x}}_n(\tau) - \boldsymbol{f}[\boldsymbol{x}_n(\tau), \tau]$, by substituting Eq. (7.8) into Eq. (7.2), we have

$$\boldsymbol{x}_{n+1}(t) = \boldsymbol{x}_n(t) + \int_{t_0}^{t} \{\boldsymbol{T}_0[\boldsymbol{\lambda}] + \boldsymbol{T}_1[\boldsymbol{\lambda}](\tau - t) + \ldots + \boldsymbol{T}_K[\boldsymbol{\lambda}](\tau - t)^K\}\mathbf{G}\mathrm{d}\tau \tag{7.11}$$

Considering that $\boldsymbol{T}_k[\boldsymbol{\lambda}]$, $k = 0, 1, \ldots K$, are functions of $\boldsymbol{x}_n(t)$ and t, this expression can be rewritten in another form:

$$\boldsymbol{x}_{n+1}(t) = \boldsymbol{x}_n(t) + \boldsymbol{A}_0(t)\int_{t_0}^{t} \mathbf{G}\mathrm{d}\tau + \boldsymbol{A}_1(t)\int_{t_0}^{t} \tau\mathbf{G}\mathrm{d}\tau + \ldots + \boldsymbol{A}_K(t)\int_{t_0}^{t} \tau^K\mathbf{G}\mathrm{d}\tau \tag{7.12}$$

where the coefficient matrices $\boldsymbol{A}_K(t)$, $k = 0, 1, \ldots K$, are combinations of $\boldsymbol{T}_k[\boldsymbol{\lambda}]$ and t.

Obviously, if $\boldsymbol{\lambda}(\tau)$ is approximated very simply with $\boldsymbol{T}_0[\boldsymbol{\lambda}]$, the correctional formula of VIM will degrade to the Picard iteration formula, which is written as follows:

$$\boldsymbol{x}_{n+1}(t) = \boldsymbol{x}_n(t) + \int_{t0}^{t} \boldsymbol{f}[\boldsymbol{x}_n(\tau), \tau]\mathrm{d}\tau. \tag{7.13}$$

Since Eq. (7.12) can be regarded as a higher order Picard iteration that converges much faster than conventional Picard iteration, we call it Feedback-Accelerated Picard Iteration (FAPI) in the following contents of this chapter.

7.1.2 Quasi-linearization method

The proposed LVIM is both efficient and accurate for solving IVPs. In order to take this advantage we designed a quasi-linearization and LVIM (QLVIM) to solve TPBVPs. The derivation of QLVIM involves three stages: the quasi-linearization of TPBVP, the transformation of linear TPBVPs into IVPs, and the integration of IVPs by using the LVIM.

7.1.2.1 Quasi-linearization of two-point boundary value problem

Generally, we consider nonlinear second-order differential equations

$$r'' = f(t, r, r') \tag{7.14}$$

subjected to the boundary conditions

$$r(t_0) = r_0, r(t_f) = r_f \tag{7.15}$$

where r' and r'' represent $\mathrm{d}r/\mathrm{d}t$ and $\mathrm{d}^2r/\mathrm{d}t^2$, respectively.

Eq. (7.14) can be rewritten as

$$\phi(t, r, r', r'') = r'' - f(t, r, r') = 0 \tag{7.16}$$

Denote the nth and $(n+1)$th iterations by r_n and r_{n+1}, respectively, and require that for both iterations, $\phi = 0$. For the nth iteration this gives

$$r''_n = f\left(t, r_n, r'_n\right) \tag{7.17}$$

For the $(n+1)$th iteration, we get

$$\phi\left(t, \boldsymbol{r}_{n+1}, \boldsymbol{r}'_{n+1}, \boldsymbol{r}''_{n+1}\right) = \phi\left(t, \boldsymbol{r}_n, \boldsymbol{r}'_n, \boldsymbol{r}''_n\right) + \left(\frac{\partial \phi}{\partial \boldsymbol{r}}\right)_n (\boldsymbol{r}_{n+1} - \boldsymbol{r}_n) + \left(\frac{\partial \phi}{\partial \boldsymbol{r}'}\right)_n \left(\boldsymbol{r}'_{n+1} - \boldsymbol{r}'_n\right)$$

$$+ \left(\frac{\partial \phi}{\partial \boldsymbol{r}''}\right)_n (\boldsymbol{r}''_{n+1} - \boldsymbol{r}''_n) + \ldots = 0 \tag{7.18}$$

$$-\left(\frac{\partial \boldsymbol{f}}{\partial \boldsymbol{r}}\right)_n (\boldsymbol{r}_{n+1} - \boldsymbol{r}_n) - \left(\frac{\partial \boldsymbol{f}}{\partial \boldsymbol{r}'}\right)_n \left(\boldsymbol{r}'_{n+1} - \boldsymbol{r}'_n\right) + \left(\boldsymbol{r}''_{n+1} - f\left(t, \boldsymbol{r}_n, \boldsymbol{r}'_n\right)\right) + \ldots = 0 \tag{7.19}$$

Substituting $\boldsymbol{r}''_n$ from Eq. (7.17) into Eq. (7.19) and omitting second- and higher-order terms, we get

$$\boldsymbol{r}''_{n+1} - \boldsymbol{K}_n \boldsymbol{r}'_{n+1} - \boldsymbol{J}_n \boldsymbol{r}_{n+1} = \boldsymbol{f}_n(t) - \boldsymbol{K}_n \boldsymbol{r}'_n - \boldsymbol{J}_n \boldsymbol{r}_n \tag{7.20}$$

where $\boldsymbol{K}_n = \left(\frac{\partial \boldsymbol{f}}{\partial \boldsymbol{r}'}\right)_n$, $\boldsymbol{J}_n = \left(\frac{\partial \boldsymbol{f}}{\partial \boldsymbol{r}}\right)_n$, $\boldsymbol{f}_n(t) = \boldsymbol{f}\left(t, \boldsymbol{r}_n, \boldsymbol{r}'_n, \boldsymbol{r}''_n\right)$. The boundary conditions for Eq. (7.20) are

$$\boldsymbol{r}_{n+1}(t_0) = \boldsymbol{r}_0 \tag{7.21a}$$

$$\boldsymbol{r}_{n+1}(t_f) = \boldsymbol{r}_f \tag{7.21b}$$

Thus the nonlinear TPBVP in Eqs. (7.14) and (7.15) degenerate to an iterative sequence of linear TPBVPs in Eqs. (7.20) and (7.21), which are much easier to solve.

7.1.2.2 Transformation of linear two-point boundary value problem to initial value problems

To solve the linear TPBVPs in Eqs. (7.20) and (7.21), we use the method of superposition to transform linear TPBVP into IVPs. Assume that

$$\boldsymbol{r}_{n+1} = \boldsymbol{V} + s_x \cdot \boldsymbol{W}_x + s_y \cdot \boldsymbol{W}_y + s_z \cdot \boldsymbol{W}_z = \boldsymbol{V} + \boldsymbol{W} \cdot \boldsymbol{s} \tag{7.22}$$

where $\boldsymbol{W} = \left[\boldsymbol{W}_x, \boldsymbol{W}_y, \boldsymbol{W}_z\right]$ and

$$\boldsymbol{s} = \left(s_x, s_y, s_z\right)^{\mathrm{T}} = \boldsymbol{r}'(t_0) \tag{7.23}$$

The unknown constant $\boldsymbol{s}$ is identified as the missing initial slope. $\boldsymbol{V}, \boldsymbol{W}_x, \boldsymbol{W}_y, \boldsymbol{W}_z$ are column vectors of 3×1. Their physical meanings will be discussed later. Substituting Eq. (7.22) into Eq. (7.20), we obtain

$$(\boldsymbol{V}'' + \boldsymbol{W}'' \cdot \boldsymbol{s}) - \boldsymbol{K}_n(\boldsymbol{V}' + \boldsymbol{W}' \cdot \boldsymbol{s}) - \boldsymbol{J}_n(\boldsymbol{V} + \boldsymbol{W} \cdot \boldsymbol{s}) = \boldsymbol{f}_n(t) - \boldsymbol{K}_n \boldsymbol{r}'_n - \boldsymbol{J}_n \boldsymbol{r}_n \tag{7.24}$$

Particularly, Eq. (7.24) can be separated into two differential equations:

$$\boldsymbol{V}'' - \boldsymbol{K}_n \boldsymbol{V}' - \boldsymbol{J}_n \boldsymbol{V} = \boldsymbol{f}_n(t) - \boldsymbol{K}_n \boldsymbol{r}'_n - \boldsymbol{J}_n \boldsymbol{r}_n \tag{7.25}$$

$$\boldsymbol{W}'' - \boldsymbol{K}_n \boldsymbol{W}' - \boldsymbol{J}_n \boldsymbol{W} = 0 \tag{7.26}$$

Similarly, substituting Eq. (7.22) into Eqs. (7.21a) and (7.23), we obtain

$$\boldsymbol{r}(t_0) = \boldsymbol{V}(t_0) + s_x \cdot \boldsymbol{W}_x(t_0) + s_y \cdot \boldsymbol{W}_y(t_0) + s_z \cdot \boldsymbol{W}_z(t_0) = \boldsymbol{r}_0 \tag{7.27}$$

$$\boldsymbol{r}'(t_0) = \boldsymbol{V}'(t_0) + s_x . \boldsymbol{W}'_x(t_0) + s_y \cdot \boldsymbol{W}'_y(t_0) + s_z \cdot \boldsymbol{W}'_z(t_0) = \left(s_x, s_y, s_z\right)^{\mathrm{T}} \tag{7.28}$$

As a special case, Eqs. (7.27) and (7.28) can be decomposed as follows:

$$\boldsymbol{V}(t_0) = \boldsymbol{r}_0, \boldsymbol{W}_x(t_0) = \boldsymbol{W}_y(t_0) = \boldsymbol{W}_z(t_0) = (0, 0, 0)^{\mathrm{T}} \tag{7.29}$$

$$\boldsymbol{V}'(t_0) = (0,0,0)^{\mathrm{T}}, \boldsymbol{W}'_x(t_0) = (1,0,0)^{\mathrm{T}}, \boldsymbol{W}'_y(t_0) = (0,1,0)^{\mathrm{T}}, \boldsymbol{W}'_z(t_0) = (0,0,1)^{\mathrm{T}} \tag{7.30}$$

Thus Eqs. (7.20), (7.21a), and (7.23) are separated into four IVPs.

IVP1:

$$\boldsymbol{V}'' - \boldsymbol{K}_n \boldsymbol{V}' - \boldsymbol{J}_n \boldsymbol{V} = \boldsymbol{f}_n(t) - \boldsymbol{K}_n \boldsymbol{r}'_n - \boldsymbol{J}_n \boldsymbol{r}_n, \boldsymbol{V}(t_0) = r_0, \boldsymbol{V}'(t_0) = (0,0,0)^{\mathrm{T}} \tag{7.31}$$

IVP2:

$$\boldsymbol{W}''_x - \boldsymbol{K}_n \boldsymbol{W}'_x - \boldsymbol{J}_n \boldsymbol{W}_x = 0, \boldsymbol{W}_x(t_0) = (0,0,0)^{\mathrm{T}}, \boldsymbol{W}'_x(t_0) = (1,0,0)^{\mathrm{T}} \tag{7.32}$$

IVP3:

$$\boldsymbol{W}''_y - \boldsymbol{K}_n \boldsymbol{W}''_y - \boldsymbol{J}_n \boldsymbol{W}_y = 0, \boldsymbol{W}_y(t_0) = (0,0,0)^{\mathrm{T}}, \boldsymbol{W}'_y(t_0) = (0,1,0)^{\mathrm{T}} \tag{7.33}$$

IVP4:

$$\boldsymbol{W}''_z - \boldsymbol{K}_n \boldsymbol{W}'_z - \boldsymbol{J}_n \boldsymbol{W}_z = 0, \boldsymbol{W}_z(t_0) = (0,0,0)^{\mathrm{T}}, \boldsymbol{W}'_z(t_0) = (0,0,1)^{\mathrm{T}} \tag{7.34}$$

Now if we examine the form of Eq. (7.22) again, its physical meaning will be obvious. $\boldsymbol{V}$ is the state vector of the linear system Eq. (7.25) with initial values $\boldsymbol{r}_0$ and zero initial velocity. $\boldsymbol{W}_x, \boldsymbol{W}_y, \boldsymbol{W}_z$ are the state vectors of the homogeneous system Eq. (7.26) starting from the original point with unit velocity in x direction, y direction, and z direction, respectively. Since we are solving a linear system in this step, the real trajectory of the system is the linear combination of these four component trajectories, and the coefficients (s_x, s_y, s_z) are components of initial velocity in three directions. On the other hand, from the perspective of linear differential equation theory, the form of Eq. (7.22) is also very easy to understand. As a matter of fact, $\boldsymbol{V}$ is a particular solution of Eq. (7.25), and $\boldsymbol{W}_x, \boldsymbol{W}_y, \boldsymbol{W}_z$ are three nontrivial and distinct solutions that are obtained from the homogeneous form, Eq. (7.26), of the original Eq. (7.25). According to the superposition principle, any solution of Eq. (7.25) can be represented as the linear combination of $\boldsymbol{V}, \boldsymbol{W}_x, \boldsymbol{W}_y, \boldsymbol{W}_z$, that is, Eq. (7.22). However, the value of $\boldsymbol{s}$ is still unknown. By substituting Eq. (7.22) into Eq. (7.23), we obtain

$$\boldsymbol{s}_{new} = \boldsymbol{W}(tf)^{-1} \cdot \left[\boldsymbol{r}_f - \boldsymbol{V}(t_f)\right] \tag{7.35}$$

where $\boldsymbol{W}(t_f) = \left[\boldsymbol{W}_x(t_f), \boldsymbol{W}_y(t_f), \boldsymbol{W}_z(t_f)\right]$. Eq. (7.35) gives the iterative formula of initial slope $\boldsymbol{s}$.

7.2 Perturbed orbit propagation

In the following, three scenarios, including a low earth orbit (LEO), a high eccentric orbit (HEO), and a geosynchronous earth orbit (GEO), are used to test the proposed method. These three orbit regimes represent most cases in orbit determination and propagation and thus make it possible to evaluate the proposed method comprehensively. The parameters and initial conditions of the three orbit regimes are listed in Table 7.1. The same configurations are used by Berry and Healy [9], where the fourth-order Runge-Kutta method and the Gauss-Jackson method are compared with each other by using several different accuracy assessment techniques.

Table 7.1 Parameters and initial conditions.

Orbit type	$\boldsymbol{r}_0$ (m)	$\dot{\boldsymbol{r}}_0$ (m/s)	Eccentricity	Inclination (rad)
LEO	−0.3889e6 7.7388e6 0.6736e6	−3.5794e3 0 6.1997e3	0.1	π/3
HEO	4.05e6 0 −7.0148e6	0 9.1464e3 0	0.7	π/3
GEO	4.2164172e7 0 0	0 3.074660237e3 0	0	0

7.2.1 Comparison of local variational iteration method with the modified Chebyshev picard iteration method

The proposed method and the regular MCPI method are tested in a spherical harmonic gravity field of 40 degree (EGM2008). The effect of air drag is not considered herein so that the computational error of Hamiltonian in these two methods can be obtained. Although the Hamiltonian cannot capture all the computational errors, as Berry and Healy point out [9], it provides some insights into the highest accuracy that these two methods may achieve. The relative error of Hamiltonian presented in the numerical results is calculated by $\varepsilon = \Delta H / H_0$, where H_0 is the Hamiltonian of the system at the initial instant and ΔH is the discrepancy of the Hamiltonian in computation.

To demonstrate the performance of these two methods without bias, the number of Chebyshev-Gauss-Lobatto nodes M and the tolerance criterion $tol = norm(\boldsymbol{U}_{n+1}) - norm(\boldsymbol{U}_n)$ in iteration process are tuned to ensure that the MCPI method works optimally, and then the same tuned parameters are used in the proposed method. First, the step size Δt adopted in these two methods is selected as one orbit period T_p, with which the integration during one orbit revolution is completed in a single step for all the three orbit regimes. Then much smaller step sizes are used, with which the orbits are integrated successively. The accuracy and convergence speed of these two methods are evaluated and presented in Figs. 7.1 and 7.2 for the cases of large and small step sizes, respectively. In all the cases considered, the initial approximation in each step is selected as a straight line, which is used as a "cold-start" for the methods.

Overall, in the case where the step size $\Delta t \approx 1$ *orbit period*, the accuracy of the feedback-accelerated Picard iteration (FAPI) method and the MCPI method are almost the same. As is shown in Fig. 7.1, the solutions that are obtained by these two methods are both of very high accuracy. The relative computational errors of the Hamiltonian in these three orbits are kept near to or less than 10^{-13}, which is close to the machine precision. In the case of HEO, the accuracy deteriorates a little bit. The main reason for this is that the HEO is affected more by the gravity force near perigee than near apogee, which means that many more nodes are needed to precisely approximate the trajectory near perigee. To solve this problem, Macomber et al. [10] suggest that true anomaly segmentations may be used to improve the computational accuracy and efficiency.

As shown in Fig. 7.1, the main difference between the proposed method and the MCPI method lies in the convergence speed. Since the computation time in each iteration is roughly the same for these two methods, the convergence speed can be used to evaluate the computational efficiency. As shown, the number of iterations that are needed to converge is halved by using the proposed method. In practical applications, the evaluations of the force model consume most of the computational time, thus the reduction of iterations is significant in reducing the computational cost.

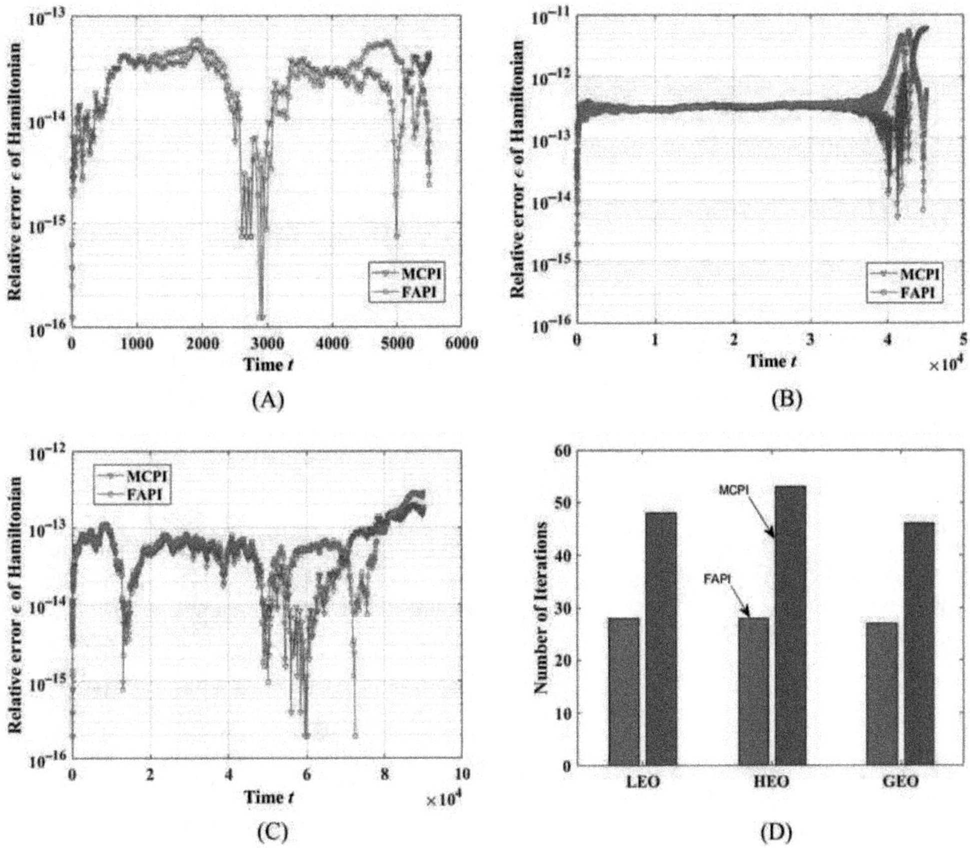

Figure 7.1 Performance of LVIM and MCPI with (**Δt** ≈ 1 ***orbit period***).

By using a very large time step size, lots of nodes are need to accurately approximate the solution in each step, as is shown in Table 7.2. A large number of nodes could occupy a lot of memory and cause a heavy burden on the computer processor, which will in turn slow down the computation. By using a smaller time step size, the number of nodes and the iteration steps in the FAPI method and the MCPI method can be much reduced. Moreover, it makes the algorithms more stable, since the convergence in small time intervals is guaranteed and the convergence speed is relatively fast.

With smaller time step sizes as shown in Table 7.3, the relative computational error of Hamiltonian is further reduced in the proposed FAPI method. As shown in Fig. 7.2, the error ε is kept well under 10^{-13} in the first orbit revolution for the LEO, HEO (3 orbit revolutions are presented in Fig. 7.2B), and GEO. The accuracy of the two methods concerned herein is still almost the same except for the case of HEO that is presented in Fig. 7.2B, which is mainly due to the computational error in the

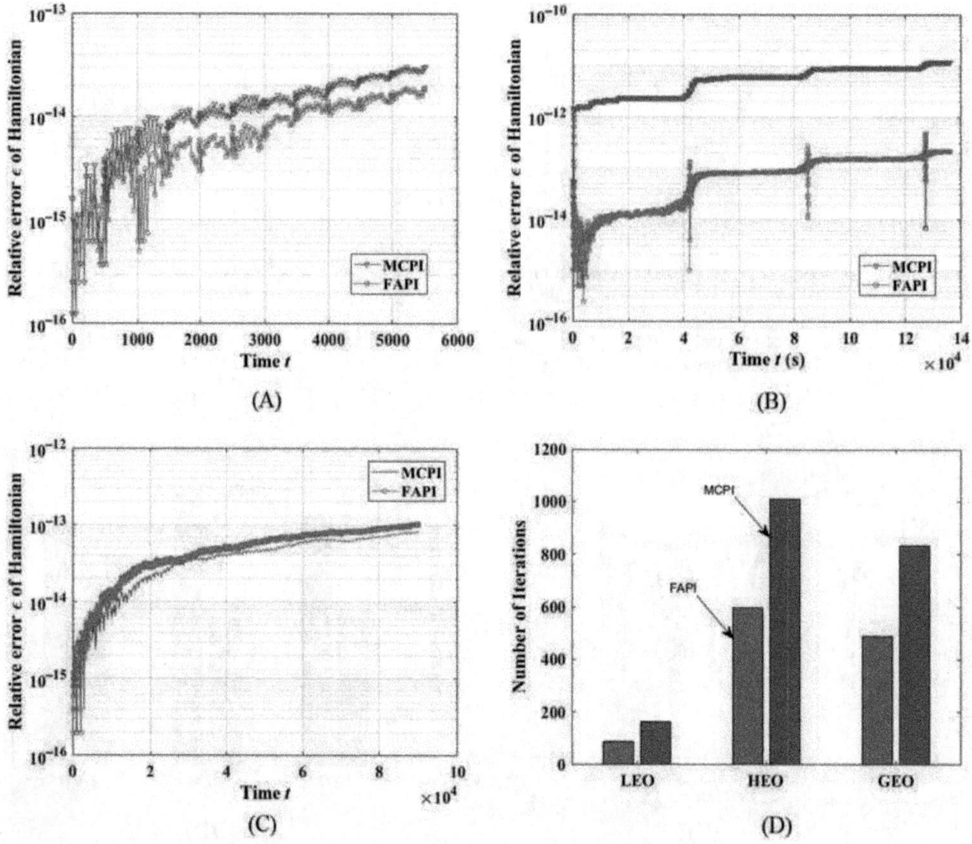

Figure 7.2 Performance of LVIM and MCPI by using small integration steps.

Table 7.2 Tuning parameters ($\boldsymbol{\Delta t} \approx 1$ ***orbit period***).

Case	Number of nodes	Step size (1 orbit period) (s)	Error tolerance, *tol*
LEO	251	8000	1e − 5
HEO	1001	5e4	2e − 5
GEO	1001	9e4	4e − 5

Table 7.3 Tuning parameters.

Case	Number of nodes	Step size (s)	Error tolerance
LEO	19	500	1e − 7
HEO (3 orbits)	31	500	1e − 7
GEO	25	1000	1e − 7

first few steps. In Fig. 7.2B, it is also observed that the computational error of both methods accumulates relatively fast in the steps near the perigee.

In Fig. 7.2D it can be seen that the total number of iterations is halved by the proposed method, which is consistent with the previous result in Fig. 7.1D. Specifically, the iteration numbers in each time step is about 5−7 for the FAPI method and 10−15 for the MCPI method.

7.2.2 Comparison of FAPI with Runge-Kutta 12(10)

To investigate the performance of the proposed method in solving real-life problems, the air drag force is included in the perturbed forced model in the comparison with Runge-Kutta (RK) 12(10) method [11]. To evaluate the drag force, a basic cannon ball drag model is used, with the ballistic coefficient set as $0.01\text{m}^2/\text{kg}$. The MSIS-E-90 atmosphere model is used to provide the information of air density, which is supported online by NASA. When the air drag is included, the force evaluation in LEO and HEO becomes more difficult, and the atmospheric reentry in the case of HEO may further stress the integrators.

To get reasonable assessments of the integration accuracy, we used an error ratio ρ in terms of the root-mean-square error of the integration, as is defined by Berry and Healy [9]. Take position error for instance, ρ is calculated from

$$\rho = \frac{\Delta r_{RMS}}{r_A N_{orbits}}, \Delta r_{RMS} = \sqrt{\frac{1}{N}\sum_{i=1}^{N}(\Delta r_i)2}$$

where Δr_i is the position error on each sampling point, r_A is the distance of apogee, and N_{orbit} is the number of orbits. Similarly, one may define the error ratio of velocity. Note that the exact value of the computational error in the proposed method and the RK 12 (10) method is unlikely to be obtained, since there is no accurate solution for reference. Using other integration methods as reference is also unreliable, because these two methods have relatively high accuracy in comparison with other methods in the literature. Considering that, the position errors of these two methods is evaluated by using the step size halving technique, as it is verified to be reliable by Berry and Healy [9]. The parameters used in the FAPI method are listed in Table 7.4. The step sizes of RK 12 (10) are set as 50s, 50s, and 100 s for the cases of LEO, HEO, and GEO, respectively.

The evaluations of error ratios for the RK12 (10) and the proposed method are obtained and presented in Table 7.5. The accuracy of these two methods is shown to be similarly high in the cases of LEO and GEO. However, in the integration of HEO, the accuracy of RK 12 (10) deteriorates significantly, while the proposed method still

Table 7.4 Parameters for FAPI.

Case	Number of nodes per step	Step size (s)	Error tolerance
LEO (3 orbits)	25	500	1e − 7
HEO (3 orbits)	501	1.5e4	2e − 6
GEO (3 orbits)	301	3e4	2e − 6

Table 7.5 Evaluation of error ratios.

Case	RK12(10)		FAPI	
	Position	Velocity	Position	Velocity
LEO (3 orbits)	3e − 14	3e − 14	3e − 14	3e − 14
HEO (3 orbits)	2e − 9	2e − 9	6e − 12	6e − 12
GEO (3 orbits)	5e − 13	5e − 13	1e − 13	1e − 13

achieves relatively high accuracy. A more concrete comparison is made in Fig. 7.3, where the position errors in three orbit revolutions of LEO, HEO, and GEO are plotted separately.

The results in Fig. 7.3A–C further verify the results in Table 7.5. As shown, the maximum position error in the proposed methods is about 10^{-6} m in the case of LEO. This value increases to about 10^{-3} and 10^{-4} in the cases of HEO and GEO, respectively. The computational error of velocity in these two methods is similar to the position error except that the value is three magnitudes smaller. In Fig. 7.3D the computational cost is plotted, which reveals the computational time that is consumed in evaluating the force model. The calculation follows.

First, the time that is needed to evaluate the perturbed force on one node is defined as the unit time t_u. Then the computational time for one iteration in the FAPI method is determined as t_{FAPI}. Thereby, the computational cost for the FAPI method is calculated as $Cost_{FAPI} = N_{it}t_{FAPI}/t_u$, where N_{it} is the total number of iterations. As in the RK 12 (10) method, the computational time for one step, which involves 25 successive iterations, is determined as $25t_u$. Thus the corresponding computational cost is $Cost_{RK} = 25N_{st}$, where N_{st} is the number of integration steps. As can be seen in Fig. 7.3D, the proposed method is far more efficient than the RK 12 (10) method. Regarding the three kinds of orbit regimes that are considered herein, the cost saving by using the proposed method is about 95%. If the parallel computing technique is adopted, the efficiency of the proposed method could be further improved.

In the above simulations, the step size of the FAPI method is fixed. It has a drawback that some steps are over calculated, while some others may not be calculated very well, as is observed in Fig. 7.2B. This leads to the wasting of computational resources and faster accumulation of errors. Being aware of that, one may use adaptive

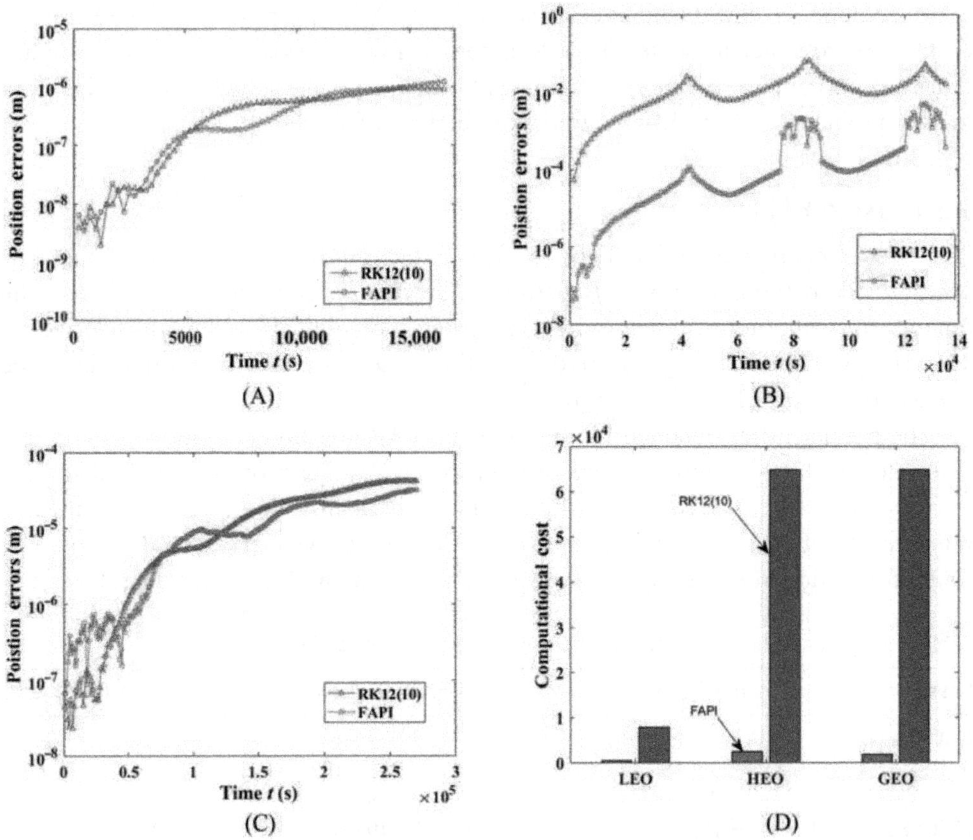

Figure 7.3 Comparison of FAPI and RK 12 (10).

steps in practice to further improve the computational efficiency and accuracy of the proposed method, as well as the MCPI method. A possible approach to achieve that is to divide the orbit into multiple true anomaly segments and make approximations in each segment, as is proposed by Macomber et al. [10].

7.3 Perturbed Lambert's problem

The solution of Lambert's problem is fundamental to the targeting and orbit determination applications. For nonperturbed Keplerian motion, there has already been a variety of elegant methods for solving this problem [12]. However, in the more general situations, such as perturbed two-body motion, three-body motion, and relative motion, the solvers of the classical Lambert problem can be useless. Unlike these solvers, the FAPI method is not limited to unperturbed motion or any specific TPBVP. Using the perturbed Lambert's problem as an example, the effectiveness of the proposed method is verified.

In addition, a new technique labeled the FSGM is proposed in this section. It enables one to generally solve TPBVPs in conservative systems with limited computational resource. This technique is demonstrated by solving a high eccentricity orbit transfer problem and a multi-revolution Lambert transfer problem.

7.3.1 Using FAPI

The two Lambert transfer problems listed in Table 7.6 are solved by using the proposed FAPI method. A simple J_2 perturbed gravity force model is used here. The number of collocation points and Chebyshev polynomials are both selected as $N = M = 64$. The initial guess for the solution is selected as a uniform straight line connecting the initial and final position.

The transfer orbits obtained by the proposed FAPI method are plotted in Fig. 7.4. The velocities at the boundaries are also determined by the proposed method and are listed in Table 7.7. As a reference, the MATLAB built-in ODE45 function is used to numerically integrate the transfer orbit with the initial velocity in Table 7.7. The ODE45 function uses an explicit Runge-Kutta4(5) method. An introduction to this

Table 7.6 Initial and final positions for the Lambert transfer problems.

Cases	Initial position, r_0 (m)	Final position, r_f (m)	Transfer time (s)
LEO transfer	− 0.3889e6 7.7388e6 0.6736e6	− 3.6515e6 − 4.2152e6 6.3103e6	2,500
HEO transfer	− 1.4e7 2.1e7 2.4249e7	− 3.1497e7 − 0.0462e7 5.4554e7	25,000

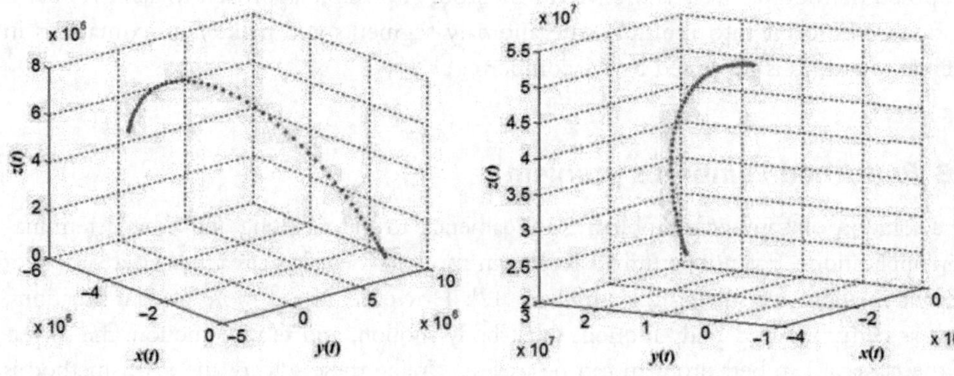

Figure 7.4 (A) Low earth transfer orbit. (B) High eccentricity transfer orbit.

Table 7.7 Two general transfer orbits obtained using the FAPI method.

Cases	Boundary velocities (m/s)		Errors	
	Initial	Final	Position (m)	Velocity (m/s)
Low eccentricity	−3579.396550	1798.099253	−3.2143e − 05	−2.5436e − 07
	0.008964	−5510.306049	1.1292e − 05	−7.5450e − 06
	6199.705320	−3124.253368	9.1111e − 06	−2.4637e − 05
High eccentricity	−1687.308996	20.559111	−2.6535e − 05	−1.2676e − 06
	−0.025078	−1124.689391	1.0044e − 05	1.0028e − 06
	2922.606386	−35.693697	−1.2435e − 05	−2.5208e − 07

function can be found in the literature [13]. The relative tolerance and the absolute tolerance of ODE45 are both set to 10^{-15}. The discrepancies in final positions and velocities between the results obtained by ODE45 and the proposed method are then measured and are listed in Table 7.7. As is shown, with 64 CGL nodes as collocation points, the Lambert's problems considering J_2 perturbation are solved with very high accuracy. The MCPI method is also used, but somehow it appears not to be convergent in these two cases.

7.3.2 Using the fish-scale-growing method

In the area of solving TPBVPs, all the methods, including various shooting methods, finite difference methods, and collocation methods, are complicated by the long duration or large interval between the initial and final states, especially when the dynamical system is complex and strongly nonlinear. In the simple shooting method, for example, as the shooting distance increases, the method becomes more and more sensitive to the initial guess, and at the same time, the selection of a proper initial guess become more and more difficult. This leads to the development of the multiple shooting method and the modified simple shooting method, both of which have larger convergence domains than the simple shooting method. However, these two methods are much more laborious than the simple shooting method and can still become hard to implement if a proper initial guess cannot be obtained.

Here, we propose a FSGM for solving long-duration TPBVPs. As will be shown, the initialization and the implementation of this method are rarely simple. Moreover, as an iterative method, the FSGM is very robust and converges fast. It can also be conveniently coded for parallel computation. The iterative procedure of this method is as follows.

An initial approximation is provided by a reference trajectory, which could be a nominal solution obtained from the linearized problem or unperturbed problem. Then divide the domain (t_0, t_f) of the TPBVP into multiple isometric intervals $[t_i, t_{i+1}], 0 \leq i < N, t_{i+1} - t_i = (t_f - t_0)/N$. For each interval, the points $\boldsymbol{x}(t_i)$ and

$\boldsymbol{x}(t_{i+1})$ on the reference trajectory are set as boundaries. In each iteration we need to solve the N TPBVPs defined by the boundaries $\boldsymbol{x}(t_i)$ and $\boldsymbol{x}(t_{i+1})$ in the corresponding interval $[t_i, t_{i+1}]$. The points $\tilde{\boldsymbol{x}}(t_j)$, $t_j = [t_i, t_{i+1}]/2$ on the solutions of these $N+1$ TPBVPs are collected for the next step. Then we solve the $N-1$ TPBVPs that are defined by the boundaries $\tilde{\boldsymbol{x}}(t_j)$ and $\tilde{\boldsymbol{x}}(t_{j+1})$, $0 \le j < N-1$, in the corresponding interval $[t_j, t_{j+1}]$. After that, the points $\overline{\boldsymbol{x}}(t_i)$ on the solutions of these $N-1$ TPBVPs are used to replace $\boldsymbol{x}(t_i)$ and $\boldsymbol{x}(t_{i+1})$. If $\|\overline{\boldsymbol{x}}(t_i) - \boldsymbol{x}(t_i)\| \le \varepsilon$, the iteration ends. Otherwise, we should replace $\boldsymbol{x}(t_i)$ with $\overline{\boldsymbol{x}}(t_i)$ and restart the iteration.

According to the principle of least action, the solution of a TPBVP in a conservative system is the trajectory that has the least action among any nearby trajectories that connect the initial and final position. Thus it can be concluded that in each iteration, the action along the corrected piecewise solution connecting $\boldsymbol{x}(t_0)$, $\tilde{\boldsymbol{x}}(t_{j=0})$, ..., $\tilde{\boldsymbol{x}}(t_{j=N-1})$, $\boldsymbol{x}(t_f)$ is less than that along the trajectory connecting $\boldsymbol{x}(t_0)$, $\boldsymbol{x}(t_{i=1})$, ..., $\boldsymbol{x}(t_{i=N-1})$, $\boldsymbol{x}(t_f)$. Therefore it is guaranteed for a conservative system that the total action along the piecewise trajectory is monotonically decreasing as the iteration goes on. Therefore this method is absolutely convergent for conservative systems, provided that the sub-TPBVPs are solvable. Although there is no mathematical method to estimate the convergence speed, the numerical examples show that the FSGM converges fast within the first 15 iterations.

The FSGM is illustrated in Fig. 7.5. The domain $[t_0, t_f]$ is divided into two smaller intervals $[t_0, t_1]$ and $[t_0, t_f]$, where $t_1 = (t_0, t_f)/2$. In each iteration, three boundary value problems are defined and need to be solved to correct the position values $\boldsymbol{x}_1$ at

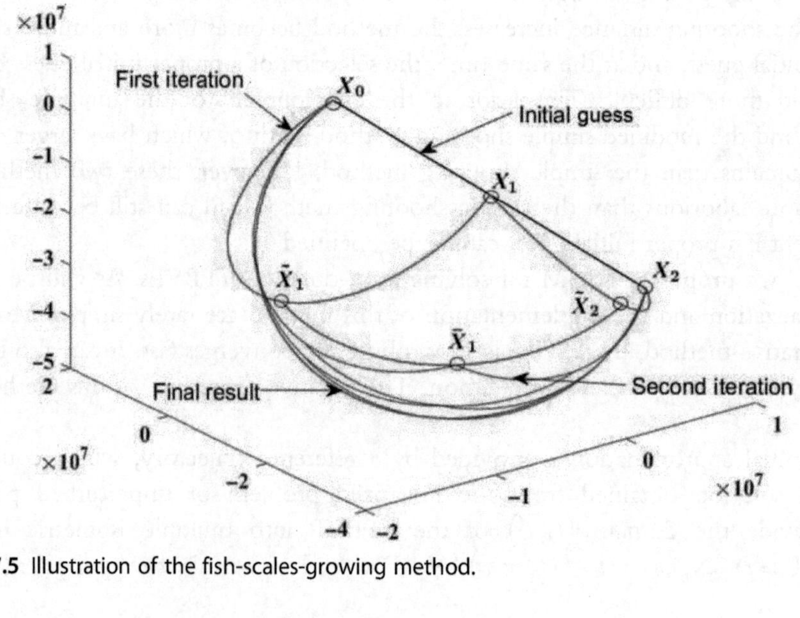

Figure 7.5 Illustration of the fish-scales-growing method.

the time instant t_1. As the iteration goes on, $\boldsymbol{x}_1$ approaches to the real value. Thus the piecewise trajectory evolves to the true solution of the original TPBVP.

First, a high eccentricity orbit transfer problem is solved by using the FSGM. The boundary conditions are listed in Table 7.8. A very rough initial guess is selected as follows:

$$\boldsymbol{r}(t) = \boldsymbol{r}_0 + (\boldsymbol{r}_f - \boldsymbol{r}_0)/(t_f - t_0),$$

where the initial and final time of orbit transfer are set as $t_0 = 0s$, $t_f = 50,000s$. The entire domain is divided into two equal time interval. The TPBVPs that are defined in the smaller intervals are solved by using the FAPI method. The number of collocation points are selected to be $M = 13$. The results are plotted in Fig. 7.6 It can be seen that the piecewise solution of the FSGM converges to the true solution in a few iterations. In this example it is shown that the original problem is divided into multiple sub-TPBVPs, each of which can be solved easily with 13 collocation points. Fig. 7.6D indicates that the rate of convergence is logarithmic for the first 15 iterations.

In practice, the FSGM can also be combined with other TPBVP solvers to enlarge the application area of these solvers in long-duration transfer problems and strongly nonlinear boundary value problems.

In solving the multi-revolution orbit transfer problem, the whole time interval $[0s, 180,000s]$ is divided into 10 segments, while in each segments the solution is approximated by using 13 collocation points. Unlike the preceding case, the Keplerian orbit is used as an initial approximation. To evaluate the improvement in the solution that is obtained by using the proposed method, the discrepancies δr of position and δv of velocity between the exact solution and the iteratively corrected solution are plotted in Fig. 7.7. It is shown that the proposed method can further improve the Keplerian solution, but the convergence speed is relatively low.

According to the numerical results in solving the HEO transfer and multi-revolution orbit transfer problems, it is found that the proposed FSGM is quite efficient in solving Lambert problems within one orbit revolution, while for the multi-revolution problem the convergence speed could become very slow. This drawback is not addressed in this chapter, but we plan to address it in our future research to improve the performance of the FSGM in solving multi-revolution orbit transfer problems.

Table 7.8 Boundary conditions for orbit transfer problems.

Cases	Initial position, $\boldsymbol{r}_0$ (m)	Final position, $\boldsymbol{r}_f$ (m)	Transfer time (s)
HEO transfer	− 1.4e7 2.1e7 2.4249e7	− 1.2572e7 − 2.0930e7 2.1772e7	50,000
Multirevolution transfer	4.2164172e7 0 0	3.5737110e7 2.2375813e7 0	180,000

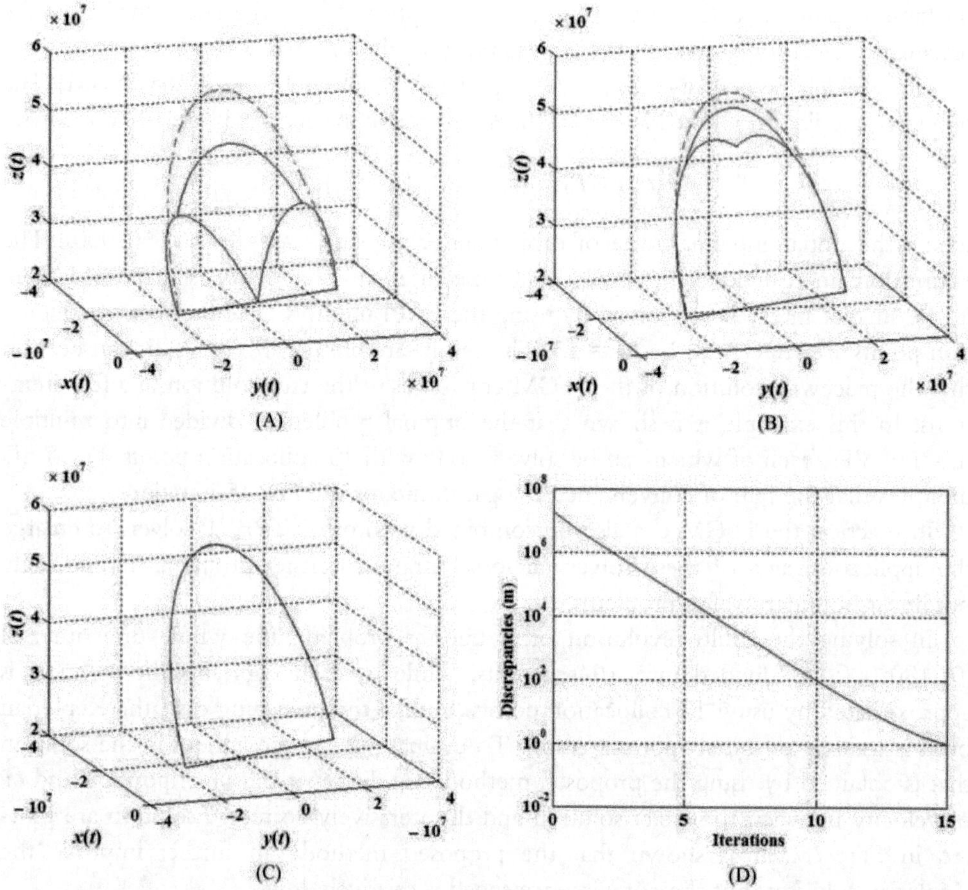

Figure 7.6 Piecewise solution obtained with FSGM.

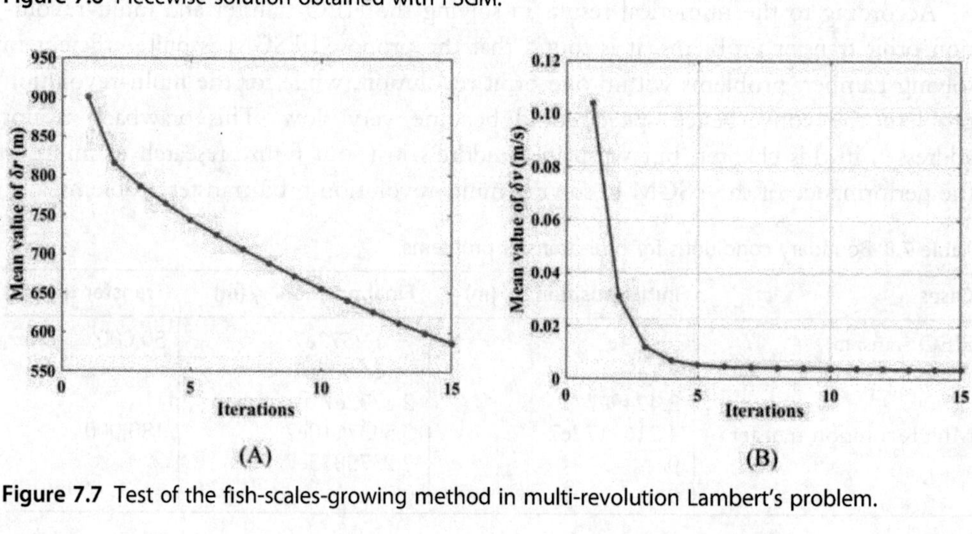

Figure 7.7 Test of the fish-scales-growing method in multi-revolution Lambert's problem.

7.3.3 Using quasilinearization and local variational iteration method

While a spacecraft is orbiting the earth, it is disturbed by nonspherical earth perturbation, atmospheric drag perturbation, lunisolar gravitational perturbation, and solar radiation pressure perturbation. In LEO and HEO cases, the first two kinds of perturbations are the main perturbations to the spacecraft. So we consider only nonspherical earth perturbation and atmospheric drag perturbations in the dynamic equation here. In the equatorial inertial coordinate system, the orbital dynamic equation of the spacecraft is

$$\ddot{\boldsymbol{r}} = -\frac{\mu}{r^3}\boldsymbol{r} + \boldsymbol{a}_J + \boldsymbol{a}_d \tag{7.36}$$

where $\boldsymbol{r} = [x, y, z]^{\mathrm{T}}$ is the radius vector, $\mu = GM = 3986004.418 \times 10^8 m^3/s^2$ is the geocentric gravitational constant, and $r = \|\boldsymbol{r}\|$ is the geocentric distance. Here, $\boldsymbol{a}_J$ is the acceleration caused by nonspherical earth perturbation, and we consider a spherical harmonic gravity field of 10 degrees (EGM2008); $\boldsymbol{a}_d$ is the acceleration caused by atmospheric drag force. To evaluate the drag force, a basic cannonball drag modal is used, with the ballistic coefficient set as $0.01\mathrm{m}^2/\mathrm{kg}$. The MSIS-E-90 atmosphere model is used to provide the information about air density, which is supported online by NASA. Based on the radius and velocity vectors, $\boldsymbol{a}_J$,$\boldsymbol{a}_d$, and their Jacobian matrices can be determined by the two models above. Besides the proposed QLVIM, two other solvers are also applied here to solve the Lambert's problem. The first solver is the Newton-RK4 method, which combines the classical Newton-type shooting method with the fourth-order Runge-Kutta method. The other solver is the FAPI method, which is highly precise and efficient. In practice, these three solvers all need initial guesses. For the QLVIM and the FAPI method the initial guess is a trajectory. Here, we choose the straight line connecting the two boundary points as an initial approximation, which is also called "cold start". For the Newton-RK4 method the initial guess is a velocity, here we choose $s = [-3100, 100, 5900]$ as the initial guess for the LEO case, and $s = [-1200, 10, 2500]$ for the HEO case. The boundary conditions for Lambert's problem are listed in Table 7.6. The parameters for these methods are listed in Table 7.9. M is the number of time nodes in one time step, N is the degree of basis function. Δt is the time step, and $tol_{\tilde{x}}$ is the iteration tolerance of $\tilde{\boldsymbol{x}}_n$. In Eq. (7.13), we denote $\varepsilon_{\tilde{x}} = norm(\tilde{\boldsymbol{x}}_{n+1} - \tilde{\boldsymbol{x}}_n)$. When $\varepsilon_{\tilde{x}} < tol_{\tilde{x}}$ is satisfied, the iteration stops. Similarly, tol_s is the iteration tolerance of $\boldsymbol{s}$.

The transfer orbits that are obtained by the proposed QLVIM are plotted in Fig. 7.8. The MATLAB built-in ODE45 function is used to numerically integrate the transfer orbit with the initial velocity that was determined by the proposed method. The ODE45 function uses an explicit RK4(5) method. The relative tolerance and the absolute tolerance are set to 1e − 13 and 1e − 15, respectively. Fig. 7.8 shows that the results of the QLVIM and ODE45 are in good agreement.

Table 7.9 Tuning parameters.

Algorithm	Cases	M	N	Δt	$tol_{\tilde{x}}$	tol_s
QLVIM	LEO	10	10	250	1e − 7	1e − 5
	HEO	10	10	2500	1e − 7	1e − 5
Newton-RK4	LEO			3.125		1e − 5
	HEO			12.5		1e − 5
FAPI	LEO	32	32	2500	1e − 5	
	HEO	32	32	25000	4e − 6	

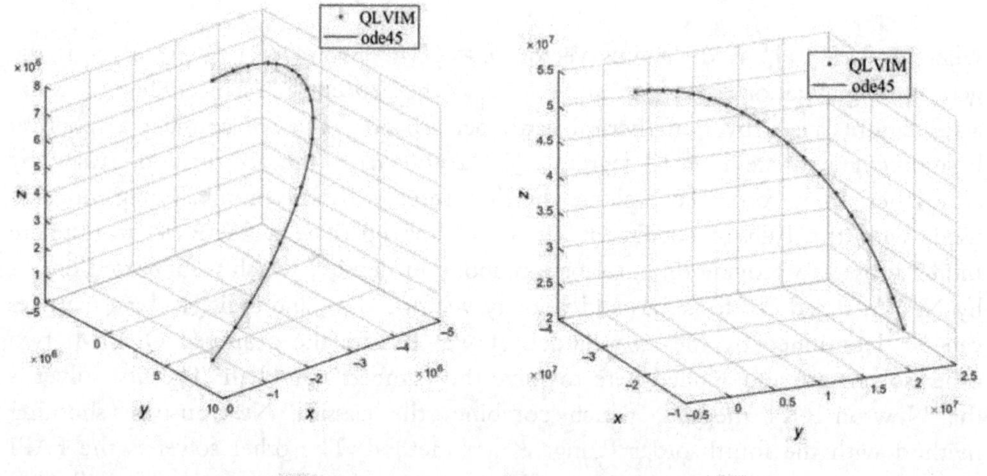

(A) (B)

Figure 7.8 (A) Low earth transfer orbit (B) High-eccentricity transfer orbit.

The initial velocities that were calculated by these methods are listed in Table 7.10, and the computational errors and computational times are listed in Table 7.11. Before the analysis, we give some explanations about the computational errors here. As we know, the key to solving the TPBVP is to find the accurate initial velocity of the transfer orbit. The precision of the initial velocity manifests the precision of the algorithm. The more accurate the calculated initial velocity is, the closer is the final position propagated with it to the boundary value $\boldsymbol{r}_f$, so we define the difference between them as computational error.

Table 7.10 shows that in the LEO and HEO cases, the six decimal places of the velocities calculated by the three algorithms are exactly the same, so the validity of the algorithms and the accuracy of the initial velocities are proved. Table 7.11 shows that the QLVIM can solve these two problems accurately with very low computational cost. While all the computational errors reach the magnitude of 1e-6, the computational speed of the QLVIM is more than 10 times higher than the Newton-RK4 method and 2 times higher than the FAPI method.

Table 7.10 Initial velocities calculated by three algorithms.

Algorithm	LEO (m/s)	HEO (m/s)
QLVIM	− 3583.94021700243; 0.428692017154153; 6194.3206904754	− 1687.37254745742; 0.033017745670880; 2922.5938559553
Newton-RK4	− 3583.94021718111 0.428692011011873 6194.3206903777	− 1687.37254745767 0.033017746529212 2922.5938559549
FAPI	− 3583.94021700365 0.428692017691296 6194.3206904743	− 1687.37254745773 0.033017745704072 2922.5938559552

Table 7.11 Calculation precision and efficiency of three algorithms.

Algorithm	LEO		HEO	
	Position error (m)	Computing time (s)	Position error (m)	Computing time (s)
QLVIM	−1.50e − 6 6.04e − 7 2.60e − 6	0.81	−1.68e − 6 −4.85e − 7 2.97e − 6	0.65
Newton-RK4	−2.7e − 6 9.38e − 6 4.63e − 6	45.79	−2.09e − 6 2.52e − 6 3.48e − 6	119.76
FAPI	−2.63e − 6 1.44e − 6 1.22e − 6	1.97	−8.74e − 6 7.60e − 7 3.40e − 6	1.65

Both the QLVIM and the FAPI method apply the VIM, but the computational speed of the former is faster. The reason is that in solving the TPBVP with the FAPI method, the iteration must be implemented over the entire time interval. Since the time interval is relatively large, we have to set more basis functions and collocation nodes for the method, which increases the computational amounts and the number of iterations. In using the LVIM for solving IVPs, the entire time interval is divided into several subintervals, and the iteration is implemented within each subinterval. The solutions could reach the same precision with much less collocation nodes and basis functions.

In comparison with the traditional Newton-RK4 method, the QLVIM takes much less time to derive the solution. Generally, both the QL method and the Newton's method converge quadratically to the solution of the original equation. Therefore the high computational efficiency of the QLVIM mainly originates from the advantage of the LVIM. While the RK4 method propagates the solution from one node to another, the LVIM approximates the solution by iterating series of collocation points. Furthermore, the time step of the LVIM can be chosen to be much larger than the RK4 method without losing computational accuracy. This comparison also proves that the LVIM is more efficient than RK4 for solving IVPs.

7.4 Conclusion

In this chapter we studied the orbit propagation and the Lambert's problem for a perturbed two-body system using the LVIM method. With the use of Chebyshev polynomials of the first kind as the basis functions, a time integrator was developed to solve IVPs in orbit propagation. The examples showed that the proposed LVIM method is very efficient and highly accurate for the long-term integration of the perturbed orbital equations. The proposed method was shown to be theoretically correct and practically effective. Then we solved the Lambert's problem for both LEO transfer and HEO transfer, using the proposed method. The solution achieves very high accuracy in both the final position and the final velocity.

For solving TPBVPs in conservative systems, we proposed a FSGM. An example was provided to illustrate that the FSGM converges to an accurate solution in a very simple and straightforward approach. Among the merits of the FSGM are that it is insensitive to the initial guess and that only the sub-TPBVPs need to be solved, which provides much convenience in manipulation.

References

[1] H.H. Dai, X.K. Yue, J.P. Yuan, S.N. Atluri, A time domain collocation method for studying the aeroelasticity of a two dimensional airfoil with a structural nonlinearity, Journal of Computational Physics 270 (2014) 214–237. Available from: https://doi.org/10.1016/j.jcp.2014.03.063.0021-9991.

[2] R.M. Woollands, A.B. Younes, J.L. Junkins, New solutions for the perturbed lambert problem using regularization and picard iteration, Journal of Guidance, Control, and Dynamics 38 (9) (2015) 1548–1562. Available from: https://doi.org/10.2514/1.G001028.

[3] X. Bai, J.L. Junkins, Modified Chebyshev-Picard iteration methods for solution of boundary value problems, The Journal of Astronautical Sciences 58 (4) (2011) 615–642. Available from: https://doi.org/10.1007/BF03321534.

[4] X. Bai, J.L. Junkins, Modified Chebyshev-Picard iteration methods for orbit propagation, The Journal of Astronautical Sciences 58 (4) (2011) 583–613. Available from: https://doi.org/10.1007/BF03321533.

[5] X. Bai, Modified Chebyshev-Picard iteration methods for solution of initial value and boundary value problems. Ph.D. Dissertation, Texas A&M University, 2010.

[6] H.Y. Feng, X.K. Yue, X.C. Wang, A quasi-linear local variational iteration method for orbit transfer problems[J], Aerospace Science and Technology 119 (2021) 107222. Available from: https://doi.org/10.1016/j.ast.2021.107222.

[7] X.C. Wang, S.N. Atluri, A novel class of highly efficient & accurate time-integrators in nonlinear computational mechanics, Computational Mechanics, (published online). Available from: https://doi.org/10.1007/s00466-017-1377-4.

[8] M. Jang, C. Chen, Y. Liy, On solving the initial-value problems using the differential transformation method, Applied Mathematics and Computation 115 (2–3) (2000) 145–160. Available from: https://doi.org/10.1016/S0096-3003(99)00137-X.
[9] M. Berry, L. Healy, Comparison of accuracy assessment techniques for numerical integration, in: 13th Annual AAS/AIAA Space Flight Mechanics Meeting, AAS 03–171, Ponce, Puerto Rico, February 9–13, 2003.
[10] B. Macomber, A.B. Probe, R. Woollands, J. Read, J.L. Junkins, Enhancements to modified Chebyshev-Picard iteration efficiency for perturbed orbit propagation, Computer Modelling in Engineering & Sciences 1 (1) (2015) 1–36.
[11] E. Hairer, S.P. Norsett, G. Wanner, Solving Ordinary Differential Equations I: Nonstiff Problems, second revised edition, Springer-Verlag, 1993, pp. 173–185. Available from: http://doi.org/10.1007/978-3-540-78862-1.
[12] R.H. Battin, An Introduction to the Mathematics and Methods of Astrodynamics, AIAA, New York, 1999, pp. 237–242. Available from: http://doi.org/10.2514/4.861543.
[13] C.B. Moler, Numerical Computing with MATLAB, SIAM, Philadelphia, PA, 2008, pp. 185–234, Chaps. 7. Available from: http://doi.org/10.1137/1.9780898717952.

CHAPTER 8

Applications of the Local Variational Iteration Method in Structural Dynamics

SUMMARY

Nonlinearities in the mechanics of structures could from various sources such as large deformations, large rotations, nonlinearities of the material, damping, and boundary conditions, and so on [1]. They are so common in real life that the nonlinear behaviors can be found everywhere ranging from civil engineering, automobile manufactory, aircraft design, to robotic dynamics and on-orbit maneuvers of spacecraft. With the development of lighter, more flexible and multifunctional materials and structures, the effect of nonlinearity will become more significant in these areas.

The investigation of structural vibration helps to enhance the reliability and performance of structures, as well as reveal potential failures. With proper structural models, one can use numerical simulation as an alternative to experiments, and to save a lot of time and expense. In chapter 7 the Local Variational Iteration Method (LVIM) was applied to orbital mechanics problems [2]. It is proved that the LVIM algorithms were much superior to traditional methods, in terms of computational time, accuracy, and speed of convergence. In this chapter, the LVIM algorithms are reintroduced with new interpretations and applications in nonlinear ordinary differential equations arising out of semi-discrete spatial discretizations of structures in structural dynamics.

The nonlinear vibration of a buckled beam is very representative in nonlinear structural dynamics and has attracted a number of researchers [3,4]. Commonly, the dynamics of buckled beam can be well modeled by finite element, boundary element, or meshless methods. However, it is shown in that the clamped-clamped buckled beam model can also be well discretized into a four-mode approximation using the spatial Galerkin method [5]. In this chapter, both the Hilber-Hughes-Taylor-α (HHT-α) and LVIM algorithms are used to investigate the nonlinear vibrations of a buckled beam which may exhibit bifurcation, jump phenomena, and chaos. The validity of the four-mode buckled beam model is examined, and the nonlinear dynamical behaviors such as period doubling process and chaotic motion are investigated. The numerical results obtained by the LVIM method are compared with those obtained by the HHT-α method to evaluate the performance of these two methods.

Computational Methods for Nonlinear Dynamical Systems
DOI: https://doi.org/10.1016/B978-0-323-99113-1.00008-X

It is known that the LVIM algorithms are far more superior to the HHT-α algorithms in terms of accuracy, computational time, convergence speed in problems involving bifurcation and chaos in structural dynamics.

8.1 Elucidation of LVIM in structural dynamics

In structural dynamics modeling, the vibration of a structure is often described by a system of second-order nonlinear ordinary differential equations when spatial discretization has already been carried out, where $\boldsymbol{x}$ is the vector of generalized displacements, and $\boldsymbol{x}''$ are the accelerations, $\boldsymbol{M}$ is the mass matrix, $\boldsymbol{C}$ is the damping matrix, and $\boldsymbol{N}$ is the vector of nonlinear restoring forces which may depend on $\boldsymbol{x}$ and the velocity vector $\boldsymbol{x}''$:

$$\boldsymbol{M}\boldsymbol{x}''+\boldsymbol{C}\boldsymbol{x}'+\boldsymbol{N}(\boldsymbol{x}',\boldsymbol{x})=0 \tag{8.1}$$

Eq. (8.1) can be further rewritten as a system of nonlinear first-order ODEs:

$$\begin{cases}\boldsymbol{x}_1'=\boldsymbol{x}_2\\ \boldsymbol{x}_2'=\boldsymbol{M}^{-1}\boldsymbol{C}\boldsymbol{x}_2-\boldsymbol{M}^{-1}\boldsymbol{N}(\boldsymbol{x}_1,\boldsymbol{x}_2)\end{cases}, \tag{8.2}$$

where $\boldsymbol{N}$ denotes the nonlinear terms and $\boldsymbol{x}_1$ and $\boldsymbol{x}_2$ represent the displacement and velocity vector of the motion.

8.1.1 Formulas of the local variational iteration method

Let $\boldsymbol{u}_1$ and $\boldsymbol{u}_2$ be the trial functions for $\boldsymbol{x}_1$ and $\boldsymbol{x}_2$ in a finite large local time interval $t\in[t_i,t_{i+1}]$. By substituting them into Eq. (8.2), the error residual function is obtained as follows:

$$\boldsymbol{R}(t)=\begin{bmatrix}\boldsymbol{u}_1'\\ \boldsymbol{u}_2'\end{bmatrix}+\begin{bmatrix}-\boldsymbol{u}_2\\ \boldsymbol{M}^{-1}\boldsymbol{C}\boldsymbol{u}_2+\boldsymbol{M}^{-1}\boldsymbol{N}(\boldsymbol{u}_1,\boldsymbol{u}_2)\end{bmatrix},\quad t\in[t_i,t_{i+1}] \tag{8.3}$$

To optimally correct the solution, a simple mechanism of feedback of the error is adopted herein, which has the following expression:

$$\begin{bmatrix}\boldsymbol{u}_1(t)\\ \boldsymbol{u}_2(t)\end{bmatrix}_c=\begin{bmatrix}\boldsymbol{u}_1(t)\\ \boldsymbol{u}_2(t)\end{bmatrix}+\int_{t_i}^{t}\boldsymbol{\lambda}(\tau)\left\{\begin{bmatrix}\boldsymbol{u}_1'\\ \boldsymbol{u}_2'\end{bmatrix}+\begin{bmatrix}-\boldsymbol{u}_2\\ \boldsymbol{M}^{-1}\boldsymbol{C}\boldsymbol{u}_2+\boldsymbol{M}^{-1}\boldsymbol{N}(\boldsymbol{u}_1,\boldsymbol{u}_2)\end{bmatrix}\right\}\mathrm{d}\tau,\quad t\in[t_i,t_{i+1}] \tag{8.4}$$

In Eq. (8.4), $\boldsymbol{\lambda}(\tau)$, a matrix, is the set of weighting functions for the feedback of the error residual $\boldsymbol{R}(t)$ to be optimized. Eq. (8.4) indicates that the solution vector $[\boldsymbol{u}_1(t);\boldsymbol{u}_2(t)]$ at any time t in the interval $t_i\le t\le t_{i+1}$ is corrected by an optimally weighted error residual from time t_i to t.

The iteration formula of Picard iteration method [6] can be regarded as a special case of Eq. (8.4), where $\boldsymbol{\lambda}(\tau)$ is simply selected as the negative unit matrix $\boldsymbol{\lambda}(\tau) = -\boldsymbol{I}$. Although the Picard iteration may converge, it is not the most efficient approach, since $\boldsymbol{\lambda}(\tau)$ is selected too roughly. The derivation of the optimal $\boldsymbol{\lambda}(\tau)$ is as follows.

For $\boldsymbol{u}_1$ and $\boldsymbol{u}_2$ in the neighborhood of the true solutions, that is, $\boldsymbol{u}_1 = \boldsymbol{u}_1^{true} + \delta\boldsymbol{u}_1$, $\boldsymbol{u}_2 = \boldsymbol{u}_2^{true} + \delta\boldsymbol{u}_2$, $\boldsymbol{\lambda}$ is expected to be optimally determined so that

$$\begin{bmatrix} \boldsymbol{u}_1(t) \\ \boldsymbol{u}_2(t) \end{bmatrix}_c = \begin{bmatrix} \boldsymbol{u}_1^{true}(t) \\ \boldsymbol{u}_2^{true}(t) \end{bmatrix}_c \tag{8.5}$$

It means that the variation of Eq. (8.4) should be equals to zero if $\boldsymbol{u}_1 = \boldsymbol{u}_1^{true}$, $\boldsymbol{u}_2 = \boldsymbol{u}_2^{true}$, which leads to

$$\begin{aligned} \delta\begin{bmatrix} \boldsymbol{u}_1 \\ \boldsymbol{u}_2 \end{bmatrix}_c &= \delta\begin{bmatrix} \boldsymbol{u}_1 \\ \boldsymbol{u}_2 \end{bmatrix} + \int_{t_i}^{t} \lambda\delta\left\{\begin{bmatrix} \boldsymbol{u}_1' \\ \boldsymbol{u}_2' \end{bmatrix} + \begin{bmatrix} -\boldsymbol{u}_2 \\ \boldsymbol{M}^{-1}\boldsymbol{C}\boldsymbol{u}_2 + \boldsymbol{M}^{-1}\mathbf{N} \end{bmatrix}\right\}\mathrm{d}\tau \\ &+ \int_{t_i}^{t} \delta\lambda\left\{\begin{bmatrix} \boldsymbol{u}_1' \\ \boldsymbol{u}_2' \end{bmatrix} + \begin{bmatrix} -\boldsymbol{u}_2 \\ \boldsymbol{M}^{-1}\boldsymbol{C}\boldsymbol{u}_2 + \boldsymbol{M}^{-1}\mathbf{N} \end{bmatrix}\right\} = 0 \end{aligned} \tag{8.6}$$

Since $\mathbf{u}_1$ and $\mathbf{u}_2$ in Eq. (8.6) are supposed to be the true solutions, we have

$$\begin{bmatrix} \boldsymbol{u}_1' \\ \boldsymbol{u}_2' \end{bmatrix} + \begin{bmatrix} -\boldsymbol{u}_2 \\ \boldsymbol{M}^{-1}\boldsymbol{C}\boldsymbol{u}_2 + \boldsymbol{M}^{-1}\mathbf{N} \end{bmatrix} = 0 \tag{8.7}$$

Therefore, Eq. (8.6) leads to

$$\begin{aligned} \delta\begin{bmatrix} \boldsymbol{u}_1 \\ \boldsymbol{u}_2 \end{bmatrix}_c &= \delta\begin{bmatrix} \boldsymbol{u}_1 \\ \boldsymbol{u}_2 \end{bmatrix} + \int_{t_i}^{t} \lambda\delta\left\{\begin{bmatrix} \boldsymbol{u}_1' \\ \boldsymbol{u}_2' \end{bmatrix} + \begin{bmatrix} -\boldsymbol{u}_2 \\ \boldsymbol{M}^{-1}\boldsymbol{C}\boldsymbol{u}_2 + \boldsymbol{M}^{-1}\mathbf{N} \end{bmatrix}\right\}\mathrm{d}\tau \\ &= (I+\lambda)\delta\begin{bmatrix} \boldsymbol{u}_1 \\ \boldsymbol{u}_2 \end{bmatrix} + \int_{t_i}^{t} \{-\lambda' + \lambda\boldsymbol{J}\}\delta\begin{bmatrix} \boldsymbol{u}_1 \\ \boldsymbol{u}_2 \end{bmatrix}\mathrm{d}\tau = 0 \end{aligned} \tag{8.8}$$

where $\boldsymbol{J} = \begin{bmatrix} 0 & 1 \\ \boldsymbol{M}^{-1}(\partial\mathbf{N}/\partial\boldsymbol{u}_1) & \boldsymbol{M}^{-1}\mathbf{C} + \boldsymbol{M}^{-1}(\partial\mathbf{N}/\partial\boldsymbol{u}_2) \end{bmatrix}$.

In Eq. (8.8) there are variations both inside and outside the integral over time, thus the variations are made to be zero separately by enforcing the weighting function matrix $\boldsymbol{\lambda}$ to satisfy the following conditions:

$$\boldsymbol{\lambda}(t) + \boldsymbol{I} = 0,\ \text{and} -\boldsymbol{\lambda}'(\tau, t) + \boldsymbol{\lambda}(\tau, t)\boldsymbol{J}(\tau) = 0,\quad \tau \in [t_i, t] \tag{8.9}$$

Obviously, $\boldsymbol{\lambda}$ is related to $\boldsymbol{u}_1$ and $\boldsymbol{u}_2$, which are supposed to be $\boldsymbol{u}_1^{true}$ and $\boldsymbol{u}_2^{true}$. However, the true solutions are not known in advance. Considering that, we calculated $\boldsymbol{\lambda}$ in this chapter by using approximated solutions instead of the true solutions in the implementation of the algorithms, which are labeled as LVIM algorithms for convenience.

8.1.2 Large time interval collocation

Further, the following weak formulation of Eq. (8.4) can be established, using a matrix of test functions $\boldsymbol{v}(t)$:

$$\int_{t_i}^{t_{i+1}} \boldsymbol{v}(t)\begin{bmatrix}\boldsymbol{u}_1(t)\\ \boldsymbol{u}_2(t)\end{bmatrix}_c \mathrm{d}t = \int_{t_i}^{t_{i+1}} \boldsymbol{v}(t)\left\{\begin{bmatrix}\boldsymbol{u}_1(t)\\ \boldsymbol{u}_2(t)\end{bmatrix} + \int_{t_i}^{t_{i+1}} \boldsymbol{\lambda}\left\{\begin{bmatrix}\boldsymbol{u}_1'\\ \boldsymbol{u}_2'\end{bmatrix} + \begin{bmatrix}-\boldsymbol{u}_2\\ \boldsymbol{M}^{-1}\boldsymbol{C}\boldsymbol{u}_2 + \boldsymbol{M}^{-1}\mathbf{N}(\boldsymbol{u}_1,\boldsymbol{u}_2)\end{bmatrix}\right\}\mathrm{d}\tau\right\}\mathrm{d}t \tag{8.10}$$

In this formula, $\boldsymbol{v}(t)$ are test functions, the same as those in the classical weighted residual methods. Let $\boldsymbol{v}(t)$ be a diagonal matrix $\boldsymbol{v} = diag[v, v, ...]$, and v be the Dirac delta function for a group of collocation points t_m in the finite large time interval t_i to t_{i+1}:

$$v = \delta(t - t_m),\quad t_m \in [t_i, t_{i+1}],\quad m = 1, 2, ..., M \tag{8.11}$$

then Eq. (8.10) becomes

$$\begin{bmatrix}\boldsymbol{u}_1(t_m)\\ \boldsymbol{u}_2(t_m)\end{bmatrix}_c = \begin{bmatrix}\boldsymbol{u}_1(t_m)\\ \boldsymbol{u}_2(t_m)\end{bmatrix} + \int_{t_i}^{t_m} \boldsymbol{\lambda}\left\{\begin{bmatrix}\boldsymbol{u}_1'\\ \boldsymbol{u}_2'\end{bmatrix} + \begin{bmatrix}-\boldsymbol{u}_2\\ \boldsymbol{M}^{-1}\boldsymbol{C}\boldsymbol{u}_2 + \boldsymbol{M}^{-1}\mathbf{N}\end{bmatrix}\right\}\mathrm{d}\tau,\quad t_m \in [t_i, t_{i+1}],\quad m = 1, \ldots, M \tag{8.12}$$

By using a set of orthogonal polynomials $\boldsymbol{\Phi} = \{\phi_0, \phi_1, \phi_2, \ldots\}^{\mathrm{T}}$ as basis functions, the trial functions $\boldsymbol{u}_1$ and $\boldsymbol{u}_2$ can be constructed as

$$u_{1,e} = \sum_{n=0}^{N} a_{1,e,n}\phi_n, \text{ and } u_{2,e} = \sum_{n=0}^{N} a_{2,e,n}\phi_n, \tag{8.13}$$

where $u_{1,e}$ and $u_{2,e}$ are elements of $\boldsymbol{u}_1$ and $\boldsymbol{u}_2$ respectively, and $a_{1,e,n}$ and $a_{2,e,n}$ are coefficients to be determined. From Eq. (8.13) we have

$$\left[u_{p,e}(t_1), u_{p,e}(t_2), ..., u_{p,e}(t_M)\right]^{\mathrm{T}} = \boldsymbol{B}\left[a_{p,e,0}, a_{p,e,1}, \ldots, a_{p,e,N}\right]^{\mathrm{T}},\quad p = 1, 2 \tag{8.14}$$

and

$$\left[u'_{p,e}(t_1), u'_{p,e}(t_2), ..., u'_{p,e}(t_M)\right]^{\mathrm{T}} = \boldsymbol{LB}\left[a_{p,e,0}, a_{p,e,1}, \ldots, a_{p,e,N}\right]^{\mathrm{T}},\quad p = 1, 2 \tag{8.15}$$

where

$$\boldsymbol{B} = \begin{bmatrix}\phi_0(t_1) & \phi_1(t_1) & \cdots & \phi_N(t_1)\\ \phi_0(t_2) & \phi_1(t_2) & \cdots & \phi_N(t_2)\\ \vdots & \vdots & \ddots & \vdots\\ \phi_0(t_M) & \phi_1(t_M) & \cdots & \phi_N(t_M)\end{bmatrix}_{(N+1)\times M},\quad \boldsymbol{LB} = \begin{bmatrix}\phi_0'(t_1) & \phi_1'(t_1) & \cdots & \phi_N'(t_1)\\ \phi_0'(t_2) & \phi_1'(t_2) & \cdots & \phi_N'(t_2)\\ \vdots & \vdots & \ddots & \vdots\\ \phi_0'(t_M) & \phi_1'(t_M) & \cdots & \phi_N'(t_1)\end{bmatrix}_{(N+1)\times M} \tag{8.16}$$

Normally, the number of colocation points M is as the same as the number of basis functions $N+1$. It can be found that the values of $u_{p,e}^{'}$ at collocation points are related with those of $u_{p,e}$ by

$$\left[u_{p,e}^{'}(t_1), u_{p,e}^{'}(t_2), ..., u_{p,e}^{'}(t_M)\right]^{\mathrm{T}} = (\boldsymbol{LB})\boldsymbol{B}^{-1}\left[a_{p,e,0}(t_1), a_{p,e,1}(t_2), \ldots, a_{p,e,N}(t_M)\right]^{\mathrm{T}} \quad (8.17)$$

In an analogous way, the values of $\int_{t_i}^{t} u_{p,e}\mathrm{d}\tau$ at collocation points t_m, $m=1,2,\ldots,M$, can be obtained from those of $u_{p,e}$ through the following transformation

$$\left[\int_{t_i}^{t_1} u_{p,e}\mathrm{d}\tau, \ldots, \int_{t_i}^{t_2} u_{p,e}\mathrm{d}\tau \int_{t_i}^{t_M} u_{p,e}\mathrm{d}\tau\right]^{\mathrm{T}} = \left(\boldsymbol{L}^{-1}\boldsymbol{B}\right)\boldsymbol{B}^{-1}\left[u_{p,e}(t_1), u_{p,e}(t_2), \ldots, u_{p,e}(t_M)\right]^{\mathrm{T}} \quad (8.18)$$

where

$$\boldsymbol{L}^{-1}\boldsymbol{B}\begin{bmatrix} \int_{t_i}^{t_1}\phi_0\mathrm{d}\tau & \int_{t_i}^{t_1}\phi_1\mathrm{d}\tau & \cdots & \int_{t_i}^{t_1}\phi_N\mathrm{d}\tau \\ \int_{t_i}^{t_2}\phi_0\mathrm{d}\tau & \int_{t_i}^{t_2}\phi_1\mathrm{d}\tau & \cdots & \int_{t_i}^{t_2}\phi_N\mathrm{d}\tau \\ \vdots & \vdots & \ddots & \vdots \\ \int_{t_i}^{t_M}\phi_0\mathrm{d}\tau & \int_{t_i}^{t_M}\phi_2\mathrm{d}\tau & \cdots & \int_{t_i}^{t_M}\phi_N\mathrm{d}\tau \end{bmatrix}_{(N+1)\times M} \quad (8.19)$$

8.1.3 LVIM algorithms for structural dynamical system

8.1.3.1 LVIM Algorithm-1

Although we have already obtained the iterative formula through collocation, the matrix of generalized Lagrange multipliers $\boldsymbol{\lambda}$ in it remain a puzzle. The possibility of directly solving the constraint equations Eq. (8.6) for $\boldsymbol{\lambda}$ is unlikely for most nonlinear cases. Fortunately, there is an approach to bypass this dilemma.

Differentiating Eq. (8.4) leads to

$$\frac{\mathrm{d}}{\mathrm{d}t}\begin{bmatrix}\boldsymbol{u}_1(t)\\ \boldsymbol{u}_2(t)\end{bmatrix}_c = \frac{\mathrm{d}}{\mathrm{d}t}\begin{bmatrix}\boldsymbol{u}_1(t)\\ \boldsymbol{u}_2(t)\end{bmatrix} + \boldsymbol{\lambda}(t)\boldsymbol{R}(t) + \int_{t_i}^{t}\frac{\partial\boldsymbol{\lambda}}{\partial t}\boldsymbol{R}(\tau)\mathrm{d}\tau, \quad (8.20)$$

where $\boldsymbol{R}(\tau)$ is the residual function:

$$\boldsymbol{R}(\tau)\begin{bmatrix}\boldsymbol{u}_1'(\tau)\\ \boldsymbol{u}_2'(\tau)\end{bmatrix} + \begin{bmatrix}-\boldsymbol{u}_2(\tau)\\ \boldsymbol{M}^{-1}\boldsymbol{C}\boldsymbol{u}_2(\tau)+\boldsymbol{M}^{-1}\boldsymbol{N}(\tau)\end{bmatrix}$$

According to Eq. (8.9) and using the theory of Magnus series, it can be proved [17] that $\boldsymbol{\lambda}(t) = -\boldsymbol{I}$, and $\frac{\partial\boldsymbol{\lambda}}{\partial t} = -\boldsymbol{J}(t)\boldsymbol{\lambda}$. Substituting them into Eq. (8.20), we have

$$\frac{\mathrm{d}}{\mathrm{d}t}\begin{bmatrix}\boldsymbol{u}_1(t)\\ \boldsymbol{u}_2(t)\end{bmatrix}_c = -\begin{bmatrix}-\boldsymbol{u}_2\\ \boldsymbol{M}^{-1}\boldsymbol{C}\boldsymbol{u}_2+\boldsymbol{M}^{-1}\boldsymbol{N}\end{bmatrix} - J(t)\int_{t_i}^{t}\lambda\left\{\begin{bmatrix}\boldsymbol{u}_1'\\ \boldsymbol{u}_2'\end{bmatrix} + \begin{bmatrix}-\boldsymbol{u}_2\\ \boldsymbol{M}^{-1}\boldsymbol{C}\boldsymbol{u}_2+\boldsymbol{M}^{-1}\boldsymbol{N}\end{bmatrix}\right\}\mathrm{d}\tau \quad (8.21)$$

Noting that $\int_{t_i}^{t} \boldsymbol{\lambda}\left\{\begin{bmatrix} \boldsymbol{u}'_1 \\ \boldsymbol{u}'_2 \end{bmatrix} + \begin{bmatrix} -\boldsymbol{u}_2 \\ \boldsymbol{M}^{-1}\boldsymbol{C}\boldsymbol{u}_2 + \boldsymbol{M}^{-1}\mathbf{N} \end{bmatrix}\right\}\mathrm{d}\tau = \begin{bmatrix} \boldsymbol{u}_1 \\ \boldsymbol{u}_2 \end{bmatrix}_c - \begin{bmatrix} \boldsymbol{u}_1 \\ \boldsymbol{u}_2 \end{bmatrix}$ according to the optimal error-feedback iteration formula (Eq. (8.4)), Eq. (8.21) is rewritten as

$$\frac{\mathrm{d}}{\mathrm{d}t}\begin{bmatrix} \boldsymbol{u}_1(t) \\ \boldsymbol{u}_2(t) \end{bmatrix}_c = -\begin{bmatrix} -\boldsymbol{u}_2 \\ \boldsymbol{M}^{-1}\boldsymbol{C}\boldsymbol{u}_2 + \boldsymbol{M}^{-1}\mathbf{N} \end{bmatrix} - \boldsymbol{J}(t)\left\{\begin{bmatrix} \boldsymbol{u}_1 \\ \boldsymbol{u}_2 \end{bmatrix}_c - \begin{bmatrix} \boldsymbol{u}_1 \\ \boldsymbol{u}_2 \end{bmatrix}\right\} \tag{8.22}$$

After rearrangement, it is rewritten as

$$\begin{bmatrix} \boldsymbol{u}_1(t) \\ \boldsymbol{u}_2(t) \end{bmatrix}_c + \boldsymbol{J}(t)\begin{bmatrix} \boldsymbol{u}_1(t) \\ \boldsymbol{u}_2(t) \end{bmatrix}_c = -\begin{bmatrix} -\boldsymbol{u}_2 \\ \boldsymbol{M}^{-1}\boldsymbol{C}\boldsymbol{u}_2 + \boldsymbol{M}^{-1}\mathbf{N} \end{bmatrix}_c + \boldsymbol{J}(t)\begin{bmatrix} \boldsymbol{u}_1(t) \\ \boldsymbol{u}_2(t) \end{bmatrix} \tag{8.23}$$

Eq. (8.23) can be regarded as being equivalent to Eq. (8.4). By collocating in the local time interval $[t_i, t_{i+1}]$, an algebraic iterative formula is obtained as

$$\begin{bmatrix} \boldsymbol{u}'_1(t_m) \\ \boldsymbol{u}'_2(t_m) \end{bmatrix}_c + \boldsymbol{J}(t_m)\begin{bmatrix} \boldsymbol{u}_1(t_m) \\ \boldsymbol{u}_2(t_m) \end{bmatrix}_c = -\begin{bmatrix} -\boldsymbol{u}_2(t_m) \\ \boldsymbol{M}^{-1}\boldsymbol{C}\boldsymbol{u}_2(t_m) + \boldsymbol{M}^{-1}\mathbf{N}(t_m) \end{bmatrix} + \boldsymbol{J}(t_m)\begin{bmatrix} \boldsymbol{u}_1(t_m) \\ \boldsymbol{u}_2(t_m) \end{bmatrix} \tag{8.24}$$

where $m = 1, 2, \ldots, M$, $t_m \in [t_i, t_{i+1}]$, and $t_{i+1} - t_i$ is a finite large time interval.

After rearranging the sequence of the collocation equations and using the relationship in Eq. (8.17), we rewrite Eq. (8.24) as

$$(\tilde{\boldsymbol{E}} + \tilde{\boldsymbol{J}})\begin{bmatrix} \boldsymbol{U}_1 \\ \boldsymbol{U}_2 \end{bmatrix}_c = -\begin{bmatrix} -\boldsymbol{U}_2 \\ \tilde{\boldsymbol{M}}^{-1}\tilde{\boldsymbol{C}}\boldsymbol{U}_2 + \tilde{\boldsymbol{M}}^{-1}\tilde{\mathbf{N}} \end{bmatrix} + \tilde{\boldsymbol{J}}\begin{bmatrix} \boldsymbol{U}_1 \\ \boldsymbol{U}_2 \end{bmatrix} \tag{8.25}$$

where

$$\boldsymbol{U}_p = \begin{bmatrix} u_{p,1}(t_1) & u_{p,1}(t_2) & \ldots & u_{p,1}(t_M) & | u_{p,2}(t_1) & u_{p,2}(t_2) & \ldots & u_{p,2}(t_M) | & \ldots \end{bmatrix}^{\mathrm{T}}, \ p = 1, 2$$

The configuration of the matrices $\tilde{\boldsymbol{E}}$, $\tilde{\boldsymbol{J}}$, $\tilde{\boldsymbol{M}}$, $\tilde{\boldsymbol{C}}$, $\tilde{\mathbf{N}}$ are provided in Appendix C.

It should be noted that the initial conditions are not incorporated in Eq. (8.25). For that, without loss of generality, we usually select the first collocation point at the initial boundary, of which the values $u_{p,e}(t_1)$ are given. By doing that, Eq. (8.25) becomes overdetermined, thus it is necessary to drop excess collocation equations at time t_1. Finally, Eq. (8.25) is modified as

$$\begin{bmatrix} \boldsymbol{U}_1 \\ \boldsymbol{U}_2 \end{bmatrix}_c^r = \begin{bmatrix} \boldsymbol{U}_1 \\ \boldsymbol{U}_2 \end{bmatrix}_c^r - (\tilde{\boldsymbol{E}}^r + \tilde{\boldsymbol{J}}^r)^{-1} - \left\{\tilde{\boldsymbol{E}}\begin{bmatrix} \boldsymbol{U}_1 \\ \boldsymbol{U}_2 \end{bmatrix} + \begin{bmatrix} -\boldsymbol{U}_2 \\ \tilde{\boldsymbol{M}}^{-1}\tilde{\boldsymbol{C}}\boldsymbol{U}_2 + \tilde{\boldsymbol{M}}^{-1}\tilde{\mathbf{N}} \end{bmatrix}^r \right. \tag{8.26}$$

The symbol $[\bullet]^r$ denotes the remained matrix after the $(lM+1)th$ rows and columns in $[\bullet]$ are dropped, $l = 0, 1, 2, \ldots$. If $[\bullet]^r$ is a vector, we just need to remove the $(lM+1)th$ rows in $[\bullet]$.

A flow diagram is presented in Fig. 8.1 to illustrate LVIM algorithm-1.

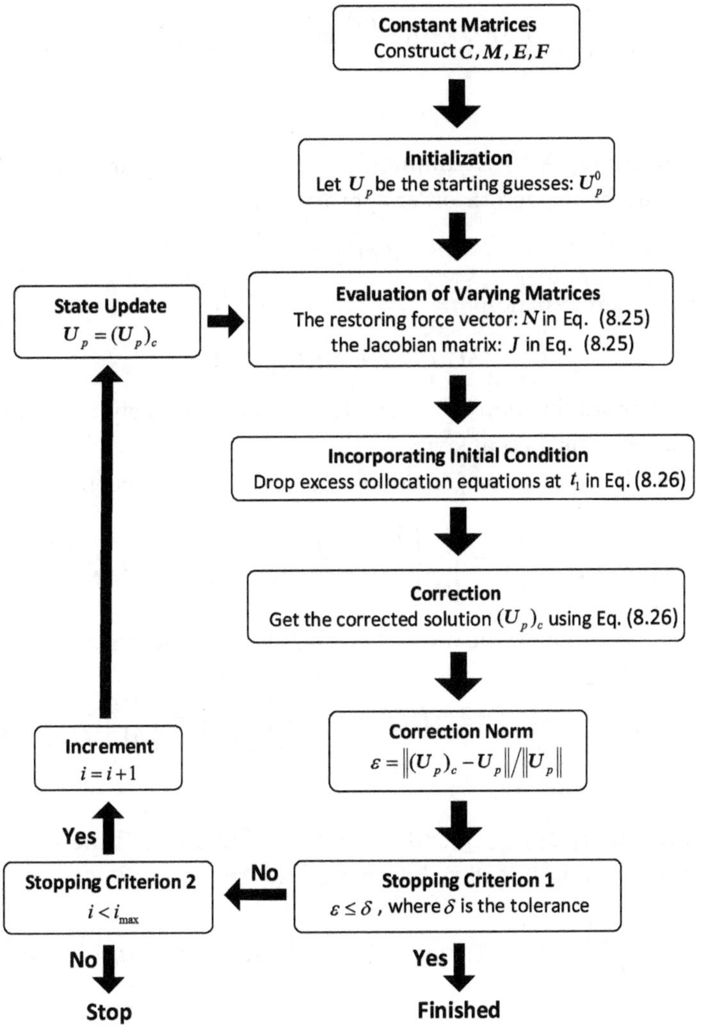

Figure 8.1 Flow chart overview of LVIM algorithm-1.

8.1.3.2 LVIM Algorithm-2

According to Eq. (8.9), the Lagrange multipliers can be approximated by Taylor series as follows:

$$\boldsymbol{\lambda}(\tau) = \boldsymbol{T}_0 + \boldsymbol{T}_1(\tau - t) + \boldsymbol{T}_2(\tau - t)^2 + \cdots \tag{8.27}$$

where $\boldsymbol{T}_0 = -1$, $\boldsymbol{T}_1 = -\boldsymbol{J}(t)$, $\boldsymbol{T}_2 = -\boldsymbol{J}(t)^2/2$, and so on. Substituting it into Eq. (8.12), the correctional formula is obtained as

$$\begin{bmatrix} \boldsymbol{u}_1(t_m) \\ \boldsymbol{u}_2(t_m) \end{bmatrix}_c = \begin{bmatrix} \boldsymbol{u}_1'(t_m) \\ \boldsymbol{u}_2'(t_m) \end{bmatrix} + \int_{t_i}^{t_m} \left[-\boldsymbol{I} - \boldsymbol{J}(t)(\tau - t) - \frac{\boldsymbol{J}(t)^2}{2}(\tau - t)^2 + \cdots \right] \left\{ \begin{bmatrix} \boldsymbol{u}_1' \\ \boldsymbol{u}_2' \end{bmatrix} + \begin{bmatrix} -\boldsymbol{u}_2 \\ \boldsymbol{M}^{-1}\boldsymbol{C}\boldsymbol{u}_2 + \boldsymbol{M}^{-1}\boldsymbol{N} \end{bmatrix} \right\} \mathrm{d}\tau \tag{8.28}$$

In implementations, $\boldsymbol{\lambda}(\tau)$ is commonly approximated by truncated Taylor series. The simplest could be the zeroth order approximation

$$\boldsymbol{\lambda}(\tau) = -\boldsymbol{I} \tag{8.29}$$

or the first order approximation

$$\boldsymbol{\lambda}(\tau) = -\boldsymbol{I} - \boldsymbol{J}(t)(\tau - t) \tag{8.30}$$

Higher order approximations are possible, but they are rarely used in practice considering the computational complexity.

With the zeroth order approximation of $\boldsymbol{\lambda}(\tau)$, Eq. (8.28) becomes

$$\begin{bmatrix} \boldsymbol{u}_1(t_m) \\ \boldsymbol{u}_2(t_m) \end{bmatrix}_c = \begin{bmatrix} \boldsymbol{u}_1(t_m) \\ \boldsymbol{u}_2(t_m) \end{bmatrix} - \int_{t_i}^{t_m} \left\{ \begin{bmatrix} \boldsymbol{u}_1' \\ \boldsymbol{u}_2' \end{bmatrix} + \begin{bmatrix} -\boldsymbol{u}_2 \\ \boldsymbol{M}^{-1}\boldsymbol{C}\boldsymbol{u}_2 + \boldsymbol{M}^{-1}\boldsymbol{N} \end{bmatrix} \right\} \mathrm{d}\tau \tag{8.31}$$

With the first order approximation of $\boldsymbol{\lambda}(\tau)$, Eq. (8.28) becomes

$$\begin{bmatrix} \boldsymbol{u}_1(t_m) \\ \boldsymbol{u}_2(t_m) \end{bmatrix}_c = \begin{bmatrix} \boldsymbol{u}_1(t_m) \\ \boldsymbol{u}_2(t_m) \end{bmatrix} + \int_{t_i}^{t_m} [-\boldsymbol{I} - \boldsymbol{J}(t_m)(\tau - t_m)] \left\{ \begin{bmatrix} \boldsymbol{u}_1' \\ \boldsymbol{u}_2' \end{bmatrix} + \begin{bmatrix} -\boldsymbol{u}_2 \\ \boldsymbol{M}^{-1}\boldsymbol{C}\boldsymbol{u}_2 + \boldsymbol{M}^{-1}\boldsymbol{N} \end{bmatrix} \right\} \mathrm{d}\tau \tag{8.32}$$

where $m = 1, 2, \ldots M$, $t_m \in [t_i, t_{i+1}]$, and $t_{i+1} - t_i$ is a finite large time interval. By separating the terms involving t_m and τ, Eq. (8.32) leads to

$$\begin{bmatrix} \boldsymbol{u}_1(t_m) \\ \boldsymbol{u}_2(t_m) \end{bmatrix}_c = \begin{bmatrix} \boldsymbol{u}_1(t_m) \\ \boldsymbol{u}_2(t_m) \end{bmatrix} - \int_{t_i}^{t_m} \boldsymbol{R}(\tau)\mathrm{d}\tau + \boldsymbol{J}(t_m)t_m \int_{t_i}^{t_m} \boldsymbol{R}(\tau)\mathrm{d}\tau - \boldsymbol{J}(t_m) \int_{t_i}^{t_m} \tau \boldsymbol{R}(\tau)\mathrm{d}\tau, \tag{8.33}$$

The Eqs. (8.31)and (8.33) are two low-order iterative formulas in LVIM algorithm-2. The integral terms in Eqs. (8.31, 8.33) are calculated using the transformation in Eq. (8.18). After some rearrangements, Eq. (8.33) can be rewritten as

$$\begin{bmatrix} \boldsymbol{U}_1 \\ \boldsymbol{U}_2 \end{bmatrix}_c = \begin{bmatrix} \boldsymbol{U}_1 \\ \boldsymbol{U}_2 \end{bmatrix} - \tilde{\boldsymbol{H}}\tilde{\boldsymbol{R}} + \tilde{\boldsymbol{J}}\tilde{\boldsymbol{T}}\tilde{\boldsymbol{H}}\tilde{\boldsymbol{R}} - \tilde{\boldsymbol{J}}\tilde{\boldsymbol{H}}\tilde{\boldsymbol{T}}\tilde{\boldsymbol{R}} \tag{8.34}$$

where $\tilde{\boldsymbol{H}}$ is the transformation matrix corresponding to integral, and $\tilde{\boldsymbol{T}}$ is the matrix related with time t. The configurations of matrices $\tilde{\boldsymbol{H}}$, $\tilde{\boldsymbol{T}}$, $\tilde{\boldsymbol{J}}$, and $\tilde{\boldsymbol{R}}$ are provided in Appendix C of this chapter. Fig. 8.2 shows the flow chart of LVIM algorithm-2.

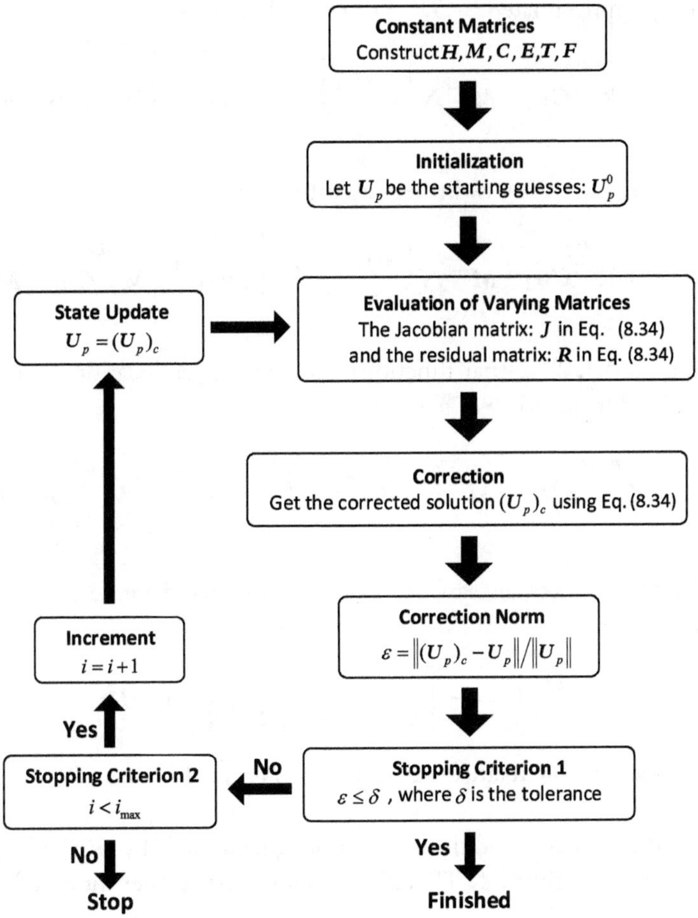

Figure 8.2 Flow chart overview of the LVIM algorithm 2.

8.1.3.3 LVIM Algorithm-3

Considering Eq. (8.21), if the Lagrange multiplier $\boldsymbol{\lambda}$ is approximated by Taylor series, we have

$$\frac{\mathrm{d}}{\mathrm{d}t}\begin{bmatrix} \boldsymbol{u}_1(t) \\ \boldsymbol{u}_2(t) \end{bmatrix}_c = -\begin{bmatrix} -\boldsymbol{u}_2 \\ \boldsymbol{M}^{-1}\boldsymbol{C}\boldsymbol{u}_2 + \boldsymbol{M}^{-1}\boldsymbol{N} \end{bmatrix} -\boldsymbol{J}(t)\int_{t_i}^{t}[\boldsymbol{T}_0 + \boldsymbol{T}_1(\tau - t) + \cdots]\left\{\begin{bmatrix} \boldsymbol{u}_1' \\ \boldsymbol{u}_2' \end{bmatrix} + \begin{bmatrix} -\boldsymbol{u}_2 \\ \boldsymbol{M}^{-1}\boldsymbol{C}\boldsymbol{u}_2 + \boldsymbol{M}^{-1}\boldsymbol{N} \end{bmatrix}\right\}\mathrm{d}\tau \tag{8.35}$$

If $\boldsymbol{\lambda}$ is simply approximated by $\boldsymbol{T}_0$, Eq. (8.35) becomes

$$\frac{\mathrm{d}}{\mathrm{d}t}\begin{bmatrix}\boldsymbol{u}_1(t)\\ \boldsymbol{u}_2(t)\end{bmatrix}_c = -\begin{bmatrix}-\boldsymbol{u}_2\\ \boldsymbol{M}^{-1}\boldsymbol{C}\boldsymbol{u}_2+\boldsymbol{M}^{-1}\boldsymbol{N}\end{bmatrix} - \boldsymbol{J}(t)\int_{t_i}^{t}\boldsymbol{T}_0\left\{\begin{bmatrix}\boldsymbol{u}_1'\\ \boldsymbol{u}_2'\end{bmatrix}+\begin{bmatrix}-\boldsymbol{u}_2\\ \boldsymbol{M}^{-1}\boldsymbol{C}\boldsymbol{u}_2+\boldsymbol{M}^{-1}\boldsymbol{N}\end{bmatrix}\right\}\mathrm{d}\tau \tag{8.36}$$

Since $\boldsymbol{T}_0 = -\boldsymbol{I}$, Eq. (8.36) is further rewritten as follows:

$$\frac{\mathrm{d}}{\mathrm{d}t}\begin{bmatrix}\boldsymbol{u}_1(t)\\ \boldsymbol{u}_2(t)\end{bmatrix}_c = -\begin{bmatrix}-\boldsymbol{u}_2\\ \boldsymbol{M}^{-1}\boldsymbol{C}\boldsymbol{u}_2+\boldsymbol{M}^{-1}\boldsymbol{N}\end{bmatrix} + \boldsymbol{J}(t)\int_{t_i}^{t}\boldsymbol{T}_0\left\{\begin{bmatrix}\boldsymbol{u}_1'\\ \boldsymbol{u}_2'\end{bmatrix}+\begin{bmatrix}-\boldsymbol{u}_2\\ \boldsymbol{M}^{-1}\boldsymbol{C}\boldsymbol{u}_2+\boldsymbol{M}^{-1}\boldsymbol{N}\end{bmatrix}\right\}\mathrm{d}\tau \tag{8.37}$$

By using $\boldsymbol{u}_p$ $(p=1,2)$ as trial functions and making collocations, the LVIM algorithm-3 is obtained from Eq. (8.37) as

$$\begin{bmatrix}\boldsymbol{u}_1'(t_m)\\ \boldsymbol{u}_2'(t_m)\end{bmatrix}_c = -\begin{bmatrix}-\boldsymbol{u}_2(t_m)\\ \boldsymbol{M}^{-1}\boldsymbol{C}\boldsymbol{u}_2(t_m)+\boldsymbol{M}^{-1}\boldsymbol{N}(t_m)\end{bmatrix} + \boldsymbol{J}(t_m)\int_{t_i}^{t}\left\{\begin{bmatrix}\boldsymbol{u}_1'\\ \boldsymbol{u}_2'\end{bmatrix}+\begin{bmatrix}-\boldsymbol{u}_2\\ \boldsymbol{M}^{-1}\boldsymbol{C}\boldsymbol{u}_2+\boldsymbol{M}^{-1}\boldsymbol{N}\end{bmatrix}\right\}\mathrm{d}\tau \tag{8.38}$$

where $\boldsymbol{u}_p'(t_m)$ and the integrals can be obtained as is stated in Eqs. (8.17) and (8.18). After rearrangements, Eq. (8.38) can be written as

$$\tilde{\boldsymbol{E}}\begin{bmatrix}\boldsymbol{U}_1\\ \boldsymbol{U}_2\end{bmatrix}_c = -\begin{bmatrix}-\boldsymbol{U}_2\\ \tilde{\boldsymbol{M}}^{-1}\tilde{\boldsymbol{C}}\boldsymbol{U}_2+\tilde{\boldsymbol{M}}^{-1}\tilde{\boldsymbol{N}}\end{bmatrix} + \tilde{\boldsymbol{J}}\tilde{\boldsymbol{H}}\tilde{\boldsymbol{R}} \tag{8.39}$$

The configurations of matrices $\tilde{\boldsymbol{E}}$, $\tilde{\boldsymbol{M}}$, $\tilde{\boldsymbol{C}}$, $\tilde{\boldsymbol{N}}$, $\tilde{\boldsymbol{J}}$, $\tilde{\boldsymbol{H}}$, and $\tilde{\boldsymbol{R}}$ are provided in Appendix C.

Note that the initial conditions are not guaranteed by Eq. (8.39), just like Eq. (8.25) in LVIM algorithm-1. Therefore it should be further modified as follows:

$$\tilde{\boldsymbol{E}}^r\begin{bmatrix}\boldsymbol{U}_1\\ \boldsymbol{U}_2\end{bmatrix}_c^r = \left\{-\begin{bmatrix}-\boldsymbol{U}_2\\ \tilde{\boldsymbol{M}}^{-1}\tilde{\boldsymbol{C}}\boldsymbol{U}_2+\tilde{\boldsymbol{M}}^{-1}\tilde{\boldsymbol{N}}\end{bmatrix} + \tilde{\boldsymbol{J}}\tilde{\boldsymbol{H}}\tilde{\boldsymbol{R}}\right\}^r - \tilde{\boldsymbol{E}}^d\begin{bmatrix}\boldsymbol{U}_1\\ \boldsymbol{U}_2\end{bmatrix}_c^d \tag{8.40}$$

The symbol $[\bullet]^r$ denotes the remained matrix after the $(lM+1)th$ rows and columns in $[\bullet]$ are dropped, $l=0,1,2,\ldots$. If $[\bullet]^r$ is a vector, we just need to remove the $(lM+1)th$ rows in $[\bullet]$. The symbol $[\bullet]^d$ denotes the parts of $[\bullet]$ that were dropped to obtain $[\bullet]^r$. The flow chart of this algorithm is provided in Fig. 8.3.

8.2 Mathematical model of a buckled beam

The governing equation of a buckled one-dimensional Bernoulli beam is written in non-dimensional form as

$$\ddot{w} + w^{iv} + Pw'' + c\dot{w} - \frac{1}{2}w''\int_0^1 w'^2\mathrm{d}x = F(x)\cos\Omega t, \tag{8.41}$$

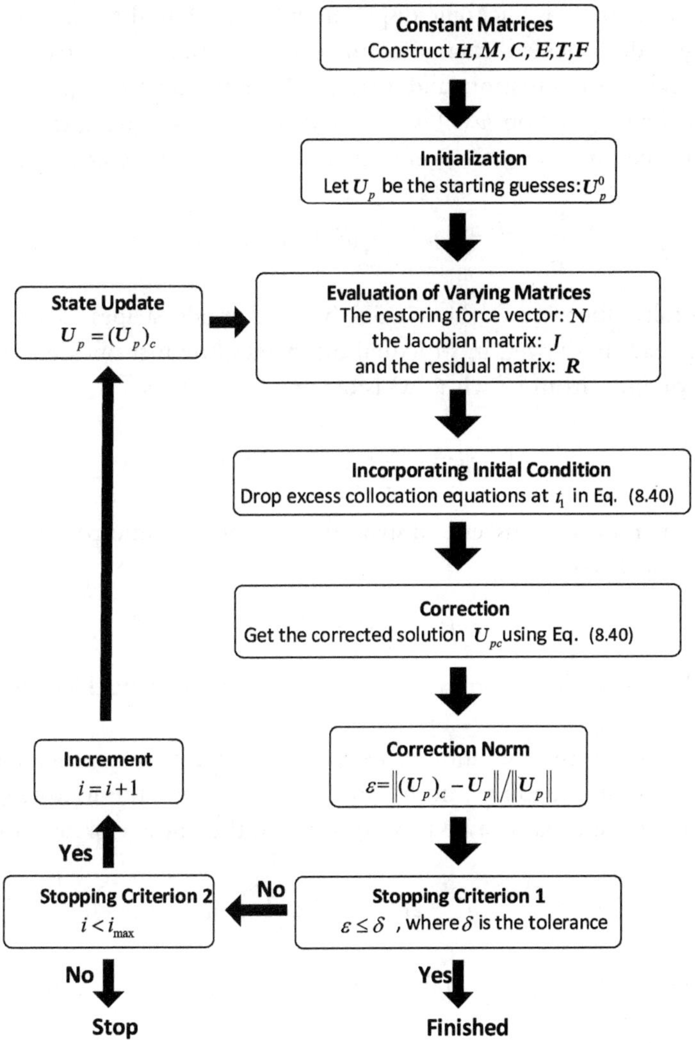

Figure 8.3 Flow chart overview of the LVIM algorithm-3.

where w is the transverse displacement of the beam and P is the axial load on the beam. $F(x)$ is the transverse distributed load on the beam, and Ω is the frequency of the applied load $F(x)$. The overdot denotes the derivative with respect to time t, while the prime denotes the derivative with respect to the spatial coordinate x.

To solve the preceding partial differential equation, we first assume the spatial modes:

$$w(x,t) = w_s(x) + v(x,t) = w_s(x) + \sum_{n=1}^{N} \phi_n(x) q_n(t), \tag{8.42}$$

where w_s is the static postbuckling displacement, $v(x,t)$ is the superposed dynamic response around the buckled configuration. $\phi_n(x)$ are the mode shapes of vibration around the buckled configuration, and $q_n(t)$ are the amplitudes of $\phi_n(x)$.

The buckled configuration $w_s(x)$ can be obtained by first solving the static buckling problem where the time derivatives and the dynamic load are removed in Eq. (8.41).

$$w^{iv} + Pw'' - \frac{1}{2}w'' \int_0^1 w'^2 \mathrm{d}x = 0 \tag{8.43}$$

Mathematically, there could be various buckled mode shapes, depending on the corresponding load. However, in structural mechanics, the first buckled mode shape is mostly of importance, from which $w_s(x)$ is obtained as follows:

$$w_s(x) = \frac{1}{2}b(1 - \cos 2\pi x) \tag{8.44}$$

The nondimensional transverse displacement b at the midspan of the beam is related to the load P via

$$b^2 = 4(P - P_c)/\pi^2 \tag{8.45}$$

where P_c is the critical load corresponding to the first Euler buckled mode, namely, $P_c = 4\pi^2$.

Substituting the assumed solution of $w(x,t)$ into the governing equation and dropping all the nonlinear, damping, and forcing terms, we have the following linear partial differential equation that can be tackled by using the linear vibration mode theory

$$\ddot{v} + v^{iv} + 4\pi^2 v'' - 2b^2\pi^3 \cos 2\pi x \int_0^1 v' \sin 2\pi x s x = 0 \tag{8.46}$$

By assuming $v(x,t) = \phi(x)e^{i\omega t}$ and substituting it into Eq. (8.46), the mode shape is obtained as follows:

$$\phi(x) = \phi_h + \phi_p = (c_1 \sin s_1 x + c_2 \cos s_1 x + c_3 \sinh s_2 x + c_4 \cosh s_2 x) + c_5 \cos 2\pi x, \tag{8.47}$$

where $s_{1,2} = \left(\pm 2\pi^2 + \sqrt{4\pi^2 + \omega^2}\right)^{1/2}$, and c_5 should satisfy the following equation:

$$\left(2b^2\pi^4 - \omega^2\right)c_5 = 2b^2\pi^3 \int_0^1 \phi_h' \sin 2\pi x \mathrm{d}x \tag{8.48}$$

Using the boundary conditions and Eq. (8.48), the mode shapes $\phi_n(x)$ can be obtained.

The resulting linear vibration mode shapes $\phi_n(x)$ are used to construct the solution $v(x,t)$. Using the multi-mode Galerkin discretization, where the weighting functions are the same as the trial functions $\phi_n(x)$, the partial differential Eq. (8.41) is then reduced to a system of coupled Duffing equations.

$$\ddot{q}_n + \omega_n^2 q_n = -c\dot{q}_n + b\sum_{i,j}^{N} A_{nij}q_iq_j + \sum_{i,j,k}^{N} B_{nijk}q_iq_jq_k + f_n\cos\Omega t, \quad n = 1, 2, ..., N \tag{8.49}$$

In this chapter, four modes are retained in the reduced model. The buckling level is selected as $b = 4$. Correspondingly the natural frequencies of the four linear vibration modes are obtained as follows: $\omega_1 = 30.7067$, $\omega_2 = 44.3627$, $\omega_3 = 108.3322$, and $\omega_4 = 182.1178$. The parameters A_{nij} and B_{nijk} are provided in Appendixes A and B.

8.3 Nonlinear vibrations of a buckled beam

Considering the discretized nonlinear system (Eq. 8.49), several types of nonlinear resonances of the buckled beam may occur due to the external harmonic excitation. The primary resonance is mostly observed when the excitation frequency Ω is close to one of the mode frequencies ω_n, which often leads to a periodic motion of large amplitude in that mode. For the existence of quadratic nonlinearities and cubic nonlinearities, the subharmonic resonances and superharmonic resonances may also occur for Ω that has an integer relationship with ω_n.

In the following, the resonance of the buckled beam under harmonic excitations is investigated. The frequency Ω is selected to be close to the natural frequency of the first vibration mode ω_1. The external force is supposed to be uniform over the length of the beam, thus $F(x)$ in Eq. (8.41) is constant. Through Galerkin discretization, the forces that are imposed on the four linear vibration modes are obtained as $f_1 = -0.850654F, f_2 = 0, f_3 = 0.309884F$, and $f_4 = 0$.

In the numerical simulations, the force-sweep and frequency-sweep processes are used to obtain an overview of the nonlinear dynamical behaviors of the buckled beam it is when subjected to a primary resonance excitation of its first vibration mode. Considering that the magnitudes of the coefficients ω_n^2, bA_{nij}, and B_{nijk} are roughly between 10^3 and 10^4 in the restoring forces, the amplitude of the external force F is set to vary between $F = 40$ and $F = 600$ in the force-sweep process. In the frequency-sweep process, F is fixed at $F = 400$, while the frequency Ω is varied between $\Omega = 30$ and $\Omega = 28$.

There are abundant types of basis functions for choice in the literature. In this chapter, the basis functions used in LVIM method is selected as Chebyshev polynomials of the first kind and the collocation points are selected as Chebyshev-Gauss-Lobatto nodes [6].

8.3.1 Bifurcations and chaos

With the discretized equations derived above, the nonlinear vibrations of a buckled beam are first investigated under uniform harmonic excitations, in which the frequency of the external load $F(x)\cos\Omega t$ is $\Omega = 30$. A force sweeping approach is adopted herein to capture the bifurcation phenomenon. For simplicity, a periodic motion is referred to as period n motion if its period is nT, where $T = 2\pi/\Omega$. It is shown in Fig. 8.4 that a period-one motion is obtained for $F = 40$. Fig. 8.4A is the Poincaré map that indicates the evolution of the displacement $q_1(t)$ and the velocity $\dot{q}_1(t)$ from the initial values to the final state, which is referred to as the sink. It can be found that the transient process of period one motion is rather simple. Since the sink is the only attractor in this area, all the points around it are attracted to it asymptotically.

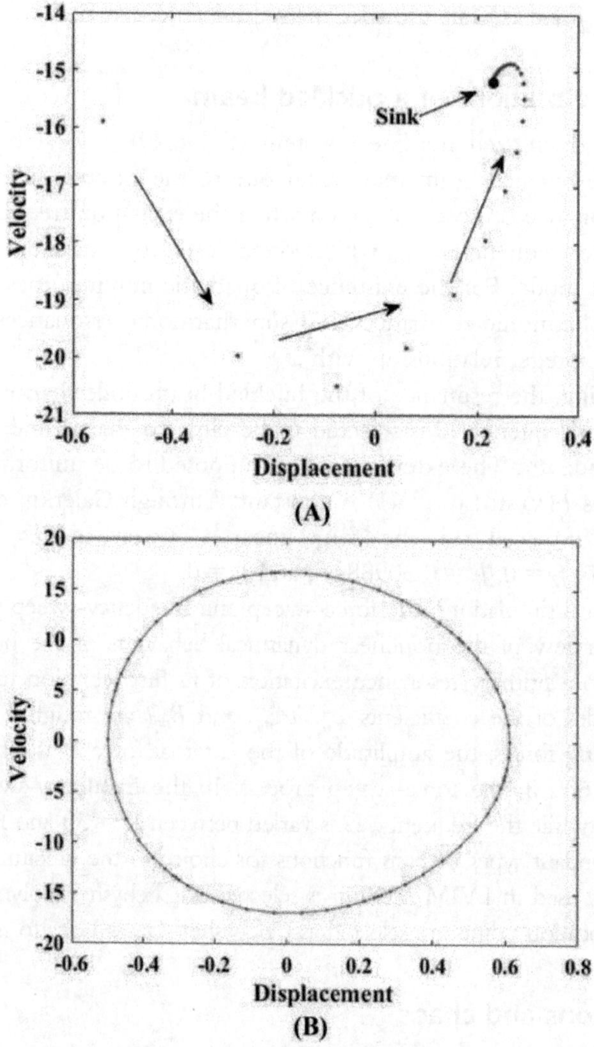

Figure 8.4 (A) The Poincaré map for $F = 40$. (B) The period 1 limit cycle oscillation. Same results for the HHT-α and LVIM algorithm, but with different computational performances.

As the excitation force F increases, the first period-doubling bifurcation occurs at about $F=400$. A typical period 2 motion is presented in Fig. 8.5 for $F=440$. In Fig. 8.5A the Poincaré maps of the transient process are obtained by using LVIM algorithms of various integrating accuracy. It is found that the two sinks standing for the

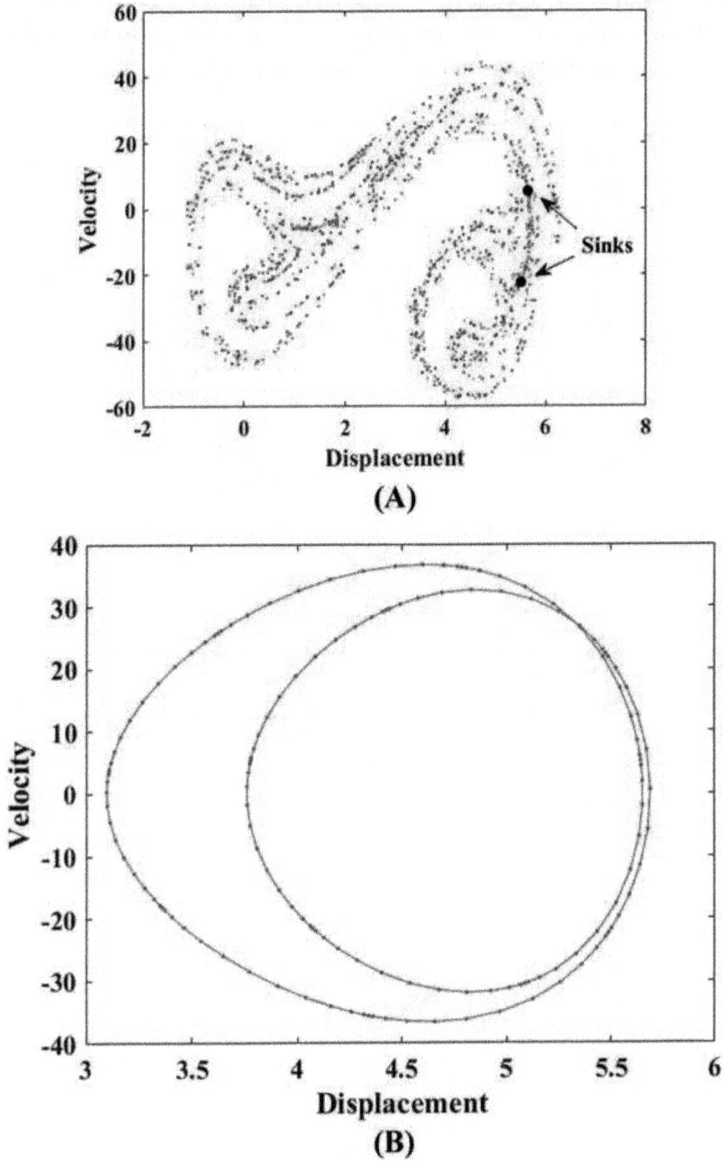

Figure 8.5 (A) The Poincaré map for $F=440$. (B) The period 2 limit cycle oscillation. Same results for the HHT-α and LVIM algorithms but with different computational performances.

period-two motion exist in a "strange attractor." This means that the transient response of this motion is chaotic-like. For the existence of sinks, the motion will eventually settle down to a steady limit cycle oscillation. However, it could be difficult to determine the exact time when the motion settles down if the transient stage is prolonged. The accurate integration of the system will also be a challenge, because the trajectory in the chaotic regime can easily diverge even with a very small integration error.

By further increasing the force amplitude, the second period-doubling occurs at about $F = 454$. A period-four motion is presented in Fig. 8.6 for $F = 460$. It can be

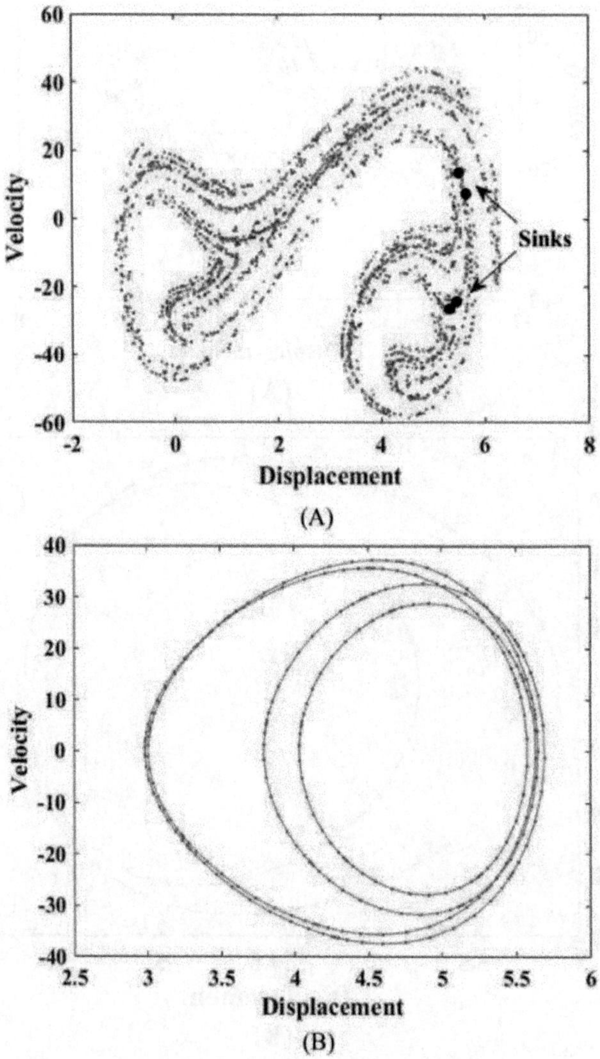

Figure 8.6 (A) The Poincaré map for $F = 460$. (B) The period 4 limit cycle oscillation. Same results for the HHT-α and LVIM algorithms but with different computational performances.

observed in Fig. 8.6A that the four sinks also coexist with the chaotic attractor, thus the transient motion is also chaotic. By comparing the chaotic regimes in Fig. 8.5A and Fig. 8.6A, there is not much difference before and after the bifurcation occurs. Therefore, it seems that the chaotic attractor is barely affected by the bifurcation of periodic motions. The next bifurcation occurs at $F = 461$, leading to the period-eight motion. The corresponding Poincaré map, phase portrait and response curve are plotted in Fig. 8.7.

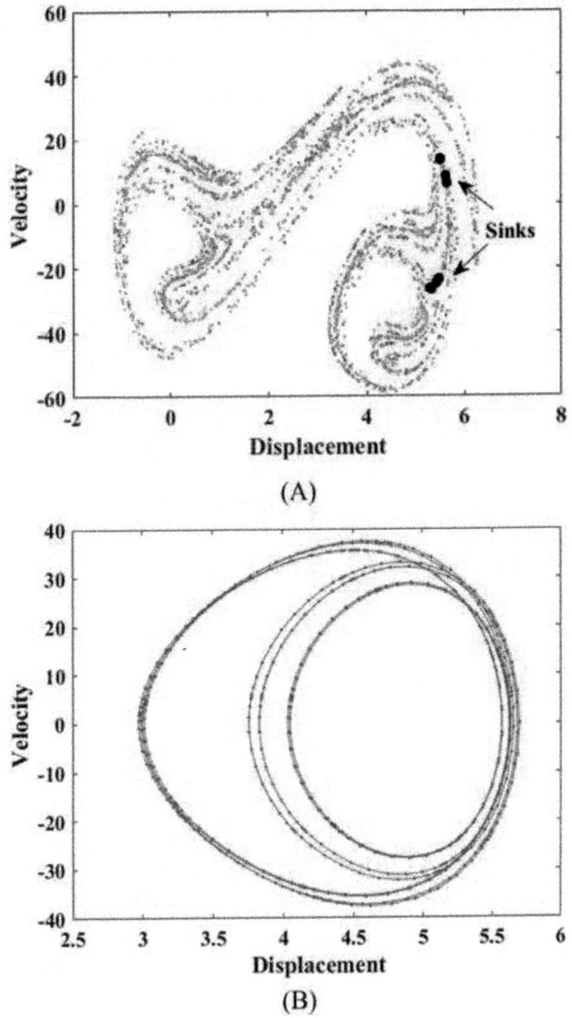

Figure 8.7 (A) The Poincaré map for $F = 461$. (B) The period 8 limit cycle oscillation. Same results for the HHT-α and LVIM algorithms but with different computational performances.

The results in Figs. 8.4–8.7 can be obtained by both the HHT-α and LVIM algorithms. However, the time step size has to be selected to be very small to accurately obtain the steady periodic motions by using the HHT-α algorithm, of which the computational time will be much prolonged. On the contrary, the step size of the LVIM algorithms can be selected relatively very large, but still easily achieve high accuracy in predicting these limit cycle oscillations. It will be shown in the next subsection that the LVIM algorithm has far better performance than the HHT-α method in terms of accuracy, computational time, and speed of convergence, in predicting the nonlinear dynamical responses of the buckled beam involving bifurcation and chaos.

As the period-doubling bifurcation proceeds, more sinks appear in the chaotic regime, and eventually the motion becomes completely chaotic. In Fig. 8.8 the chaotic motion is presented through a phase portrait, a response curve, a Poincaré map,

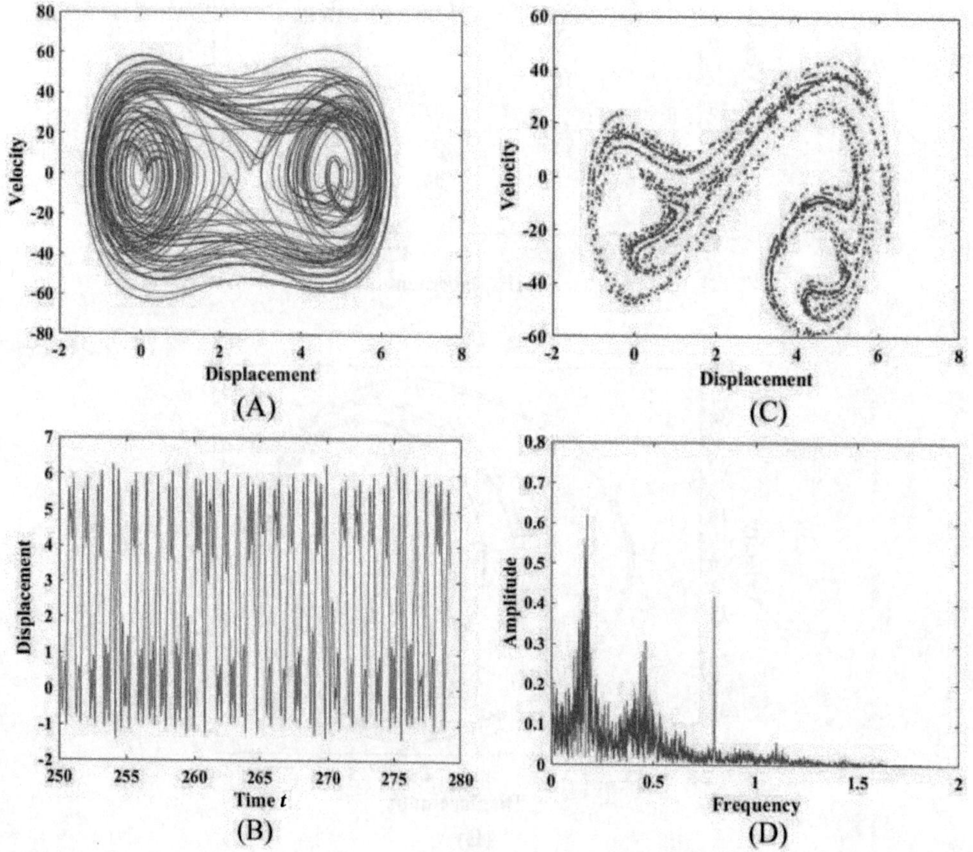

Figure 8.8 The same chaotic motion revealed by the HHT-α (with a very small step size) and LVIM algorithms. (A) Phase portrait. (B) Time responses. (C) Poincaré map. (D) Fast Fourier transform curves.

and curve. Similar period-doubling route to chaos is also observed for varied excitation frequency with the amplitude fixed at $F = 400$. The bifurcation diagrams obtained through force-sweeping process and frequency-sweeping process are plotted in Fig. 8.9.

In predicting chaotic motion, the trajectories obtained by using the considered two methods diverge after a short time. However, the Poincaré maps and the FFT curves coincides very well, which shows that the same chaotic motion is revealed by these two methods.

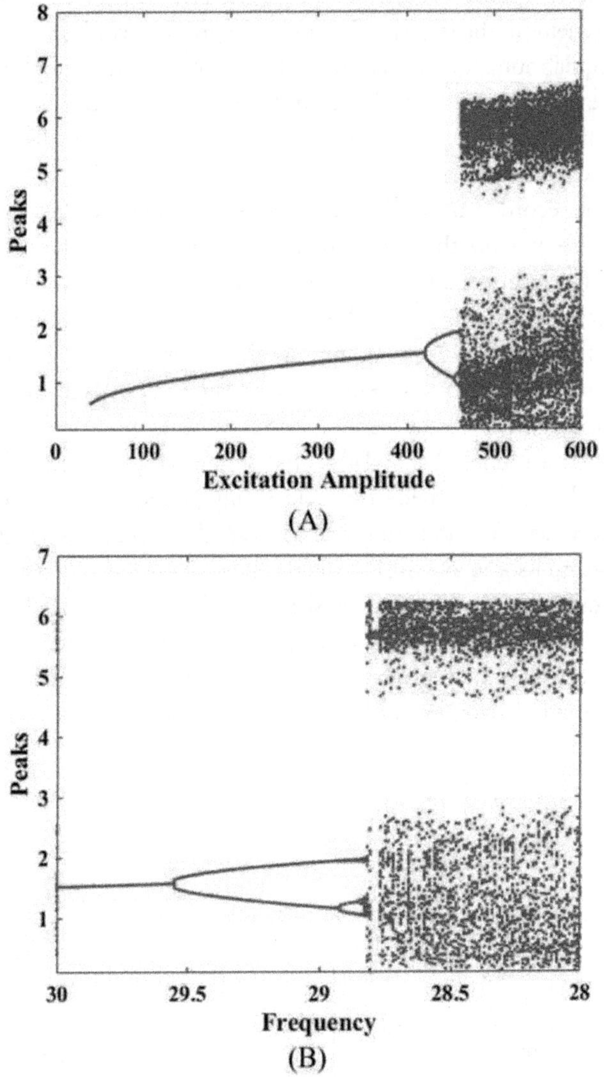

Figure 8.9 Bifurcation diagrams obtained by (A) the force-sweep process and (B) the frequency sweep process. Same results obtained for the HHT-α (with a very small step size) and LVIM algorithms.

8.3.2 Comparison between HHT and LVIM algorithms

The discretized model is solved by using both the HHT-α method and the LVIM. It is found that both methods can predict the limit cycle oscillations and chaos. However, the computational performances of these two methods are very different. In the analysis below, various step sizes are used to test the stability of the algorithms. It is found through simulations that the LVIM method converges for both periodic and chaotic motions with step sizes as large as $T/5$, where T is the period of the first vibration mode, $T = 2\pi/\omega_1$. The largest step size of the HHT-α method depends on the nonlinear algebraic equations solver. By using the Newton-Raphson method, the step size should be no greater than $T/20$; otherwise the solution of nonlinear equilibrium equations of the HHT-α method may not be found.

We tried to reveal the steady periodic motions of the buckled beam under different external excitations, using the HHT-α method and the LVIM separately. After the motion settles down, the extremes of the displacement $q_1(t)$ that are obtained by these two methods are as recorded in Table 8.1. The exact values are provided by ode45, and they are fully consistent with the results that were obtained by the LVIM, with a time step size of $T/5$. The number of collocation points that are used in the LVIM is $N = 7$. Increasing the collocation points can further improve the accuracy of this method.

The HHT-α method fails to work for $\Delta t = T/5$ and $T/10$ because the Newton-Raphson iteration scheme cannot converge for such large steps. With the step size of $T/50$, the HHT-α method provides steady periodic motions, although they are quite erroneous in comparison to the exact ones. Particularly, in the parameter region where the period-8 motion dominates, the HHT-α method provides only period-four motion, which is due to the integrating inaccuracy. By shortening the step size, the accuracy is significantly improved. However, there are still some observable discrepancies between the results of the HHT-α method and the exact ones, even with a step size as small as $\Delta t = T/1000$. In the simulations, different values of α are used. In the case of $\alpha = -0.1$, a moderate numerical damping is included, which could filter the high-frequency component of motion and avoid divergence. For $\alpha = 0$, no numerical damping is introduced, and the HHT-α method becomes the

Table 8.1 Extremes of the steady periodic motions obtained by the HHT-α method and the LVIM methods.

Methods	**Period 1**		**Period 2**		**Period 4**		**Period 8**	
LVIM ($T/5$)	**0.6006**		**3.0983**		**2.9881**		**2.9752**	
HHT-α	$\alpha = -0.1$	$\alpha = 0$	$\alpha = -0.1$	$\alpha = 0$	$\alpha = -0.1$	$\alpha = 0$	$\alpha = -0.1$	$\alpha = 0$
$T/5$	—	—	—	—	—	—	—	—
$T/10$	—	—	—	—	—	—	—	—
$T/50$	0.5626	0.5954	3.1282	3.1164	3.0015	2.9955	—	—
$T/100$	0.5826	0.5995	3.1074	3.1025	2.9912	2.9898	—	2.9828
$T/500$	0.5972	0.6006	3.0994	3.0984	2.9884	2.9881	2.9770	2.9755
$T/1000$	0.5989	0.6006	3.0988	3.0983	2.9882	2.9881	2.9760	2.9756

average acceleration method. By comparing the values of extremes with the exact ones, Table 8.1 indicates that the periodic motions are better predicted with $\alpha = 0$ than with $\alpha = -0.1$. This result is reasonable, since the numerical damping that is introduced in the HHT-α method makes the system behave as if extra damping exists in the simulation.

The limit cycles of period-four and period-eight motions are obtained by using the LVIM with $\Delta t = T/5$. They are compared with the results obtained from the HHT-α method, where $\alpha = -0.1$ and $\Delta t = T/50$. Fig. 8.10 shows that the HHT-α

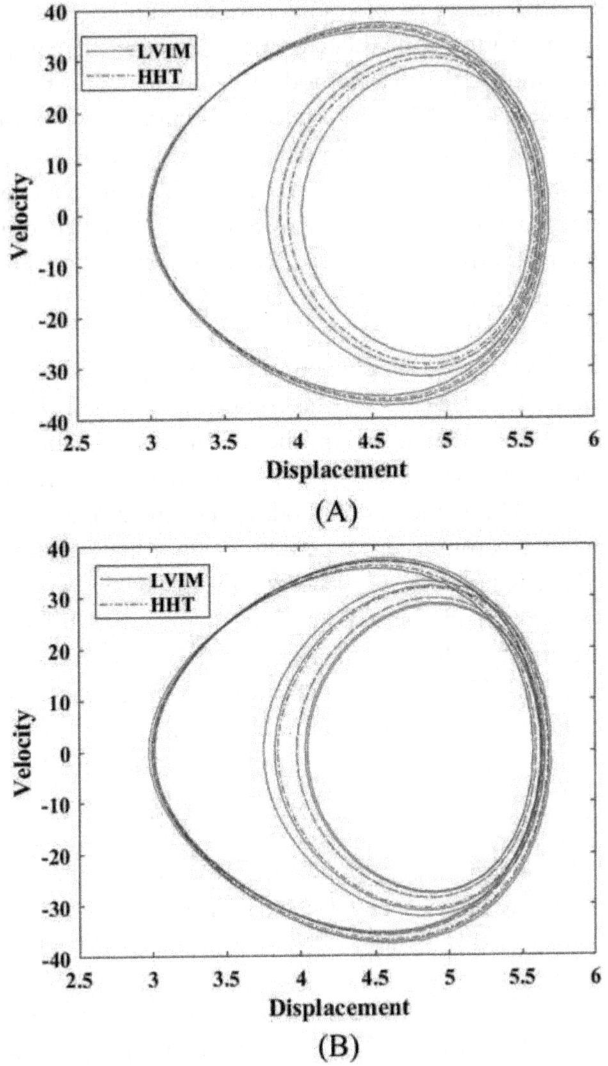

Figure 8.10 Comparisons between the limit cycle oscillations obtained by the LVIM ($\Delta t = T/5$) and HHT-α ($\Delta t = T/50$, $\alpha = -0.1$) algorithms.

method cannot predict the true dynamical behavior with such a large step size. In Fig. 8.10A an obvious discrepancy exists between the results of the LVIM and the HHT-α method, while in Fig. 8.10B the HHT-α method gives a period-four motion at where the period 8 motion should be revealed. In all, both the LVIM and the HHT-α method can stably integrate the nonlinear system of buckled beam. However, the step size of the HHT-α method should be selected very small to obtain valid solutions.

Generally, the evaluation of computational accuracy is nontrivial for numerical methods, especially in nonlinear problems. Herein, the numerical solution of MATLAB built in ode45 function is used as the benchmark to estimate the computational error of the HHT-α method and the LVIM methods. The absolute and the relative accuracies of ode45 are both set as $1E-15$. The transient responses of period-one, period-two, period-four and period-eight motions are obtained using the HHT-α method and the LVIM methods respectively. Then by comparing with those obtained by ode45, the discrepancies of the numerical results are shown in Fig. 8.11. To evaluate the highest accuracy these two methods may achieve, the steps size of the HHT-α method is selected as $\Delta t = 1e - 4$, while that of the LVIM is $\Delta t = T/6 = 0.035$ with 13 collocation points in each time step.

As is illustrated in Section 8.3, the transient response of period-one motion is relatively simple and nonchaotic. It is reflected by the consistent numerical discrepancies over the simulation time in Fig. 8.11A. As is shown, the LVIM method achieves very high accuracy with respect to ode45. The discrepancies between them is five magnitudes lower than that between the HHT-α method and ode45. For the multiple-period motions, the numerical discrepancies accumulate exponentially for both the HHT-α and LVIM algorithms before $t = 10$, as is shown in Fig. 8.11B. This can be explained by the fact that a chaotic regime exists around the periodic motion. It is shown that the accuracy of the LVIM method is still much higher than that of the HHT-α method in the integration of transient chaotic motions. Although the results of the LVIM method diverge from that of ode45 after $t = 10$, this does not mean that the LVIM fails. Actually, the chaotic regime is so sensitive to the initial state that even an error that occurred on the machine precision could blow up and a cause significant discrepancy in the final state.

Table 8.2 lists the computational time, iteration steps and step sizes of the HHT-α method and the LVIM method. It is found that the LVIM method saves up to 80% as compared to the computational time of the HHT-α method. The step size that is used in the HHT-α method is 0.0001, while that in the LVIM is 0.035. According to the performances demonstrated in Fig. 8.11 and Table 8.2, the LVIM method is much more accurate and efficient than the HHT-α method in addition to requiring only much larger step sizes. For further comparison the performance indexes of ode45 are also presented in Table 8.2. It is shown that the computational efficiency of the LVIM algorithm is far superior to that of ode45. Noting that the ode45 is a built-in function

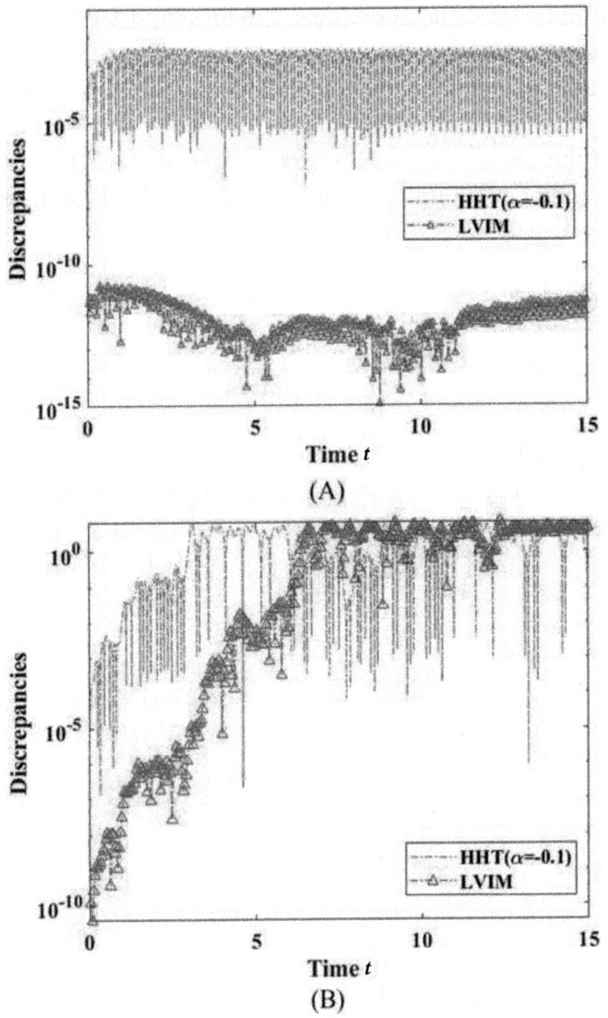

Figure 8.11 The computational error of the HHT-α and LVIM algorithms.

Table 8.2 Computational cost of the HHT-α method and the LVIM method.

Cases	Computational time(s)			Iteration steps			Step size		
	HHT-α	LVIM	ode45	HHT-α	LVIM	ode45	HHT-α	LVIM	ode45
Period-1	11.4	2.6	17.6	593468	5392	903797	0.0001	0.035	1.6e − 05
Period-2	11.8	2.5	22.5	599664	5509	1022317	0.0001	0.035	1.5e − 5
Period-4	11.6	2.6	21.2	599670	5608	1015945	0.0001	0.035	1.5e − 5
Period-8	11.7	2.4	19.6	599629	5567	1015821	0.0001	0.035	1.5e − 5

in MATLAB that has already been optimized, while the LVIM algorithms are implemented in MATLAB with a crudely designed program. Moreover, the LVIM algorithms can be easily realized by using parallel programming, which will further improve the computational efficiency.

To investigate the energy conservation properties of the HHT-α method and the LVIM, a simple undamped Duffing equation is used for demonstration:

$$x'' - x + x^3 = 0.$$

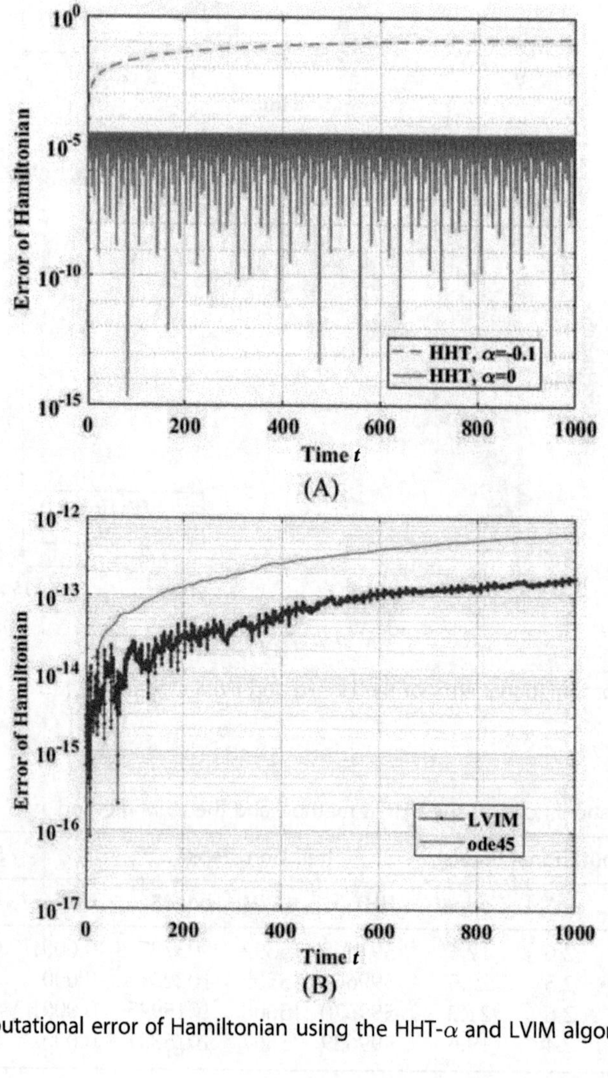

Figure 8.12 Computational error of Hamiltonian using the HHT-α and LVIM algorithms and ode45.

The Hamiltonian energy of this system is

$$H = \frac{x'^2}{2} - \frac{x^2}{2} + \frac{x^4}{4}$$

Starting from the initial state $x(0) = 1.5$, $x'(0) = 0$, the system is integrated by using the HHT-α method, the LVIM, and the ode45 method. The step size of the HHT-α method is $\Delta t = 0.01$, and the simulation is carried in the time interval $t \in [0, 1000]$. For the LVIM the step size is as $\Delta t = 2$, with 32 collocation points in each step. The absolute and relative accuracies of the ode45 method are both set to be $1E - 15$. The computational errors of the Hamiltonian are recorded and plotted in Fig. 8.12. It can be seen that both the LVIM method and ode45 are much superior to the HHT-α method in energy conservation. Notably, the LVIM method behaves even better than ode45, with a negligible error of the Hamiltonian of $1E - 13$. Among all these methods, the HHT-α method with $\alpha = -0.1$ is the worst in energy conservation, as can be seen in Fig. 8.12A. After dropping out the numerical damping by setting $\alpha = 0$, the performance of the HHT-α method is much improved, with the error of the Hamiltonian being $1E - 5$.

8.4 Conclusion

The LVIM Algorithm is used to solve a nonlinear dynamical model of the clamped-clamped buckled beam. The numerical results show that the proposed method is very accurate and efficient in predicting nonlinear dynamical behaviors including bifurcation and chaos. Through force-sweep and frequency-sweep processes, the periodic motions and the period-doubling routes to chaos are successfully captured by the LVIM algorithms. It is also found in the simulations that the steady multiple-periodic motions are actually surrounded by chaotic motions in the neighborhood. It is the reason for the chaotic transient motions that can be hardly predicted precisely.

The robustness of the Wang-Atluri method is indicated by the highly accurate integration using a large step size $\Delta t = T/5$. Compared with the Hilber-Hughes-Taylor-α method, the proposed method is superior on many aspects involving accuracy, efficiency, stability, and energy conservation. It is shown in this chapter that the Hilber-Hughes-Taylor-α method cannot accurately predict the periodic vibrations of the buckled beam unless extremely small step sizes are used. An obvious discrepancy can still be observed between the exact phase portrait and that provided by Hilber-Hughes-Taylor-α method even when a relatively small step size $\Delta t = T/100$ is used. The example of Duffing equations also indicates that the energy of the system is barely conserved by using the Hilber-Hughes-Taylor-α method after the simulation time $t = 1000$, while the error of energy in the results of the Wang-Atluri algorithm is negligible.

Appendix A

$$A_{1,1,1}=124.905,\ A_{1,1,2}=0,\ A_{1,1,3}=162.554,\ A_{1,1,4}=0,\ A_{1,2,2}=171.098,$$
$$A_{1,2,3}=0,\ A_{1,2,4}=-141.515,\ A_{1,3,3}=407.072,\ A_{1,3,4}=0,\ A_{1,4,4}=625.188;$$

$$A_{2,1,1}=0,\quad A_{2,1,2}=342.195,\quad A_{2,1,3}=0,\quad A_{2,1,4}=-141.515,\quad A_{2,2,2}=0,$$
$$A_{2,2,3}=391.398,\quad A_{2,2,4}=0,\quad A_{2,3,3}=0,\quad A_{2,3,4}=-161.863,\quad A_{2,4,4}=0$$

$$A_{3,1,1}=81.2768,\quad A_{3,1,2}=0,\quad A_{3,1,3}=814.145,\quad A_{3,1,4}=0,\quad A_{3,2,2}=195.699,$$
$$A_{3,2,3}=0,\ A_{3,2,3}=-161.863,\quad A_{3,3,3}=1264.72,\quad A_{3,3,4}=0,\quad A_{3,4,4}=715.082$$

$$A_{4,1,1}=0,\quad A_{4,1,2}=-141.515,\quad A_{4,1,3}=0,\quad A_{4,1,4}=1250.38,\quad A_{4,2,2}=0,$$
$$A_{4,2,3}=-161.863,\ A_{4,2,4}=0,\quad A_{4,3,3}=0,\quad A_{4,3,3}=0,\quad A_{4,3,4}=1430.16,\quad A_{4,4,4}=0$$

Appendix B

$$B_{1,1,1,1}=-65.3795,\quad B_{1,1,1,3}=-79.2731,\quad B_{1,1,2,2}=-268.675,$$
$$B_{1,1,2,4}=222.22,\quad B_{1,1,3,3}=-600.138,\quad B_{1,1,4,4}=-981.733,\quad B_{1,2,2,3}=-108.59,$$
$$B_{1,2,3,4}=89.8146,\quad B_{1,3,3,3}=-233.924,\quad B_{1,3,4,4}=-396.786$$

$$B_{2,1,1,2}=-268.675,\quad B_{2,2,2,2}=-1104.11,\quad B_{2,1,2,3}=-217.18,$$
$$B_{2,2,3,3}=-2378.47,\quad B_{2,1,1,4}=111.11,\quad B_{2,2,2,4}=1369.81,$$
$$B_{2,1,3,4}=89.8146,\quad B_{2,3,3,4}=983.614,\quad B_{2,2,4,4}=-4412.05,\quad B_{2,3,3,3}=1668.42$$

$$B_{3,1,1,1}=-26.4244,\quad B_{3,1,2,2}=-108.59,\quad B_{3,1,1,3}=-600.138,$$
$$B_{3,2,2,3}=-2378.47,\quad B_{3,1,3,3}=-701.773,\quad B_{3,3,3,3}=-5123.69,\quad B_{3,1,2,4}=89.8146,$$
$$B_{3,2,3,4}=1967.23,\quad B_{3,1,4,4}=-396.786,\quad B_{3,3,4,4}=-8690.89$$

$$B_{4,1,1,2}=111.11,\quad B_{4,2,2,2}=456.603,\quad B_{4,1,2,3}=89.8146,\quad B_{4,2,3,3}=983.614,$$
$$B_{4,1,1,4}=-981.733,\ B_{4,2,2,4}=-4412.05,\ B_{4,1,3,4}=-793.572,\ B_{4,3,3,4}=-8690.89,$$
$$B_{4,2,4,4}=5005.26,\quad B_{4,4,4,4}=-14741.6$$

Appendix C

(Table 8.3)

Table 8.3 The constant and varying matrices in the WA algorithms.

Constant matrices	Varying matrices
$\tilde{\boldsymbol{E}} = \boldsymbol{I} \otimes (\boldsymbol{LB})\boldsymbol{B}^{-1}$ $\tilde{\boldsymbol{M}} = \boldsymbol{M} \otimes \boldsymbol{I}$ $\tilde{\boldsymbol{C}} = \boldsymbol{C} \otimes \boldsymbol{I}$ $\boldsymbol{t} = [t_1, t_2, \ldots, t_M]^{\mathrm{T}}$ $\hat{\boldsymbol{t}} = diag(\boldsymbol{t})$ $\tilde{\boldsymbol{t}} = \boldsymbol{I} \otimes \boldsymbol{t}$ $\tilde{\boldsymbol{H}} = \boldsymbol{I} \otimes \left(\boldsymbol{L}^{-1}\boldsymbol{B}\right)\boldsymbol{B}^{-1}$	$\tilde{\boldsymbol{J}} = \boldsymbol{J}(\hat{\boldsymbol{t}})$ $\tilde{\boldsymbol{N}} = \boldsymbol{N}(\boldsymbol{t})$ $\tilde{\boldsymbol{R}} = \tilde{\boldsymbol{E}} \begin{bmatrix} \boldsymbol{U}_1 \\ \boldsymbol{U}_2 \end{bmatrix} + \begin{bmatrix} -\boldsymbol{U}_2 \\ \tilde{\boldsymbol{M}}^{-1}\tilde{\boldsymbol{C}}\boldsymbol{U}_2 + \tilde{\boldsymbol{M}}^{-1}\tilde{\boldsymbol{N}} \end{bmatrix}$

References

[1] G. Kerschen, K. Worden, A.F. Vakakis, J.C. Golinval, Past, present and future of nonlinear system identification in structural dynamics, Mechanical Systems and Signal Processing 2006 (20) (2006) 505–592.

[2] X. Wang, S.N. Atluri, A new feedback-accelerated picard iteration method for orbital propagation & Lambert's problem, J. Guid. Control Navig., online.

[3] W. Kreider, A.H. Nayfeh, Experimental investigation of single-mode responses in a fixed-fixed buckled beam, Nonlinear Dynamics 15 (2) (1998) 155–177.

[4] W.Y. Tseng, J. Dugundji, Nonlinear vibrations of a beam under harmonic excitation, Journal of Applied Mechanics 37 (2) (1970) 292–297.

[5] S.A. Emam, A.H. Nayfeh, On the nonlinear dynamics of a buckled beam subjected to a primary-resonance excitation, Nonlinear Dynamics 35 (1) (2004) 1–17.

[6] T. Fukushima, Picard iteration method, Chebyshev polynomial approximation, and global numerical integration of dynamical motions, The Astronomical Journal 113 (5) (1997) 1909–1914.

Index

Note: Page numbers followed by "*f*" and "*t*" refer to figures and tables, respectively.

N

O

Q